Forest Certification
Roots, Issues, Challenges, and Benefits

Forest Certification
Roots, Issues, Challenges, and Benefits

Kristiina A. Vogt • Bruce C. Larson
John C. Gordon • Daniel J. Vogt
Anna Fanzeres

School of Forestry and Environmental Studies
Yale University
New Haven, Connecticut

CRC Press
Boca Raton   London   New York   Washington, D.C.

Cover Photographer: Robert Perron

Library of Congress Cataloging-in-Publication Data

Forest certification : roots, issues, challenges, and benefits /
    Kristiina A. Vogt ... [et al.].
        p.   cm.
    Includes bibliographical references.
    ISBN 0-8493-1585-9 (alk. paper)
    1. Forest management--Standards.   2. Forest products-
-Certification.   3. Sustainable forestry.   I. Vogt, Kristiina A.
SD387.S69F67   1999
634.9209'18--dc21   99-36135
                                                                     CIP

No claim to original U.S. Government works
International Standard Book Number 0-8493-9469-4
Library of Congress Card Number 99-36135
Printed in the United States of America 1 2 3 4 5 6 7 8 9 0
Printed on acid-free paper

Preface

The idea to write a book and have a class on forest certification originated during a walk through Amazonian tropical forests in Camutá do Pucuruí (Gurupá, Pará State, Brazil). Several of us — Daniel Vogt, Kristiina Vogt, and Anna Fanzeres — were scouting for research sites for Anna's dissertation. While walking through several forest sites, we realized that although some of the areas had clearly different forest productivities, they would be evaluated as being similar using many of the criteria and indicators of the certification protocols. By using our own ecological indicators to assess these sites, it became obvious that the ecological characteristics of one of the forests would support timber harvesting without degradation of the forest landscape if conducted under existing social/ecological legacies; however, another site would become degraded if timber harvesting were part of management. This resulted in a lively discussion about the indicators relevant to determining whether or not forest-dweller communities could harvest their timber without causing degradation of their forests. Moreover, we thought it would be interesting to apply our findings to the situation in the United States, to determine whether small forestland owners could be certifed if they were conducting sustainable forest management. We then had the idea of having students from Yale evaluate one of its forests, using several different protocols, to help identify how we could develop protocols that would be sensitive to the ecological indicators we had observed in the forest.

We decided that we needed to have a class to scientifically examine certification, and the class structure and topics were developed. Two graduate students at the School of Forestry and Environmental Studies were instrumental in developing the course: Francis Raymond and Jessica Lawrence; their insights into certification were crucial in selecting our guest speakers. In addition, Yale Forest Forum sponsored the class and helped make it a reality. Gary Dunning, director of the forum, helped us acquire funds for guest speakers' fees and to defray some of the publishing costs for a book on forest certification. Through the generous support of the Cricket Foundation, we were able to support the speaker series during the spring of 1998.

Several experts spoke to the class about topics related to certification. They were instrumental in making us consider all the issues related to certification and we would like to acknowledge them. The guest speakers generated lively discussions and helped make the class the success that it was. The ideas presented in this book are those of its authors; however, the guest speakers did stimulate our thinking about the issues from a fresh perspective. We want to thank the following individuals for the insights that they offered so freely:

Harold Burnett, certified natural resource manager, Two Trees Forestry, Maine

Richard Donovan, director, Smartwood, Rainforest Alliance, Richmond, Vermont

Jim Drescher, ecologist, Windhorse Farm, Nova Scotia

Jamie Ervin, U.S. contact person, Forest Stewardship Council

Morgan Grove, social ecologist, U.S. Forest Service, Vermont

Bill Mankin, director, Global Forest Policy Project, Washington, D.C.

Catherine Mater, vice president, Mater Engineering, Corvallis, Oregon

Michael Northrup, Rockefeller Brothers Fund, New York

Blaine Puller, forest manager, Kane Hardwood/Collins Pine, Kane, Pennsylvania

Bob Simpson, National Tree Farm Association, Washington, D.C.

As part of the class, we also had a final panel discussion with representatives from several certification organizations. The discussion was sponsored by the Yale Forest Forum and the following individuals were on the panel:

Jamie Ervin, Forest Stewardship Council, Waterbury, Vermont

Julie Jack, Sustainable Forestry Initiative, AF&PA, Washington, D.C.

Gerry LaPointe, Sustainable Forestry Certification Coalition, Montreal, Canada

Bill Mankin, Global Forest Policy Project, Washington, D.C.

In addition to the class discussions, there was a field trip to the Vermont Family Forest and Value-Added Operation. Thanks must be given to David Bryne (Smartwood certified natural resource manager, Addison County, Vermont) and Alan Calfee (forest assessor, National Wildlife Federation [Smartwood's New England partner], Montpelier, Vermont) who hosted the trip.

We especially want to thank several individuals who spent a lot of time commenting on various drafts of this book. We are very grateful to Timothy Farnham and William Barclay who had helpful comments on the structure and development of Chapter 2. Rui S. Murrieta contributed many good thoughts on various topics in the book and we appreciate his input.

Many of the ideas in this book are the results of the Vogts' collaborative research in Puerto Rico, the eastern and western parts of the United States, Alaska, Iceland, Brazil, Belize, and Malaysia. Many of their collaborations have been with Yale graduate students who are a decidedly stimulating and fun group with whom to explore new ecosystems. We also would like to acknowledge other research collaborators who have contributed to the development of our ideas, but we do not have enough space to name all the individuals who have helped us to formalize our ideas during the past twenty years. We want to acknowledge the following individuals who discussed many ideas with us related to ecosystems: Philip Wargo (U.S. Forest Service's Northeast Center for Forest Health Research) was instrumental in our understanding of the roles of insects and pests in ecosystems, and Alan Covich (Colorado State University) and Larry Woolbright (Seneca College) were not only stimulating to interact with, but also helped us learn about animals in tropical ecosystems.

Many of the ideas in this book developed while conducting research supported by the National Science Foundation (NSF) and by the U.S. Forest Service. NSF support has been in the form of individual grants to the Vogts and in their involvement with the Long-Term Ecological Research Program in Puerto Rico's Luquillo Experimental Forest. The USDA Forest Service, the Northeast Global Change Program, and the USDA Forest Service Insect and Disease Lab have supported research that has stimulated our thinking on the assessment of forest ecosystems by addressing the role of disturbance in affecting ecosystem resilience.

These research interactions have been instrumental in opening our eyes to the difficulties in assessing ecosystems. They have also caused us to consider how we can incorporate humans as drivers of ecosystem functions in tropical, temperate, and boreal ecosystems.

In addition, the research has stimulated several of us to think about frameworks that would allow us to link the social and natural sciences. The insights from all the different locations have been crucial in understanding the differences that exist in any ecosystem.

The purpose of this book is to conduct a globally relevant scientific analysis of certification. Some of the discussions are United States-specific, but the issues transcend every country's boundaries. Because the Yale School of Forestry and Environmental Studies is located in the United States, the case study was conducted on school lands. In addition, the authors' familiarity with small non-industrial landowners and the issues they face means that information from the United States will be emphasized. However, the issues faced by U.S. landowners are not much different from those faced by people in developing or developed countries. The examples given will be useful for anyone studying these issues since they will face very similar problems in integrating the social and natural sciences in natural resource assessments. Despite the uniqueness of the forests in the United States and its societal structure, we see this book as an extremely useful tool that transcends borders to address global environmental issues. By using a new perspective, we hope the reader will also begin to think about certification and will help us further expand our thinking on this topic.

Authors

Kristiina A. Vogt is professor of forest ecosystem ecology at Yale University's School of Forestry and Environmental Studies. She is also the chairman of the Yale Forest Forum. Her expertise is in carbon and nutrient cycles at the ecosystem and landscape levels in wetland and terrestrial systems. She has conducted research in Iceland, Malaysia, Mexico, Brazil, Belize, Puerto Rico, and within the United States in Alaska, Oregon, Washington, New York, Vermont, New Hampshire, and New Mexico.

Kristiina has been intrigued by the roles of human and natural disturbances in controlling processes in ecosystems and in determining the links between species diversity and ecosystem sustainability. In her research, she attempts to understand the mechanisms that control how ecosystems function and what causes changes in natural cycles, believing that this knowledge will help us understand how to restore ecosystems. Kristiina believes that no adequate tools exist to document the impact of human activities on our landscape. An adequate, working framework would help to analyze the effects our activities are having on changes in the natural processes of a system and aid in our ability to determine when the trajectory of a system moves towards degradation. Her long-term interest in determining the best, minimum amounts of information needed to interpret our landscape resulted in a 1997 book written with several other authors, entitled *Ecosystems: Balancing Science with Management.*

Kristiina Vogt has published more than 80 peer-reviewed articles and two previous books. Her published work covers topics from the ecology of species to the productivity of deserts — from forest and wetland ecosystems to the development of frameworks to assess ecosystem health and resilience — to the development of indicators to determine whether ecosystems will degrade and how they will respond to invasive species. After obtaining her Ph.D. and M.S. in microbial ecology at New Mexico State University, she was a research faculty member at the University of Washington for 11 years. She earned her B.S. from the University of Texas, El Paso.

Bruce C. Larson is director of the forests belonging to Yale University and a lecturer in forest management at Yale's School of Forestry and Environmental Studies. He is also a member of the Scientific Advisory Board and a consultant to Mistik Management Ltd. in Meadow Lake, Saskatchewan, Canada; consultant to S.D. Warren Company; senior consultant for Interforest, LLC; president of CONNWOOD, Inc.; and a member of the National Research Committee.

Bruce Larson's expertise in forest management and silvicultural issues is presented in several important forestry academic books which he has coauthored. They include *The Ecology and Silviculture of Mixed-Species Forests*, published in 1992 with M. Kelty and C.D. Oliver, *Forest Stand Dynamics*, published in 1996 with C.D. Oliver, and *The Practice of Silviculture: Applied Forest Ecology (9th ed.)* published in 1997 with D. Smith, M. Kelty, and P.M.S. Ashton. Additionally, he has published numerous reports and articles in scientific journals. His field experience has been applied in the United States in the Northeast and the Pacific Northwest, British Columbia, and Iceland.

Before becoming a faculty member at Yale, Bruce taught at the University of Washington and Duke University. His B.S. in biology was earned *magna cum laude* from Harvard University. He earned his M.S. in forestry science from Yale School of Forestry and Environmental Studies and his Ph.D. from the University of Washington.

John C. Gordon is currently a Pinchot professor of forestry at Yale's School of Forestry and Environmental Studies, a position he has held since 1990. He has twice been the dean at the Yale School of Forestry and Environmental Studies (from 1983 to 1992 and from 1997 to 1999) and was acting director for the Yale Institute for Biospheric Studies. Before joining Yale, Gordon was a professor of forest science for more than five years before becoming head of the department at Oregon State University. Prior to that, he was professor of forestry at Iowa State University's department of forestry (College of Agriculture). In addition to his experience in academia, Gordon has worked as a plant physiologist for the United States Department of Agriculture, and as a consultant to business and not-for-profit sectors. He collaborated with overseas development initiatives, including USAID projects in India and Pakistan, where he led field teams and designed research and educational projects. Gordon has also worked in China, Australia, Brazil, Argentina, and Costa Rica. He was a member of the first official United States delegation on forestry to China in 1980, and in 1985 led tours given by U.S. foresters. Gordon was a member of the team in Costa Rica that made recommendations for restructuring CATIE, a key training center for agriculture and forestry for Central and South American countries.

His association with forest certification began in its early days while serving as chairman of the Workshop on Environmental Criteria and Indicators for Sustainable Development of Boreal and Temperate Forests, the precursor to the Montreal Process.

John Gordon's list of honors is extensive: he serves on committees and boards of renowned institutions, and is in demand as a lecturer and guest at national and international meetings. His contributions to scientific knowledge include being the author or coauthor of 29 books or chapters and 103 articles in scientific journals, and reports for government institutions.

Daniel J. Vogt is a lecturer on soils and ecosystem ecology and an associate research scientist at the Yale School of Forestry and Environmental Studies. His expertise is in the different aspects of the properties and processes of soils related to ecosystem ecology, sustainable forestry, and biological conservation. He has been involved in research in Belize, Brazil, Iceland, Malaysia, and Mexico, and his experience includes work in Alaska, Connecticut, Hawaii, New Hampshire, New Mexico, New York, Oregon, Vermont, and Washington.

Daniel's research is driven by his desire to understand how natural and human-derived disturbances affect soil quality and health, and sustainability of ecosystems. He has been funded by NSF, USDA (USFS and NRCS), and other foundations and NGOs. He is a coinvestigator of the Tropical Long-Term Ecological Research Project (LTER) conducted at the Luquillo Experimental Forest in Puerto Rico. His most recent research projects, concerning the global climate change programs of both the U.S. Forest Service and the Natural Resources Conservation Service, were developed to enhance sequestration of soil carbon and to model soil carbon relative to different types of forests and soils in New England. He has presented research at more than 50 national and international conferences and workshops and has published more than 30 peer-reviewed articles in scientific journals and books.

In 1997, Daniel coauthored *Ecosystems: Balancing Science with Management* with Dr. Kristiina Vogt. His academic achievements include a B.S. in biology, an M.S. in Agronomy from New Mexico State University, Las Cruces, and a Ph.D. in forestry from the University of Washington.

Anna Fanzeres is a candidate for a doctorate in forestry and environmental studies from Yale, where she also earned her M.S. in environmental studies. She received her B.S. in forest engineering from the Brazilian Rural Federal University, Rio de Janeiro. Anna's professional experience includes applied forest ecology research in the Atlantic Coastal Forest and Amazonia, and technical support to nongovernmental organizations. Anna has been heavily involved in the issues of forest certification, participating in many debates, and in the development of criteria and indicators for sustainable forest management. Her dissertation focuses on linking natural and social sciences during the evaluation of certified forest-management initiatives. Anna was instrumental in the development of a forestry certification class at Yale's School of Forestry and Environmental Studies and in the subsequent publication of this book.

Contributors

Alatorre, Enrique Consejo Civil Mexicano para la Silvicultura Sostenible A.C. Miguel Angel de Quevedo, # 103 Chimalistac, 01070 Mexico D.F.

Allen, Glenn School of Forestry and Environmental Studies, Yale University. allen4@compuserve.com

Asbjornsen, Heidi Associate Professor, Agricultural University of Norway, Department of Forest Science. heidi.asjbornsen@isf.nlh.no, heidi.asbjornsen@yale.edu

Beard, Karen H. School of Forestry and Environmental Studies, Yale University. karen.beard@yale.edu

Berry, Joyce K. Department of Forest Sciences, Colorado State University, Fort Collins, Colorado. joyceb@CNR.colostate.edu

Booth, Michael School of Forestry and Environmental Studies, Yale University. michael.booth@yale.edu

Brownlee, Allyson School of Forestry and Environmental Studies, Yale University. barbara.brownlee@yale.edu

Camara, Luisa Cabrales Forestry, University of Massachusetts, Amherst, Massachusetts. camaracabrales@forwild.umass.edu.

Cuadrado, Eva School of Forestry and Environmental Studies, Yale University. eva.cuadrado@yale.edu

Estey, J. Scott Jaakko Poyry Consulting, 580 White Plains Rd, Tarrytown, New York jse@poyryusa.com

Fanzeres, Anna School of Forestry and Environmental Studies, Yale University. anna.fanzeres@yale.edu

Finkral, Alex School of Forestry and Environmental Studies, Yale University. alex.finkral@yale.edu

Furnas, Brett School of Forestry and Environmental Studies, Yale University. brett.furnas@yale.edu

Gordon, John C. School of Forestry and Environmental Studies, Yale University. john.gordon@yale.edu

Griscom, Bronson W. School of Forestry and Environmental Studies, Yale University. bronson.griscom@yale.edu

Heintz, Jennifer School of Forestry and Environmental Studies, Yale University.
jennifer.heintz@yale.edu

Hiegel, Andrew School of Forestry and Environmental Studies, Yale University.
andrew.hiegel@yale.edu

Koteen, Laurie E. School of Forestry and Environmental Studies, Yale University.
laurie.koteen@yale.edu

Kretser, Heidi School of Forestry and Environmental Studies, Yale University.
heidi.kretser@yale.edu

Kusuma, Indah School of Forestry and Environmental Studies, Yale University.
indah.kusuma@yale.edu

Larson, Bruce C. School of Forestry and Environmental Studies, Yale University.
bruce.larson@yale.edu

Lawrence, Jessica School of Forestry and Environmental Studies, Yale University.
jessica.lawrence@yale.edu

Maxwell, Keely School of Forestry and Environmental Studies, Yale University.
keely.maxwell@yale.edu

Meyerson, Frederick A.B. School of Forestry and Environmental Studies, Yale University.
frederick.meyerson@yale.edu

Meyerson, Laura A. School of Forestry and Environmental Studies, Yale University.
laura.ahearn@yale.edu

O'Hara, Jennifer L. School of Forestry and Environmental Studies, Yale University.
jennifer.ohara@yale.edu

Paiz, Marie-Claire 2a Avenida 0-43, Zona 3 Colo, Guatemala.
marie-claire.paiz@yale.edu

Palmiotto, Peter A. Biology Department, Dartmouth College, New Hampshire and Paul
Smiths College, New York. peter.palmiotto@yale.edu; peter.palmiotto@dartmouth.edu

Parry, Brooke A. School of Forestry and Environmental Studies, Yale University.
brooke.parry@yale.edu

Patel-Weynand, Toral School of Forestry and Environmental Studies, Yale University.
toral.patel-weynand@yale.edu

Potts, Christie School of Forestry and Environmental Studies, Yale University.
christie.potts@yale.edu

Rod, Brian Taos Land Trust, 307B Paseo del Pueblo Sur, Taos, New Mexico 87571.
brian-rod@hotmail.com

Rojas, Manrique Araya Independent natural-resource management consultant. P.O. Box 72-4400, Ciudad Quesada, Costa Rica. manrique@sol.racsa.co.cr

Sigurdardottir, Ragnhildur Iceland Forest Research Station, 1-270 Mosfellsbaer, Iceland. raga@nett.is.

Smith, David M. School of Forestry and Environmental Studies, Yale University.

Smith, William H. School of Forestry and Environmental Studies, Yale University. william.smith@yale.edu

Taggart, Joe Weyerhaeuser Company, Springfied, Oregon. joetaggart@aol.com

Vogt, Daniel J. School of Forestry and Environmental Studies, Yale University. daniel.vogt@yale.edu

Vogt, Kristiina A. School of Forestry and Environmental Studies, Yale University. kristiina.vogt@yale.edu

Contents

1

Issues in Forest Certification

Kristiina A. Vogt, Bruce C. Larson, Daniel J. Vogt, John C. Gordon, Anna Fanzeres, Jennifer L. O'Hara, Peter A. Palmiotto

Despite the existence of laws governing forestlands, management of forest ecosystems is far from adequate. Certification of forestlands and forest products has been proposed as a means to improve management of these resources. Since its inception in the late 1980s, certification has become a central focus of forest management. It is also being considered by national and international agencies as a solution to many other environmental problems (e.g., global warming and carbon sequestration in forests; Apps and Kurz 1991, NRC 1998), and for achieving the goals of sustainable development (see 2.1, 2.2). Because it is in the early stages of development, the science behind forest certification has not been evaluated in order to determine whether forests are being assessed for the many benefits they provide. It is critical at this time to examine the strengths and weaknesses of certification approaches to ensure that the "emperor without clothes" analogy does not come back to haunt a very useful tool that can promote sustainable forest management. The human values integral to certification assessments must be recognized, and the consequences of focusing exclusively on them must be made transparent. If these values are not recognized, there is a strong likelihood that forest certification protocols will be ineffective in assessing the long-term maintenance of forest ecosystems.

Forest certification is a recent trend that has had an increasing market impact and has created growing interest among forestland owners and managers committed to good forest stewardship. Environmental groups and the public have continued to push forest management to meet multiple objectives, including watershed and water-quality protection, preserving scenic beauty, and providing recreational opportunities (Heissenbuttel et al. 1995). Certification has also been influenced by efforts to:

- Improve forest practices
- Make forest management more environmentally and socially responsible
- Develop and incorporate ecosystem management in national and state forest policies and private forestry

Certification has also been influenced by consumer and political pressure to promote "sustainable forest management" thereby attaining sustainable development (see 3.1.2, 4.3.5.3). However, few policymakers or citizens have thought specifically about the meaning of "sustainability," the underlying idea for sustainable forest management. Sustainability is more a set of ideals than a concrete, measurable concept. Therefore, many forest certification approaches require "sustainable forest management" but are vague and, in some cases, even misleading about the ramifications of pursuing this goal (see 1.2, 3.1.2).

1.1   Improvements crucial to successful implementation of forest certification

Two tenets exist in the forest certification debate: 1) integration of social, ecological and economic factors, and 2) the sustainability of forestlands can be attained if the previous integration has been successful.

> By focusing the discussions of forest certification on the identification and definition of criteria and indicators that translate the values behind social, economic, and ecological parameters, the central focus of supporters and promoters of certification became the development of a long laundry list of objectives to be fulfilled.

There are, however, many issues that have not been adequately analyzed in certification that will determine the effectiveness of criteria and indicators and how readily they can be generalized to assess sustainability. It is understandable that forest certification has reached a point where there is a strong need to reevaluate the approaches. For example, certification is quite young as an assessment approach, with the earliest certifications of forestlands being traced to the late 1980s (see 2.1, 2.5). Because of the eagerness of forest certification promoters to establish as many certified forests or initiatives on the ground as possible, the initial reasons for pursuing certification were lost. Certification became a goal in and of itself, rather than a consequence of adequate forest management practices. Forest certifiers had an urgent need to develop checklists that could be used to assess all types of forests: tropical, temperate, and boreal; and plantations of exotics or native tree species worldwide. This approach was contradictory to the initial goals that drove the forest certification debate, where the need had been identified to develop standards more applicable to the local level (FSC 1996, FSC 1999a). If the initial goals of certification had been followed, perhaps the parameters proposed would not have been so value oriented and would have been more indicative of the responses of natural systems to human interventions.

Several topics have not received serious analysis as part of the ongoing discussions about certification. The purpose of this book is to perform a scientific analysis of the different elements of certification in order to evaluate what works well and what needs to be adapted to make certification more effective.

The following topics are presented and evaluated in this book:

IMPLEMENTATION
- An analysis of certification as one of the suite of tools to be used, without forcing forestland owners to perform functions that may be the purview of state, federal or international-level organizations (Chapter 2)
- How to integrate social and natural science legacies that continue to control the response of a system to any disturbance or management activity (Chapter 5)
- How to develop approaches that can distinguish when value-based and non-value-based indicators are most appropriate for assessing a site (Chapters 3, 5, 7)

DEFINITIONS
- The differences between "good forest management," "sustainability," and "sustainable development" (Chapter 3)

DEVELOPMENT OF CERTIFICATION
- A comparison of the different approaches of certification protocols, and their strengths and weaknesses (Chapters 2, 4)

ECOLOGICAL AND SILVICULTURAL FOUNDATIONS
- How to more specifically link the management activities and their impacts on the environment by identifying indicators that best reflect the sensitive driving variables that control how resistant or resilient a forested system is (Chapter 6)
- Understanding the minimum information needs of an assessment, and how to weight the importance of individual data needed to assess a site (Chapter 5)
- The determination of what forest condition should be used to assess a system as "sustainable" (Chapters 3, 4, 7)

SOCIAL INPUT
- The amount and type of participation by the public in managing forestlands when they typically have little training in managing forest resources (Chapters 3, 7)
- Identification of the elements of an assessment protocol that are not being driven by human values; recognizing that human values may drive the use of good forest management practices, but may also drive the use of management practices that degrade ecosystems (Chapters 2, 3, 6, 7)

EFFECTIVENESS
- A scientific analysis of the structure and effectiveness of different certification protocols in evaluating different sites (Chapter 4)
- The ability to balance the values of the producers and their property rights with the values of consumers and society (Chapters 2, 3, 7)
- Tools to effectively integrate the social and natural science aspects of sustainability in certification (Chapters 3, 4, 5, 7)

STANDARDIZATION
- The determination of whether there is a need for the development of a uniform set of standards (criteria and indicators) for all sites in order for certification to be credible, and how this deals with natural ecosystems that have different constraints and driving variables controlling their resilience (Chapter 5)

BARRIERS
- Factors which limit the acceptance of certification as a broad scale approach, or are impediments to implementing certification (Chapters 2, 5, 7)
- The determination of what limits and how not to exclude small, non-industrial forestland owners from being certified (Chapters 2, 4, 7)

Several issues recur in the list above. Some of these points are serious impediments to certification because they result from the different values driving the development of the structure being used in the assessments. For example, many organizations promoting certification and assessing forestlands are embedding their values for a particular standard forest condition into their assessments, which are reflected in their choices of criteria and indicators (see Chapters 2, 3, 4 and 5).

Many of the solutions being generated by certifying organizations may require changes in the social and political institutions that manage forestlands without requiring changes in societal values. History has shown us that societal values will continue to change with time (Vogt et al. 1997a, see 2.1). Therefore, protocols should be structured to be adaptable to changing values, since what is currently acceptable may not be in ten years.

This is especially relevant, since many of the changes occurring in forest management are in response to society changing its values on natural-resource uses, and not in response to developments in the science of managing forest resources (see Chapter 6). This requires the development of flexible systems that are sensitive to new information, new understanding of our systems, and new values generated by the suppliers and consumers of natural-resource products.

Superimposed on the changing societal values, with little change happening in management practices, are the values being brought about by the organizations involved with certification (see 1.2, 3.2.2.2). Nongovernmental Organizations (NGOs) (i.e., environmental, socially oriented, industry representatives, feminist groups, First Nations, etc.) have gained influence in the last decade by having their values and perceptions of the environment included in the design of public policies. Despite disagreement on how much power they wield (Arts 1998, MacDonald 1998), the expression of societal values through these organizations has been a very effective way for individuals to have their values heard and for them to become reality (Tuan 1974). These organizations are extremely important for society as long as they continue to represent society's changing values and are flexible. They must also be aware that they represent just one of the many values represented in any society. If certification protocols are designed to reflect a narrow set of values, they will probably be effective tools for assessing good forest management and sustainability of social and natural systems. It is also important that every certification protocol be able to explicitly separate its desired values from those which naturally constrain how a system functions, and ultimately determine its sustainability (Vogt et al. 1999ab).

Another very serious problem is the failure, or lack of willingness, of most forest certification protocols to look at patterns and processes at the landscape level. This type of information would be invaluable as a mechanism for weighting the importance and types of data needed, and it is necessary to link the proposed management activities to the adequate scale or level of impact. The zones of influence relative to each type of activity must be taken into consideration if there is a desire to detect whether a system is changing in a negative manner. For example, the present focus of existing protocols on the need to maintain biodiversity fails to consider whether the abundance or absence of certain organisms is driving the regeneration of desired species (see 3.2.1). The choice of criteria and indicators must be based on the potential impact of management activities (see Chapter 6), and on external activities which might be influencing management. Another good example, showing the lack of a link between human values and those factors controlling ecosystem function, is hunting. For example, the elimination, or favoring of, certain game species can feedback to affect vegetation structure (i.e., fauna roles in browsing, dispersing, etc.). These feedbacks have to be taken into consideration when evaluating the responses of the system to further human interventions.

An inclusive approach is important if certification is to impact forestry practices around the world. If there is participation in certification only by very large industrial forest enterprises or small, non-industrial forestland owners, certification's ability to change forest management practices and to satisfy societal values will be minimal. How much each of these sectors participates in certification, of course, varies from country to country. In the United States, the importance of the small, non-industrial forestland owners is highlighted because of the significant amount of forestland they own. In 1992, nonfederal forestland comprised 66.2% of the total forest area in the United States (NRC 1998). These same lands were also where most of the timber harvesting was occurring in the United States (NRC 1998). The potential impact of forest certification could be quite large if these owners became certified, because of their size and economic importance in the forest products industry (see 7.4). The United States accounts for 25.2% of the world roundwood production (WRI 1994) and 25% of the world's solid wood products (Zinkhan 1992). In the United States, 6% of the gross national product comes from the forest products industry (Zinkhan 1992).

However, of the 66.2% of the forestland in nonfederal ownership in the United States, probably only a small fraction can satisfy the requirements for being certified (see 2.5, 7.4, 7.5). Certification must consider and attempt to incorporate the small landowners who, at least in the United States, will be an important source of timber in the future (Birch 1996). These small landowners have great difficulty in being certified or even participating in the process (see 7.4, 7.5) because of several limitations. These private nonindustrial forestlands contribute significantly to many global-scale environmental values and services. For example, nonindustrial private timberlands have been estimated to annually sequester 61 million metric tons of carbon in the woody plant species found on these lands (NRC 1998). These lands have great potential to increase the amount of carbon that can be sequestered in the United States.

Another major issue with certification is related to the types and quantity of data that are expected to be collected during an assessment; these strongly limit the ability of many forestland owners to pursue certification, because they cannot satisfy the data required to verify that they are managing sustainability, or are using good forest management. An approach must be developed that identifies which indicators are relevant in assessing forest management practices. Most indicators appear to be laundry lists of data that are not directly related to how an ecosystem functions. It is obvious that a small landowner will not have the required information needed to satisfy a certification assessment. This high need for so many indicators reflects the desire to satisfy several different values from a given unit of land. However, the requirement for so much data makes the process very subjective due to the lack of a system to weight the importance of different indicators. Presently, no framework exists that allows the values of the certifying organization to be selectively isolated from the constraints that determine how social and natural systems function.

Another major weakness of most of the protocols is that they do not present supporting data for the selected criteria or indicators chosen for the assessments of forest management operations. In addition, there are no guidelines for determining when a particular activity will cause the system to change or when a threshold of degradation may be crossed. These protocols have not really linked the specific impacts of management activities at the ecosystem level nor used them to assess a management unit (see Chapter 6).

Certification will have to consistently evaluate a site in the same manner if it is to be accepted as an assessment tool. This will require a certifier to be able to weight the importance of different indicators and to determine when a change in an indicator really means that a system has the potential to degrade. Assessments also need to be expanded to include potential legacies that exist for each site, and the spatial matrix within which the management unit is embedded (see Chapter 5). These legacies can be understood as cyclical temporal and spatial dynamics of an ecosystem in relation to natural disturbances and socio-economic institutions (Vogt et al. 1999ab, see Chapters 3 and 5).

1.2 Why certification is relevant

There are many reasons why certification is such a hot topic of conversation in the natural resource community. Timber certification is being described by some as a market-driven solution to social value-driven problems. For industrial and non-industrial forestland owners, it is perceived as an effective tool for producers of wood products to demonstrate, by using a label, that their forests are being managed in a sustainable manner. Since certification is voluntary, forestland owners can decide whether they are interested in pursuing this process and in meeting its criteria. The reasons forestland owners decide to become certified or to pursue certification are quite varied and will be discussed below.

Several factors are driving the use of certification as a tool in natural resource management. Some of the factors and aims of forest certification are to:

- Control resource management techniques
- Control resources economically
- Alleviate poverty
- Create a system that assures the public that environmental concerns and values have been addressed
- Manage resources holistically so that healthy environments are maintained
- Diminish the amount of regulation that is being imposed on a forestland owner
- Balance the need to extract resources from the environment while maintaining sustainable ecosystems
- Control the values of private forestland owners, or for private forestland owners to maintain their values in the face of society's drive to impose its values on them

Several rationales can be stated as to why certification has become a method of characterizing the sustainability of natural resource management. Before examining these rationales, it is worthwhile to consider the word *sustainability,* because of the problems it has created for certification assessments (see 3.1.2). This term was used during the late 1980s, when the tools for certification were first being developed. Sustainability was the forest condition for owners to aim for when certifying their lands (Crossley 1996, Granholm et al. 1996, Upton and Bass 1996, Viana et al. 1996). Distinction should be made between the original definition of this term within forestry (i.e., related to constant yields of timber), to the more recent paradigm shift which strives for a balance between socio-economic and ecological variables (see 3.1.3).

Presently, the term sustainability is problematic because of the inability of academic and non-academic communities to consistently define it (see 3.1.2). No consensus exists on the definition of sustainability, except at a very general level (Wijewardana et al. 1997). This level, however, has not been useful in developing criteria and indicators in certification protocols. The inability to build consensus on the definition of sustainability has resulted in the word providing no guidance in identifying which indicators should be used in its assessment. Because of this, several certifying organizations are eliminating the use of this term in order to avoid its uncertainties. Currently, it is common to hear the term "good forest management" or "good forest stewardship" as the forest condition expected with certification (Mankin 1998). However, in this book, the word "sustainability" will be used because it defines the values associated with managers and certifiers (see 3.1.2).

> However, spending too much time and energy in defining sustainability is not beneficial. It distracts developers of certification protocols from addressing other problems which are more significant, and can actually impede the usefulness of certification (see Chapters 4 and 7).

In the literature, there is no consensus on what the elements of sustainability are, but private forestland owners have specific elements they value. These values are expressed when forestland owners select a certification organization to certify their lands. For example, forestland owners who value strict conservation are more apt to select a certifier that reflects this approach, while commercial forestland owners are more likely to choose certifiers that understand industrial concerns.

The initial interest in certification was driven by concerns over the non-sustainable uses of tropical forests and the reported losses of species from these forests (see 2.1). Much of this interest was fueled by reports of excessive deforestation rates in the tropics. In 1983, the first global forest assessment reported that tropical moist forests were being cleared at an unprecedented rate of over 11 million hectares per year (FAO 1983). Environmental groups initially attempted to reduce the rate of deforestation of tropical forests by boycotting the international tropical timber trade (Ozane and Smith 1993, Cabarle 1994b, Brockmann 1996, Viana et al. 1996). However, these boycotts were not very effective in changing the deforestation rates. In fact, it has been suggested that these boycotts further devalued already undervalued tropical forest resources, thereby increasing the rate of conversion of these forests (Vincent 1990, Johnson and Cabarle 1993). This suggested that a different approach was needed. The concept of timber certification was introduced to initially deal with the problem in the tropics, but it was later expanded to include all types of climatic forests — boreal, temperate, and tropical, as well as plantations (see 2.1).

The impact of deforestation on the sustainability of tropical forests was an important factor pushing the development of certification by environmentally focused organizations. This is apparent from an examination of the structure and data needs of certification protocols whose development has been strongly influenced by these organizations (see Chapter 4). These protocols reflect the present ethical and philosophical beliefs of society, and include those factors which are believed to be necessary to achieve sustainability. These protocols reflect the worldwide concern about deforestation and the losses of species in areas with high biological diversity (Guruswamy and McNeely 1998). All these factors are strongly reflected in protocols used by organizations under the umbrella of the Forest Stewardship Council (see 3.2).

Combined with these natural resource issues were concerns for the human rights of indigenous groups, and the high poverty levels of people living close to forestlands, especially in the tropics. In response, forest certification also became a vehicle to ensure land security and to alleviate the poverty of communities dependent on these ecosystems for their livelihoods. The criteria and indicators used in forest certification have been chosen to address the need to respect legislation and human rights, to recognize the utilitarian role of nature, to promote alternatives for the continuous consumption of forest products, and to obtain economic benefits (see Box 1.1).

Box 1.1
Categories of value-laden parameters of forest certification protocols

Parameters with emphasis on rules, legislation, and documentation

This category of data deals with destructive forest-use practices due to non-fulfillment of existing legislation, and/or rules designed locally to ensure the maintenance of the forest cover. These groups of parameters require a strong stand on what would be considered an alternative adequate model for using forest resources (environmentally and socially). Due to an inability to trust some claims from the timber industry, there was a need to require written documentation to justify all claims. This paperwork would allow for independent verification of all claims.

Parameters to ensure the protection of biodiversity, soil, and water

This category recognizes the existence of other forest values besides timber. It also deals with the new paradigm of using natural resources in the present, while also allowing future generations to enjoy the same benefits (and those benefits not yet recognized). This group of parameters adds terms to the certification vocabulary that are considered relevant for the evaluation of forest sustainability.

Parameters for empowering the poor and/or less-favored, labor rights, and other social benefits

This category formalizes the new paradigm elaborated in the *Brutland Report*. It is thought to incorporate concerns about existing inequities in the access and distribution of benefits of development to those segments of society contributing most to the production of goods. In addition, threats to the biological and cultural survivorship of indigenous groups and traditional communities dependent on forests and their resources were to be addressed under this topic. Moreover, concerns with slave and child labor had to be addressed, as identified by the International Labor Organization (ILO).

Parameters for ensuring financial profits and economic returns

This category deals with profits from the exploitation and commercialization of forest resources. For this group of parameters, certification initiatives should involve the business sectors, and should introduce language supporting financial profits and the measurement of sustainability though traditional mechanisms (i.e., constant volume of timber).

However, these parameters, proposed as measures to curtail destructive practices to maintain biodiversity and maintain the well-being of local populations, do not address the mechanistic links between socio-economic and ecological parameters. It is important to recognize that the values for species and healthy social and natural ecosystems appear to have driven the structure of existing forest certification protocols and are not based on objective data needs.

The point is not that human value-laden parameters are wrong or useless in this process, but rather to understand when they are, and when they are not, relevant to include in forest certification. There is a need to evaluate whether all parameters in a protocol of forest certification have to be included in an assessment of sustainable forest management (see 3.1.3, 4.2). This current approach in forest certification has been useful in identifying all the components integral to an ecosystem, but has not been very effective in determining how to weight the importance of the different information (see 5.1, 7.2). An approach that strives for uniformity in standards is not effective in (see 5.2):

- Analyzing and incorporating the heterogeneity that exists in ecosystems
- Assessing the legacies that may change how a system responds to a perturbation
- Determining how it recovers from management activities

Others see certification as a tool to improve forest-harvesting practices and forest management around the world. Their philosophy is that a certification label will result in increased market share for a private forestland owner. Their higher financial return should compensate for the increased costs incurred while improving harvesting practices and, at the same time, increase the environmental services obtained from managed forests (Brockmann 1996, Viana et al. 1996). The certification label is useful only if it results in a more-informed consumer who will selectively buy those forest products that reflect his environmental values. For the producer, the ability to obtain a higher price, or to capture a market niche, should increase the financial returns from the marketplace. If this were to occur, forestland owners would definitely pursue certification.

For some, the need for certification has arisen because the distribution and marketing of forest products is so broad that the direct link between managing resources and the market has been lost. When consumers buy products, they have no way of knowing the quality of the natural resource management. The conflicts in forest harvesting methods and the public perceptions of abuses by the industry have resulted in a loss of trust between the public and the industry involved in managing and marketing forest products (see 3.2.2). Most of the certifiers see certification as a vehicle for decreasing conflicts concerning forest uses, and as a means of rebuilding the trust between the public and the forest industry (see 2.1, 2.2). This approach also assumes that only an independent third party can assess with any credibility the sustainability of natural resource management (see 2.3).

For some, certification is seen as a tool for easing the social friction between the public and the commercial forest industry (see 2.1). It is perceived that the certification label will allow the public to feel that their environmental values have been integrated into commercial forestry practices (Mater 1997). It is suggested that third-party certification systems might broker a new "social contract" between the suppliers of wood products and the consumers who buy them (Mater 1997).

Many of the differences between existing certification approaches occur at the level of what the standard forest condition should be on the ground, and how many other societal values should be obtained from these "working" forests (see 3.1). A number of the accredited certifying organizations, under the umbrella of the Forest Stewardship Council, have added many values they would like to achieve from privately held lands (see 3.2.2.2, 4.3). This places additional demands on private forestland owners and explains some of the difficulties small forestland owners have in pursuing certification (see 3.1, 7.4). The type of social agreement or contract that would allow forestland owners to manage their land in a productive and economical way, while still providing the public with the confidence that resources are being sustainably managed, is still an ongoing discussion.

Other certification protocols attempt to ensure that the values held by private forestland owners are not lost in the process of satisfying societal needs for a third-party assessment of forest management. Protocols, such as the Sustainable Forestry Initiative of the American Forests & Paper Association, reflect the values held by for-profit commercial entities and others who manage private forestlands. The primary goal of certification for "working" forestlands is to be able to extract products from forests while maintaining them in healthy conditions into the future (see 2.1, 2.2). When certification is pursued for working forestlands, its primary goal becomes the pursuit of good forest management on the ground, and the continued sustainable extraction of wood products (see 3.1.2).

Forestland owners may also pursue certification as a mechanism to decrease the amount of regulation they must satisfy (see 5.3.2, 5.3.3). Certification is perceived as a model that would replace the need to apply additional layers of strict regulations on forestland owners. For some consumers, the certification label would be sufficient to give them confidence that forests are being managed in a socially adequate and environmentally friendly manner. The idea behind this approach is that the imposition of more regulations on forestland owners would not result in a better-informed consumer. Past regulations have not produced an informed consumer, because the diversity of regulations has been difficult for a consumer to synthesize and evaluate whether forest sustainability has been achieved on the ground (see 3.1.2, 3.1.3).

Certification is also being provided as a tool that will more effectively stimulate forestland owners to change their field management practices. The idea is that people are more likely to be motivated to change their behavior (e.g., their field practices) by using the "carrot" instead of the "stick" approach. The carrot approach would work if certification encouraged people to pay a higher price for wood products or increased the producers' ability to obtain a higher share of the marketplace, thereby increasing producer profits. The carrot approach uses an economic incentive (economic sustainability) to obtain ecological sustainability and socio-political sustainability. Higher profits would be the carrot that would induce forestland owners to modify their field practices. If wood prices were high enough, many practical improvements could occur in the management of forest resources that would satisfy the human values associated with these resources (however, see 7.9). For example, some management practices driven by economic factors that could be easily modified are:

- Longer rotations
- More coarse woody debris left after a forest harvest
- Better management of riparian habitats to improve wildlife habitat on commercial timberlands

It is too early to know whether people will be willing to pay a price premium for certified wood or how much of the wood-fiber needs around the world will be satisfied from certified wood sources (see 7.4, 7.6). Since the beginning of the forest certification debate, there has been the assumption that public outrage over environmental problems was powerful enough to produce a willingness for them to pay premium prices for wood products harvested from sustainable sources (Cabarle et al. 1995, Adamowicz et al. 1996, Brockmann 1996, Jenkins 1997). More recently, however, questions have been raised regarding whether consumers really are willing to pay a premium price for sustainably produced goods (see 2.1, 7.6). According to Drescher (1998), although consumers want standards to be imposed on how forest management is conducted, they do not want prices for wood products to increase (see 7.6, 7.7). If this is correct, certification's main leverage in the marketplace will be its ability to determine where people buy their goods. There is concern that the public will become confused by the high number of labels and certification initiatives. There is the possibility that leverage in the marketplace may not materialize if the increased number of labels overwhelms the public, especially if the differences between labels are unclear (see 2.3).

In addition to the past reasons for pursuing certification, new uses are being developed. Certification is beginning to be recognized as an important tool to assess the trade-offs of global-level impacts of human activities. For example, certification is thought of as a tool to assess projects attempting to reverse large-scale impacts of humans on the global climate by sequestering carbon in forests (NRC 1998). This is an initiative to motivate industrial companies producing and releasing CO_2 into the atmosphere to balance their industrial outputs of C by sequestering an equivalent amount of C in forest biomass. Certification is the tool to ensure that these goals could be satisfied at the ground level, since companies would be allowed to continue to emit CO_2 at the industrial-plant level.

2

Roots of Forest Certification: Its Developmental History, Types of Approaches, and Statistics

Anna Fanzeres, Kristiina A. Vogt

2.1 Origins of the concept of forest certification

To establish a basis for the ongoing debate on forest certification, it is important to review some of the historical factors related to the growth of public environmental awareness. In order to make sense of the tremendous amount of information on this topic, the material will not be discussed in a continuous chronological sequence. It is hoped that the format used in this analysis will provide a more useful understanding of the issues relevant to this discussion. This analysis should be useful to those individuals who are already familiar with forest certification and for those who have more recently been confronted with this topic.

Forest certification is a remarkable social, economic, and historical phenomenon. Within a relatively short time, forest certification has solidified its place as an integral tool for addressing a diversity of forestry issues and has generated considerable controversy and debate as to its role in forestry. In fact, the fast pace at which it has become part of the dialogue and an integral part of numerous initiatives dealing with forests has been overwhelming and confusing at the same time (Wijewardana et al. 1997, Bruce 1998). The combination of media attention, public pressure, and market forces has created a web of interacting influences that can be quite complex. However, the goal of certification is simple: its purpose is to ensure that forests are managed in accordance with a set of standards considered environmentally appropriate, socially beneficial, and economically viable (Cabarle 1994, SGS 1994, Upton and Bass 1996, Viana et al. 1996, Elliott 1997). Certification recognizes that these values have been achieved in a forest by presenting a seal of approval that can be recognized by the public.

The concept of sustainable forest management can be traced back to the Middle Ages in Germany and France (see 3.1.3). Not all forestlands were set aside to be managed (Winters 1974, Shabecoff 1993, Jordan 1995), however, many forests were managed for specific products (e.g., wildlife, water, etc.). Besides the primary need to ensure the supply of timber, the practice of forest management was driven by a need to guarantee the availability of other resources (e.g., water) associated with, or originating in, forests (Hays 1959, Gottlieb 1993, Grove 1997). In some cases, concern for game availability or the desire for exclusive hunting grounds determined that forested areas were not converted to agriculture or other uses (Winters 1974, Gilbert 1979, McCormick 1989). Forests were also historically used for supplying firewood, timber, and for raising domestic animals; all activities depended on being granted permission from kings, nobility, or locally ruling religious leaders (Gilbert 1979, Linnard 1982, Neeson 1991, Elliott 1996).

Recently, the traditional paradigm of sustainable forest management has been questioned (Cabarle 1994, Colfer 1995, Elliott 1996, Evans 1996, Gordon 1996, Granholm 1996, Merino 1996, Toman and Ashton 1996, Prabhu et al. 1996, Wijewardana et al. 1997, Barthod 1998). These past approaches to sustainable forest management, which focused exclusively on assuring constant yields, did not help to reduce environmental degradation (Winters 1974, McCormick 1989, Wilman 1990, Ludwig 1993). The new trend in forest management explicitly links forest ecosystem health and the achievement of sustainable development objectives (Waring 1985, Norton 1991, Constanza et al. 1992, Hammond 1992, Haskell et al. 1992, Kessler 1992, Norton 1992, Schaeffer 1992, Belsky 1995, Margules 1996, O'Laughlin 1996, Smith 1996, Drengson and Taylor 1997, Vogt et al. 1997). Scientists and managers are trying to move beyond ensuring constant yields of timber, or any other forest product of interest, to including other values held by the public (see 3.2.2). One factor contributing to this changed management strategy appears to be society's perception that the availability of forestlands, forest resources, and species might be threatened by increased deforestation rates worldwide.

The use of forest certification as a potential tool to deal with deforestation rates can be traced back to the late 1980s. However, the 1992 United Nations Conference on Environment and Development (UNCED) can be identified as the watershed event (Princen and Finger 1994) for the environmental movement as a whole, and the point at which certification became a reality. The United Nations Conference on Environment and Development (UNCED), the Rio Earth Summit, held in June 1992, was the first worldwide attempt to reach consensus on forest issues. Forestry issues became part of the international political agenda with ECO 92, when efforts were increased to determine how forests should be managed worldwide. After UNCED, the first sets of principles, criteria, and indicators were released on how sustainable forest management should be conducted and evaluated for forest certification (see 4.2.1, Heuveldop 1994, Ervin 1996, Crossley 1996, Bruce 1998, SFF 1998).

However, the driving forces behind the development of the idea of certification can be traced to when society started changing its views on how it perceived its environment (especially in developed countries). Some authors suggest that forest certification is a consequence of the growth of environmental awareness in the late 1960s and early 1970s (Granholm 1996, Hansen 1997). A pivotal book forcing the public to recognize the consequences of industrial growth and the harmful legacy it could produce for the present and future generations was Rachel Carson's *Silent Spring* (Carson 1965). For many, this book launched worldwide the concept of an environmental movement (McCormick 1989, Dobson 1990, Gottlieb 1993, Shabecoff 1993, Jordan 1995, MacDonald 1998, Cuomo 1998). Nonetheless, most writers in forest certification prefer to attribute the development of this idea to the late 1980s (Cabarle 1994, SGS 1994, Baharuddin 1995, Brockmann 1996, Lyke 1996, Upton and Bass 1996, Viana et al. 1996, Kiekens 1997, Bruce 1998).

In the 1980s, the high awareness of the loss of tropical forests played a pivotal role in stimulating the development of forest certification as a tool to control deforestation rates. The media attention given to the impact of deforestation and forest fires in the tropics made people aware of what was being potentially lost from these forests. This helped to create awareness of these activities and the need to stop them. For people living in developed countries, deforestation in the tropics was linked to destroying potential cures for diseases, food crops and numerous other valuable products, and the livelihood of native peoples. On one hand, this stimulated an ethical drive by some to preserve nature (and all natural things, including native peoples) because of its intrinsic value (Wilson 1988, Callicott 1989, Burks 1994, Guruswamy and McNeely 1998). For others, the issue was more utilitarian. For those, the quality of life in the developed world would be endangered if this pace of forest loss was allowed to continue (Dobson 1990, Crossley 1996, Guruswamy and McNeely 1998). The public in developed countries was outraged by the high rates of deforestation that were occurring and demanded immediate solutions to this problem (Dobson 1990, Shabecoff 1993). The public perception was that this unacceptable model of land use was taking place because regulatory mechanisms were either non-existent or not enforced. For most people, this problem was recognized to be more acute in developing countries where most tropical forests are located. In developed countries, the loss of forests was also linked to a serious disrespect for the human rights of indigenous groups, which were perceived to exist in a harmonious relationship with nature (Posey and Balée 1989, Posey and Dutfield 1997).

All these viewpoints were important catalysts for environmental groups (more strongly in Europe) to use in their campaign to boycott the purchase of tropical forest products (Ozane and Smith 1993, Brockmann 1996, Viana et al. 1996). In response to these campaigns, the solutions proposed were regulatory in nature (unilaterally or not), and were delivered as policy statements or laws that focused on controlling the trade of imported tropical timber (Box 2.1). Those governments, where society strongly participated or influenced decision making, were quick to respond to the pressures being exerted by their citizens. Their response was to begin proactive measures to deal with these issues. Industrialized, tropical timber-consuming countries issued national or local ordinances to curb the importation of tropical wood altogether, or only accepted those supply sources that were certified as sustainable (Crossley 1996, Viana et al. 1996).

Box 2.1
Some governmental initiatives to respond to public pressure against deforestation

The Muntingh Proposal was launched in 1988. Under this framework, the European Community (EC)-member countries would "only import tropical hardwood products produced under forest management and protection programs, and that such products be certified" (Crossley 1996).

The Austrian government in 1992, responding to growing pressure from local NGOs, passed legislation prohibiting the importation of tropical woods produced unsustainably. Indonesia, the major supplier to Austria at the time, filed a protest with the World Trade Organization (WTO) using the argument that this legislation was a barrier for international trade and an unfair practice. The Austrian government suspended the measure in 1996 (Crossley 1996, Viana et al. 1996).

Dutch Working Group of Experts — Deskundigenwerkgroep Duurzaam Bosbeheer Standards (DDB). This initiative was an effort by the Dutch government to determine what the standards for sustainably managed tropical forests would be. Once these standards were accepted by the Parliament, legislation was to be enacted forbidding any importation of tropical timber (starting in 1995) unless it could be shown to have been harvested from sustainable sources. The legislation was suspended, however, before it could be enacted, due to strong denunciations at the General Agreement for Tariffs and Trade and the European Union (Prabhu et al. 1996, SF&CW 1999).

Initiative Tropenwald (Germany) — In 1993, a group of concerned scientists, some NGOs, and representatives of governmental agencies initiated a process to develop criteria for an evaluation of sustainable management of tropical forests. These guidelines would guide German trade on tropical timber (Heuveldop 1994).

Indonesian Lembaga Ekolabel — This initiative developed in response to growing pressure from timber-buying countries (e.g., Austria). In 1992–93, a well-connected ex-government official, Emil Salim, called for a working group to design the framework for what would be called the Indonesian Ecolabeling Institute, and to develop the standards by which sustainable forest management should be conducted. The Institute opened in 1998 and has already developed a protocol for evaluating production in natural forests. A number of concessions in Indonesia are being tested under these criteria and indicators. This process has received much criticism from environmentalists as a mechanism to ensure that the large concessionaires would continue to have the freedom to conduct their usual practices.

During the time that some governments were attempting to regulate sources of timber from tropical countries, several international governmental organizations also began to be quite active in this arena. In the 1980s, the number of multilateral or bilateral projects began to proliferate, which promoted the preservation, or more efficient utilization, of forest resources. In 1983, the United Nations Food and Agriculture Organization (FAO), through its World Forest Appraisal Programme, expanded its inventory of forests worldwide by systematically gathering quantitative and qualitative data on forests. As part of this program, human impacts on forest conditions became officially incorporated in these international datasheets (FAO 1983). This new approach led to the release of FAO's statistics on deforestation rates. This report generated a worldwide debate on the extent of forestland loss and on tools for implementing sustainable forest management on the ground.

That same year, the FAO report triggered the International Tropical Timber Organization (ITTO) to launch an agreement for its member countries to voluntarily comply with a cooperative system between tropical timber producing and consuming countries (ITTO 1997). As described by Upton and Bass (1996), ITTO "launched the first commodity agreement with political commitments to ensure that all trade in tropical timber would only come from sustainably managed sources by the year 2000." In May 1990, ITTO published the *ITTO Guidelines for the Sustainable Management of Tropical Forests*. This was followed by two other publications: *Guidelines for the Establishment and Sustainable Management of Planted Tropical Forests,* and *Guidelines on the Conservation of Biological Diversity in Tropical Production Forests* (ITTO 1997).

In 1985, the Tropical Forestry Action Plan (TFAP) was promoted by the World Bank as part of a joint effort with the United Nations Development Program (UNDP) and the World Resources Institute of Washington. The main goal of TFAP was to respond quickly and efficiently to the escalating rate of tropical deforestation (WRI and UNDP 1985). The preamble to TFAP expressed their philosophy of a utilitarian purpose for forests (i.e., they protect soil and water resources and provide habitat to half the world's species of plants and animals). This initiative aimed at identifying successful enterprises that were using forest resources sustainably (and also unsuccessful ones to learn from past mistakes), in order to promote worldwide projects that protected tropical forests and to improve the quality of life for local populations.

All these efforts by environmental organizations and governmental agencies were severely criticized by opposing groups. For example, the use of boycotts was denounced as a mechanism for controlling trade (e.g., GATT — General Agreement in Trade and Tariffs) using discriminatory and protectionist measures. Despite the fact that boycotts were voluntary and were being aimed at individual consumers (Carbale 1995), this perception occurred because they were adopted by governmental authorities. Many economists helped to fuel the dissatisfaction with boycotts that was being expressed by traders of tropical timber products, and other consumer countries. They alleged that the lack of market value for rainforests would lead to further and faster conversion of forests to alternative land uses (Carbale 1995, Viana 1994). The Southern-Northern equity trade/resource-use issue was a strong argument that was reinforced by the corroboration that deforestation was also taking place in temperate and boreal regions (Viana et al. 1996). The lack of trustworthy mechanisms to ensure the effectiveness of claims and the obvious self-interests that appeared to be inherent in many of these programs discredited them with nongovernmental organizations (NGOs).

In the case of ITTO and TFAP, the public and a growing number of NGOs shared a deep distrust of claims made by large-scale businesses and Third World governments regarding environmental and human rights issues. Gabus et al. (1993) revealed a weakness of the ITTO scheme: "Once a member country shows efforts towards implementing policies, regulations, and management plans that ensure substantial progress towards the year-2000 target...all tropical timber products of that country will be certified as sustainably produced." Thus, the ITTO scheme almost became irrelevant to the current ongoing debate on forest certification (Crossley 1996, Prabhu et al. 1996). The main reason for this was the lack of trustworthy mechanisms to ensure the effectiveness of claims and the obvious self-interest invested in the process (Crossley 1996, Prabhu et al. 1996). The TFAP also received considerable criticism. NGOs and some funding governments saw that TFAP was, in fact, increasing deforestation rates due to its promotion of logging projects in primary forest areas (Colchester and Lohmann 1990).

All these initiatives and the counter-pressure movements they generated led to the development of an alternative compromise proposition (i.e., forest certification). However, the agenda behind the proposition for certifying forests was still being driven by developed nations attempting to ensure access to forest resources.

To satisfy the goals of forest certification, there was a need to develop evaluation tools that would be credible and widely accepted globally. A set of principles and criteria was needed that could be used to independently assess sustainable forest management (see 3.1.2, 3.1.3, Cabarle 1994, Baharuddin 1995, Prabhu et al. 1996, Upton and Bass 1996, Viana et al. 1996, FAO 1997, Wijewardana et al. 1997). The proposed format used to define principles and criteria of sustainable forest management included ecological, social, and economic parameters. This approach closely tracked the sustainable development concept articulated by the United Nations report, *Our Common Future*, also known as the *Brutland Report* (WCED 1987, Viana 1994). The theoretical perspective for this developmental model was based on the need to address environmental and social problems, while satisfying the need for continued economic growth (McCormick 1989, Dobson 1990, Simonis 1990, Shabecoff 1993, Jacob 1994, Smith 1994, Williams and Haughton 1994, Dahl 1996, Henderson 1996, Doorman 1998).

As previously mentioned, society in developed countries became more aware of environmental problems, starting in the late 1960s. During the same time, social issues were also becoming a growing public concern (McCormick 1989, Dobson 1990, Gottlieb 1993, Shabecoff 1993). The younger generation began to question the status-quo — despite being middle-class, having university degrees, and being the biggest beneficiaries of the social and material security achieved by their parents (McCormick 1989, Dobson 1990, Shabecoff 1993). Moreover, health conditions and *social pollution*, or degradation, was being linked to the state of the environment; although, for many, these latter issues were still considered applicable only to poor, less-developed countries (Goldsmith 1972, Shabecoff 1993, Jamison 1996). At this time, there was a renewed interest in the population debate, which focused on poor and technologically underdeveloped countries (Ehrlich 1968, McCormick 1989). Other social issues also became prominent, starting in the 1950s. For example, discussions on poverty, racism, and gender started to mobilize an increasing number of people. These discussions continued into the 1970s with antiwar and nuclear protests (McCormick 1989, Dobson 1990, Gottlieb 1993, Shabecoff 1993, Jamison 1996).

Demonstrations of civil resistance and discontentment spilled into the environmental arena, as well. This is illustrated by the commemoration in the United States of the first Earth Day in April 1970, that had hundreds of thousands of participants (McCormick 1989, Gottlieb 1993, Shabecoff 1993, Cuomo 1998).

Even the United Nations' agencies reflected the changes that were happening in society during this period. In 1968, the Biosphere Conference, held in Paris, produced recommendations regarding the need for more and better research on ecosystems, human ecology, pollution, genetic and natural resources, and also on conducting inventories and on monitoring resources. However, more importantly, this meeting introduced the concept that "deterioration of the environment was the fault of rapid population growth, urbanization, and industrialization" (McCormick 1989). The United Nations Economic and Social Council (ECOSOC), part of the United Nations original structure, was created in cooperation with FAO to address the "social and economic rehabilitation" of the postwar world.

In response to many governments' fears of social unrest during the Cold War era, a United Nations Conference on the Human Environment was held in Stockholm in 1972 (McCormick 1989, MacDonald 1998). At this conference, the paradigm of sustainable development (later published in the *Brutland Report*) was officially introduced. It called for "the management and conservation of the natural resource base and the orientation of technological and institutional changes in such a manner as to ensure the attainment and continued satisfaction of human needs for the present and future generations" (WCED 1987). That same year, growing pressure from society to bring environmental issues more to the forefront of international attention prompted the creation of the United Nations Environmental Agency (UNEP). An important reason for the creation of UNEP may have been the need to enforce the UN system's legitimacy in dealing with international issues (McCormick 1989). This UN agency aimed to bring the design of environmental-related policies to an international framework, and to provide assistance for developing countries through information sharing and technology transfer.

Therefore, this new level of public consciousness resulted in the development of many new efforts in both the social and environmental arenas. Along with the social concerns described above, the unprecedented rate of perceived plant and animal species extinction resulted in this becoming an important topic in ecology (Myers 1984, Wilson 1988, Primack 1993, Pearce 1994, Raven 1998). Buzzwords such as *biodiversity maintenance* (and sometimes–enhancement) increasingly appeared in governmental documents and media publications. The development of the discipline of *conservation biology* has been heavily influenced by another discipline, *environmental ethics*, which was inspired by the same forces pushing for social ethics (Jordan 1995). Primack (1993) gave evidence of these early links when he outlined the principles of the former discipline:

- The diversity of organisms is good because humans enjoy watching biodiversity and a variety of potential products increases human survivorship rates;
- The untimely extinction of populations and species is bad unless it is part of a natural process of evolution;
- Ecological complexity is good and only in natural environments it is fully expressed;
- Evolution is good because it leads to new species and, thus, more biological diversity; and
- Biological diversity has an intrinsic value, despite whether humans make use of it or not *(and because it could always gain value in the future)*.

The rationale behind the discipline of conservation biology and ecology is that a more diverse complex of plants and animals will contribute to a more stable and healthy system

(Dobson 1990, Pearce 1994). However, many authors have presented arguments which contest the theory that a direct correlation existed between species diversity and ecosystems stability (see 3.2.1, May 1972, Calow 1992, Johnson et al. 1996, Bengtsson 1997). This unresolved controversy between species and ecosystem function does not, however, diminish the importance of biodiversity as a value for human societies (Guruswamy and McNeely 1998). In general, most of society and policy-makers are still responding to value-based drivers that have developed during the environmental movement, where the priority is on the maintenance of valued species. The culmination of this viewpoint can be highlighted by the United Nations Convention on Biological Diversity (CBD), one of the products of the 1992 UNCED (Box 2.2).

Box 2.2

A brief overview of the Convention on Biological Diversity (CBD)

Summary: The CBD was negotiated by the United Nations Environment Programme (UNEP). It was signed at UNCED and made active on December 29, 1993, ninety days after its ratification. As of October 1998, more than 170 countries had become parties to this convention. The three goals of the CBD are: to promote the conservation of biodiversity; the sustainable use of its components; and the fair and equitable sharing of benefits arising out of the utilization of genetic resources. The CBD secretariat is located in Montreal, Canada. The Subsidiary Body on Scientific, Technical, and Technological Advice (SBSTTA), which advises the Conference of the Parties (COP), promoted four meetings:

COP-1 took place in Nassau, the Bahamas, from November 28–December 9, 1994. Its decisions were: adoption of the medium-term work programme; designation of the Permanent Secretariat; establishment of the Clearing House Mechanism (CHM) and the SBSTTA; and designation of the Global Environment Facility (GEF) as the interim institutional structure for the financial mechanism.

COP-2 met in Jakarta, Indonesia from November 6–17, 1995. Its decisions were: designation of the permanent location of the Secretariat in Montreal, Canada; agreement to develop a protocol on biosafety; operation of the CHM; designation of the GEF as the continuing interim institutional structure for the financial mechanism; consideration on marine and coastal biodiversity; issues of Plant Genetic Resources for Food and Agriculture (PGRFA); and agreement to address forests and biodiversity, including the development of a statement from the CBD to the Intergovernmental Panel on Forests (IPF) of the Commission on Sustainable Development.

COP-3 met in Buenos Aires, Argentina, from November 4–15, 1996. Its decisions were: a work programme on agricultural biodiversity and a more limited one on forest biodiversity; agreement to hold a workshop on traditional knowledge (Article 8(j)); application by the executive secretary for observer status to the World Trade Organization (WTO) Committee on Trade and the Environment; and a statement from the CBD to the Special Session of the UN General Assembly (UNGASS) to review implementation of Agenda 21.

COP-4 took place from May 4–15, 1998 in Bratislava, Slovakia. Delegates addressed, inter alia: inland water, marine and coastal, agricultural and forest biodiversity; the clearing-house mechanism; biosafety; implementation of Article 8(j) (traditional and indigenous knowledge); access and benefit sharing; a review of the operations of the convention; and national reports. Delegates also conducted a review of the financial mechanism.

Source: *http://www.iisd.ca/linkages/biodiv/cbdintro.html*

Those searching for a new paradigm for sustainable forest management were very aware of the importance to society of conserving plant and animal species. This can be shown by their use of *biodiversity maintenance* as one of the key components in all protocols being developed (see 3.2.1, 4.2.1). However, the only problem that has arisen is the expectation that societal values can be heavily relegated to private landowners (see 7.2). This expected social role clashes with another strong social and economic value (at least in most developed countries): private property rights (Westman 1990). Furthermore, this scenario is even more complicated for the small-scale forest management initiatives being pursued in developing countries characterized by common use of land and resources. *The Tragedy of the Commons*, by Garrett Hardin, has been extensively cited as the rationale for the need for private landowners to contribute to achieving societal values (Feeny 1990, Williams and Haughton 1994, Jordan 1995, Henderson 1996, Doorman 1998). For these authors, only privatization of the land would lead to the level of commitment necessary to ensure sustainability. Others prefer to defend the idea that the recognition of intellectual property rights of traditional communities will generate the necessary mechanisms for sustainable development (McCay 1987, Posey and Dutfield 1997).

The growing public awareness of environmental problems and concerns over societal inequalities set the stage for the solutions proposed by the *Brutland Report* and later reinforced at UNCED. This entire societal shift was also echoed in the economic arena. The discipline of *environmental economics* or *ecological economics* started growing exponentially and became the backbone of the sustainable-development model (Daly 1980, Constanza 1991ab, Constanza et al. 1991, Hardin 1991, Dahl 1996, Adamowicz et al. 1996). Numerous indexes of environmental and human welfare started to appear that contemplated this social dissatisfaction but ensured that business would continue as usual (Jacob 1994, Williams and Haughton 1994, Henderson 1996, Borgström 1997).

In the 1950s, because of the technological achievements that had occurred during WWII in many areas of human knowledge (e.g., medicine, transportation, communication, and information gathering), the world was geared to accept a technocratic model of development. The Allied nations wanted to ensure that development would be based on fast industrial growth and that the main beneficiaries would be the winners of the war. However, many less-developed nations would also be allowed to participate in this vision of a good future. For them, their good future would be the fulfillment of their roles as suppliers of natural resources and/or consumers of industrial goods (Keynes 1980a, Keynes 1980b, McCormick 1989, Gottlieb 1993, Schild 1995). Developing industrial capability was seen as fundamentally important in improving the present living conditions and in securing power and independence, even if it meant depleting resources and generating pollution (MPCG 1968). The Gross National Product (GNP) and the Gross Domestic Product (GDP), units for measuring industrial output, became the favorable indexes for the evaluation of development. Thus, the quality of life was linked to consumption (Clark 1991, Farber 1991, Jordan 1995). The world began to be ruled by trade values (Goldsmith 1997, Chichilnisky 1998).

In this context, the belief among politicians and decision-makers was that to become sustainable, the world economy just needed to more clearly incorporate social and ecological factors. The re-casting of the National Accounts System and continuous expansion of the economy towards more environmentally correct and socially equitable activities would bring everybody to sustainability (Williams and Haughton 1994, Henderson 1996). The United Nations Development Programme (UNDP) has pursued an example of this solution formula, not directly related to forest certification, but to forestry in general. In the late 1980s, this agency made a great effort to develop the methodology and to convince member governments to adopt *environmental accounting*. This new approach incorporated the role of the environment in economic activities, both as a resource base and as receptacle of the residues from production and consumption processes. In developing countries, indicators (i.e., GNP and GDP) incorporating the side effects of production and consumption activities were perceived by UNDP to be essential to incorporate into any environmental accounting approach (Landerfield 1985, Bartelmus 1987, Repetto 1987, Peskin 1989, Tongeren 1990). This need was identified because developing countries frequently have growth produced at the expense of future incomes. More recently, starting in 1990, UNDP has annually published the *Human Development Index*, which measures the level of development in different countries in accordance with their population's access to education, health, freedom, and other value-oriented parameters (UNDP 1990).

Forest certification can be interpreted as an exercise to make this model operational within the forestry sector, where human-laden values dictate the rules by which forest resources are accessed, used, and preserved (see 3.2.2). This alternative does not contradict or conflict with the neoliberal dogma of economic growth (Simonis 1990, Constanza 1991, Jacob 1994, Williams and Haughton 1994, Adamowicz et al. 1996, Henderson 1996, Goldsmith 1997, Chichilnisky 1998, Doorman 1998, Reid 1998, Sagoff 1998). Since questions about patterns of natural resource uses and consumerism of processed goods were not alternatives that society (especially in developed nations) was willing to hear, forest certification has become a very palatable option.

Thus, the role of the economic sector cannot be ignored in the ongoing debates on sustainable forestry. The market issue for some supporters of sustainable forest management is that financial rewards are a consequence of achieving better management practices. For others, the ultimate pursuit of new market niches or higher profit margins are the driving forces behind implementing such an approach (Crossley et al. 1994, Cabarle 1995, Adamowicz et al. 1996, Brockmann 1996, Jenkins 1997).

The pressures imposed by environmental NGOs also contributed to the rise in forest certification among the economic sector. Some of the impetus resulted from a lack of credibility when verifying claims that timber sources were being obtained from sustainably managed forests. For example, in the early 1990s, World Wildlife Fund – U.K. found that most claims of sustainable timber sources could not be supported. The results of a survey they conducted showed that only three companies out of 80 working with tropical timber could back their claims that their wood sources were from sustainably managed forests (Read 1994). In response to the reactions to this survey, some large lumber and furniture businesses (e.g., U.K.-based DIY retail chain B&Q) became subscribers to the World Wildlife Fund program, *1995 Buyers Group* (Elliott 1997, Bruce 1998). The participation of the U.K. Timber Trade Federation and the Scottish Hardwoods Charter helped to further push the idea of creating a system to assess claims of sustainable practices. The representatives from these two forest industries, encompassing nearly 100 companies, quickly adopted the WWF target for achieving sustainable consumption of wood. For the subscribers to this program, the target date for achieving sustainable consumption of wood was later changed to the year 2000.

To conclude this section, it is important to mention that there have been numerous worldwide initiatives involving the definition of sustainable forest management (Palmer et al. 1996, Wijewardana et al. 1997). These initiatives can be grouped into those driven by national or international governments or by NGOs and/or commercial enterprises (Crossley 1996, Evans 1996, see 2.2). The new paradigm of sustainable forest management assumes that forests will supply a wide array of environmental and socio-economic benefits to society

(Baharuddin 1995, Crossley 1996 Elliott 1996, Granholm 1996, Merino 1996, Prabhu et al. 1996, Wijewardana et al. 1997, Clark et al. 1998). Despite the many efforts to define criteria and indicators of sustainable forest management, certification of forest management operations is not an aim for all the protocols. Distinction should be made between the criteria and indicators that aim to evaluate the adequacy/sustainability of forest management operations and their use in intergovernmental initiatives (Prabhu et al. 1996). In the latter efforts, the criteria and indicators are common methodologies that will be used by governmental agencies to collect data on the current state of and future trends in forests. All initiatives to define criteria and indicators in forest certification are supposed to "provide a common framework for describing, monitoring, and evaluating, over time, progress towards sustainable forest management." (Wijewardana et al. 1997).

2.2 Other relevant initiatives in forest certification

Anna Fanzeres, Kristiina A. Vogt

Many different mechanisms and institutional arrangements have addressed the problems perceived by society to exist in the forestry sector. "Institution" is defined here as any organization that officially or informally addresses and pursues solutions to problems identified by society (Southgate 1990, Colfer 1995, Gunderson 1995, Colfer et al. 1996, Crossley 1996, Richards 1996, Viana et al. 1996). Within this context, the terms "stakeholders" or "actors" have been extensively used in the forest certification debate to address interested parties and the structures they represent, or are represented by them (Colfer 1995, Colfer et al. 1995, Merino 1996).

Recently, nongovernmental organizations (NGOs) have gained more prominence as institutions which attempt to lead the discussions on how to solve societal problems (McCormick 1989, Arnt and Schwartzman 1992, Princen and Finger 1994, Eder 1996, Marks 1996, Elliott 1997, Arts 1998, Hendricks 1998, MacDonald 1998). Since the Stockholm Conference, the number of NGOs around the world has grown (McCormick 1989, Princen and Finger 1994). The contributions of NGOs to addressing social, economic, and environmental issues has boomed, despite disagreements on their real power and ability to influence decisions (Arts 1998, MacDonald 1998). This segment of society, with its diversity of mandates (environmental, social, feminist, indigenous, etc.), has included its values and perceptions of life when attempting to formulate public policy (Box 2.3).

Box 2.3
Some nongovernmental organizations with interests and/or initiatives on forests

Conservation International

CI is a field- and lobby-oriented organization with headquarters in Washington, D.C. that promotes policies by working with organizations such as the World Bank and the USAID. The focus of CI is on biodiversity maintenance, identifying regions of megadiversity and hotspots for conservation. In addition, it promotes a set of best practices for industries affecting the environment.

Friends of the Earth International (FoE)

FoE is a federation of autonomous environmental organizations from around the world. It has members in more than fifty countries who campaign on the most urgent environmental and social issues they have identified, while simultaneously working towards a shift to sustainable societies. Since its headquarters moved to the Netherlands from its original location in San Francisco, FoE has produced numerous reports on environmental problems and conducts public education and policy campaigns.

Greenpeace International

They have produced campaigns to bring attention to environmental problems that include climate, toxins, nuclear energy, oceans, genetic engineering, and forests. In the forest category, this organization has placed a strong focus on the preservation of primary forests around the world. Their headquarters are located in the Netherlands and they have offices in more than thirty countries, as well as a fleet of ships.

International Institute for Environment and Development (IIED)

The institute was founded in 1973. It was formulated in response to activities of the International Institute for Environmental Affairs (IIEA). IIEA's report, *Only One Earth,* was the intellectual and philosophical foundation for the UN Stockholm Conference. With headquarters in London, this organization produces information for NGOs and the media, regarding UN activities in environmental areas.

The World Conservation Union (IUCN)

IUCN is one of the oldest international conservation organizations. It was established in Fontainebleau, France on October 5, 1948 as the "International Union for the Protection of Nature" (IUPN). IUCN is comprised of more than 820 secretariat staff members who work in IUCN's offices throughout the world. The organization's work is divided into six global commissions: species survival; protected areas; education and communication; environmental law; ecosystem management; and environmental economics and social policy.

   According to Posey and Dutfield (1997): "The concept of sustainable development had its origins in the 1980 *World Conservation Strategy,* released by IUCN."

   IUCN's objectives are to maintain essential ecological processes and life-support systems; to preserve genetic diversity; and to ensure the sustainable utilization of species and ecosystems. A well-known example of IUCN's work is the elaboration of the "Red List of Threatened Species."

The Nature Conservancy

The mission of TNC is to preserve plants, animals, and natural communities that represent the diversity found on earth by protecting the terrestrial and aquatic habitats they need for survival. The Nature Conservancy operates the largest private system of nature sanctuaries in the world where imperiled species of plants and animals have been found. Their headquarters are located in Washington, D.C., and they work in collaboration with other organizations in the U.S., Caribbean, Latin America, and the Pacific.

World Wildlife Fund International

With campaigns on the subjects of saving endangered species, setting aside areas for protection of nature, and promoting change in the human pattern of resource uses, WWF's "mission is to conserve nature and ecological processes." They focus on three specific areas: climate, oceans, and forests. With headquarters in Switzerland, this organization is comprised of twenty-seven national organizations, five associates, and twenty-one program offices.

The World Resources Institute (WRI)

WRI is "an independent center for policy research and technical assistance on global environmental and development issues." Created in 1982, WRI is dedicated to "helping governments and private organizations of all types cope with environmental, resource, and development challenges of global significance." Based in Washington, D.C., this organization has contributed to the policy arena, focusing mainly on forests and climate. WRI annually publishes the report, *World Resources*, whose database is the most widely used tool in the debate on changes in forest cover worldwide.

Source: *http://www.webdirectory.com/, www.tnc.org, www.iucn.org,* and *www.wri.org*

The role of environmental NGOs in launching the debate on forest certification has already been discussed (see 2.1). Environmental NGOs continue to be at the forefront of this debate and have strongly influenced the rules of the game. The Forest Stewardship Council (FSC) is the strongest representative of this group (see 2.4). The economic sector also has strong participation within FSC, but the forestry industry has also developed initiatives of its own (see 2.4.1). However, some environmental NGOs do not give credibility to these industry efforts (see 2.1).

In 1993, the ISO 14000 Series on Environmental Management Systems was announced by the International Organization for Standardization (ISO). It received strong support from the industry (especially pulp and paper companies), which saw this initiative as a potential counterbalance against the Forest Stewardship Council (FSC). Differences in the dominant approaches being taken by both groups resulted in protocols being classified as *performance standards approach* or an *environmental management system's approach* (see 2.4.1). The former focuses on measuring previously specified conditions on the ground, while the latter assesses whether certain management procedures should be implemented and assumes that the result will be an environmentally desirable one (sustainable outcome) (Prabhu et al. 1996).

In 1995, an initiative was orchestrated by the Canadian pulp and paper industry and the Australian wood industry to boost ISO 14000 as the industry's set of standards as a counterbalance to FSC. The Canadian Standards Association (CSA) submitted a proposal to ISO to adapt the ISO 14000 standards for forest management. However, this attempt resulted in a strong reaction from some environmental NGOs. These environmental NGOs saw ISO 14000 as an initiative that would promote weaker standards for sustainable forest management (Benchmark 1995, Hauselmann 1997, Kiekens 1997). Surprisingly, some ISO member countries (e.g., U.S., Scandinavian countries, and Brazil) did support the NGO movement because of their perception that their sovereignty was threatened and that the standards promoted by Canada would lead to unfair trade practices. The proposal was withdrawn. However, to appease Canadian and Australian representatives, ISO called for the formation of an informal study group on sustainable forest management, to be chaired by Standards New Zealand (Technical Committee 207). The participation was limited to ISO members. Eventually some NGOs were invited to participate; however, radical environmental groups were denied attendance (Hauselmann 1997).

Given the unprecedented and continuous growth of the FSC, ISO decided to acknowledge their existence and relevance in the forest management arena. The final draft version of the TC 207 working-group report has a specific section written on the "relationship between Sustainable Forest Management (SFM) Principles, Criteria, and Indicators (PC&I), and a forestry organization's Environmental Management System (EMS)" (ISO 1997). Their considerations are "organizations that wish to make a commitment to the goal of

SFM can incorporate PC&I, appropriate to the scope of the EMS, into their policy, objectives, and targets" (ISO 1997).

Neither the ISO nor the FSC approach is clearly distinct from the other. Representatives of the two major initiatives have recognized that both approaches are complementary to one another (Crossley 1996, ISO 1997). These initiatives are two of the major approaches that have survived from the early days of the forest certification debate (Hayward 1999). The trend appears to be that a combination of both initiatives will ultimately be used to assess for sustainable forest management (see 2.4.2).

Governments and their agencies are also institutions that are actively involved in forest certification. In addition, several international institutions have been important players in this debate (i.e., the United Nations, The European Community, NAFTA, etc.). For example, The United Nations Development Programme (UNDP), jointly with The World Bank and UNEP, developed a program called "Global Environmental Facility" (GEF). This program has provided the necessary financial and operational resources on issues such as forest biodiversity conservation; watershed management; combating desertification; and tree planting and/or forest preservation for the purposes of carbon sequestration. Development agencies, such as the World Bank, are increasingly participating in forest certification. Currently, the Bank's forest policy is undergoing revision after receiving heavy criticism from NGOs. Additionally, the World Bank is forming a partnership with the World Wildlife Fund. This partnership will support the establishment of 50 million hectares of new protected areas by the year 2000, and will promote the independent forest certification of an additional 200 million hectares of sustainably managed forests (Lampman 1999a, 1999b).

Several governmental agencies (national and international) have been involved in developing initiatives for the management, conservation, and sustainable development of all types of forests. Several of these initiatives use as a working framework the following: The Forest Principles, Agenda 21 (The Action Plan for Sustainable Development), and the work of the United Nations Commission on Sustainable Development (CSD). Similar to other UNCED activities, these initiatives are not designed to challenge consumption or trade mechanisms (see 2.1).

The Forest Principles (Appendix 1) is a legally nonbinding authoritative document "constructed of clauses which were either already part of national policies, or were ineffectively worded so as to provide escape clauses, or were without clear links to operational procedures" (Upton and Bass 1996). Moreover, forests have been tied to the model of sustainable development (see 3.1.2). Therefore, forests are interpreted as having an environmental role of maintaining ecological processes, a cultural and spiritual role satisfying the needs of urban and forest-dweller communities (present and future), and a socio-economic role of providing income opportunities. Included in the preamble to the Forest Principles are concepts such as:

- Multiple use of resources and the ultimate utilitarian purpose of forests that includes that they be the repositories of the survival and future of humanity (genetic basis for biotechnology)
- Conducting environmental impact assessments
- Balancing and harmonizing the relationship of small traditional communities with nature
- Using and preserving resources that are driven by economic and political interests

An interesting component in this document is the call for women to be considered as their own interest group parallel to other recognized stakeholders (e.g., indigenous peoples,

NGOs, industry, governments). The need to fulfill obligations to the less-favored segments of society (e.g., indigenous groups, poor [traditional] communities, women, etc.) is apparent in most initiatives pursuing sustainability. Many of the solutions being proposed to solve social problems involve increasing financial returns for these groups. However, the interrelations and interdependencies between society and the environment are not adequately addressed (see 3.1.2, 3.2.2).

Since the Stockholm Conference, there has been an attempt to set "an agenda for global action in response to environment and development problems" (MacDonald 1998). This is also reflected in the *Agenda for the 21st Century* which verbalizes the ideas that we are a worldwide connected community, must deal with environmental problems at a universal scale, and must base solutions on global markets (MacDonald 1998). Chapter 11 of *Agenda 21* (Appendix 2) deals specifically with forest issues. This chapter, entitled *Combating Deforestation*, places the primary role of forests as "the source of timber, firewood and other goods…also play[ing] an important role in soil and water conservation, maintaining a healthy atmosphere…and biological diversity of plants and animals" (Appendix 2). Throughout this document, poverty is characterized as the major cause of forest degradation. The document also calls for industrialized nations to be responsible for generating pressure over forest resources and land. The proposed solution is based on a "technological fix" model, to be provided and supported by rich nations. Several activities were proposed as part of this document: intensive data collection (utilizing highly technological approaches, such as satellite images); international cooperation for the purchase of modern equipment (for logging and processing); and identifying valuable genetic resources that can be transformed into biotechnology products. It also mentions the need for the expansion of the system of protected areas where the social and spiritual values of indigenous peoples, forest dwellers, and local communities could be fulfilled.

The United Nations Commission on Sustainable Development (CSD) was created after UNCED. Its mandate was to monitor the progress, review the implementation of Agenda 21 and make recommendations on necessary further steps. The United Nations' vision for CSD was for it to be the political ground for sharing experiences, proposals, and ideas. All of these were seen as advocacy functions to be largely exercised during meetings involving ministers of state and other governmental officials (IISD 1997). The aspect of CSD that was more directly related to forests was the establishment in 1995 of an Intergovernmental Panel on Forests (IPF), following a decision by the United Nations' Economic and Social Council (ECOSOC). This panel convened five sessions, with the last session held in April 1997. During the fifth session, conclusions and policy recommendations were submitted to the CSD General Assembly. These conclusions were presented with all other issues that were part of Agenda 21 during the 19th United Nations General Assembly Special Session (UNGASS). This July 1997 special session was held in New York and is commonly known as the Rio + 5 Meeting.

UNGASS, occurring five years after UNCED, demonstrated that government commitments were exclusively rhetorical. Discussions in forestry did not support implementing changes to management practices that were considered negative. Most participants concluded that this meeting was "a sobering reminder that little progress has been made over the past five years in implementing key components of Agenda 21 and moving toward sustainable development" (IISD 1997). At the conclusion of this meeting, it was decided that the work of the IPF would be extended. Now renamed Intergovernmental Forum on Forests (IFF), the mandate of this group is to deal with matters unresolved by the former IPF. Some of these issues are trade regulations; transfer of technology and financial resources; and developing mechanisms to implement a legally binding instrument (e.g., Forest Convention). The forum will report at the CSD's eighth session in 2000.

The creation of a Forest Convention has been supported by many UN member governments. However, international NGOs are approaching the Forest Convention with suspicion and many reservations. Some of the issues brought up by international NGOs can be placed into five main categories (Crampton and Ozinga 1997):

- Lack of evaluation of existing approaches because global attempts to address forest problems have not succeeded and their causes have not been analyzed
- Low standards for criteria and indicators being the only possible way of achieving international consensus on sustainable forest management
- Lack of participation of local communities and indigenous peoples because of insufficient mechanisms to ensure that they have a voice in the decision-making process
- No means of regulating transnational corporations in the mining and forestry sectors, whose practices are one of the major forces behind deforestation and forest degradation
- The right issues are not being addressed because two other important causes of deforestation (i.e., land rights for local communities in the South and unsustainable consumption of wood and wood products in the North) have not been included

Thus, the influence of NGOs can be found in all these governmental responses to forestry issues following UNCED. A list of the most relevant intergovernmental initiatives following UNCED is presented in Appendix 3. There has been a change in how forests are monitored by focusing primarily on productivity for the development of criteria and indicators of sustainability as "…tools for assessing national trends in forest conditions and forest management" (Wijewardana et al. 1997). Governmental initiatives are primarily being developed as frameworks for policy-making. They focus heavily on data collection. However, at least two initiatives — *Helsinki* and *Tapapoto* — have tried to produce indicators to be used at the Forest Management Unit Level (FMU).

Those United Nations organizations that were formed or became part of this debate continue to try to fulfill their roles as key players in this process. The Intergovernmental Forum on Forests (IFF), following the lead of FAO, continues to pursue the task of producing, or at least contributing to, the development of Principles, Criteria, and Indicators for sustainable forest management. In its final moments, IFF released a series of recommendations on criteria and indicators for sustainable forest management. This effort was coordinated jointly with the secretariat of the Inter-Governmental Seminar on Criteria and Indicators for Sustainable Forest Management (held in Helsinki, Finland from August 19–22, 1996), the Center for International Forest Research (CIFOR), the United Nations Environment Programme (UNEP), and the Economic Commission for Europe (ECE). Based on existing initiatives on the development of criteria and indicators for sustainable forest management, the panel suggested that the following parameters be minimally used to characterize sustainable management of all types of forests (DPCSD 1996):

- Extent of forest resources
- Biological diversity
- Productive, protective, environmental and socio-economic functions and conditions of forests
- Forest health and vitality

- Global carbon cycles
- Policy and legal framework, including the capacity to implement sustainable forest management

In the face of CSD's recommendation of criteria and indicators of forests' sustainability to be adopted by all UN agencies, FAO began incorporating these indicators into national and international forest inventories. The FAO/Finland "Kotka III" Expert Consultation on the Global Forest Resources Assessment 2000 was held in Finland in June 1996. The conclusion of this meeting was that 16 out of the 80 national level indicators being proposed by different initiatives could "in principle, be considered for inclusion in future global Forest Resources Assessments" starting at FRA 2000 (DPCSD 1996):

- Extent of forest resources
- Health and vitality of the forests
- Production of wood and non-wood products
- Biological diversity of forests
- Soil and water conservation
- Social and economic functions of forests

These common indicators are expected to be used by all country reports to be produced under the requirements of the Convention on Biological Diversity, Framework Convention on Climate Change, Convention on Combating Desertification and for the Convention on International Trade in Endangered Species of Wild Fauna and Flora (CITES). It was hoped that this would facilitate reports progressing towards documenting sustainable forest management practices and to better determine international trends (DPCSD 1996). Another task under FAO's responsibility is the need for harmonizing concepts and terminology. FAO is collaborating with the International Union of Forestry Research Organizations (IUFRO) to review forestry concepts and twenty core terms. These terms were originally defined in the FAO's Forest Resources Assessment 1990 and have to be examined within the context of a range of language groups in some twenty-five countries (DPCSD 1996, FAO 1997). The outcome of this harmonization effort was the FAO/ITTO Expert Meeting on the Harmonization of Criteria and Indicators for Sustainable Forest Management that was held in February 1995 (DPCSD 1996).

At the 13th session of the FAO Committee on Forestry (COFO), held in March 1997, member countries requested the development of a strategic plan for forestry. This draft document has been circulating among the six regional forestry commissions of the FAO. It was also made available to the participants of the Eleventh World Forestry Congress (Box 2.4) and to NGOs for further discussion and input. It was hoped that a final version would be approved in early 1999. This strategic plan is based on FAO's mission in forestry "to enhance human well-being through the sustainable management of the world's trees and forests" (FAO 1997). The three goals of this document encompass environmental, economic, and social aspects. It calls for the maintenance of biological diversity and ecosystem health, the full realization of forests' economic potential, and providing benefits for both the local level and the private sector. FAO intends to be "recognized for leadership and partnership in promoting sustainable management of the world's trees and forests" (FAO 1997). FAO wants to be "widely perceived as an effective and technically competent service organization, alert to new trends, and a catalyst of action in current and emerging areas of need in forestry" (FAO 1997). FAO believes that this strong role-playing will allow the organization to fulfill the objectives of slowing the rate of deforestation in the tropics and decreasing forest degradation in the rest of the world.

Box 2.4
Recent meetings related to the debate on forest certification

The UBC-UPM Conference on Ecological, Social and Political Issues in Certification of Forest Management (Malaysia, May 12–16, 1996)

Sponsored by the University of British Columbia (Canada) and the University of Pertanian (Malaysia), this meeting focused on the following topic areas: stand-level concepts and indicators; landscape-level concepts and indicators; economic concepts and indicators; social dimensions of certification; monitoring of forest practices; impacts of certification; national and sub-national institutional arrangements; and international institutional arrangements.

International Conference on Certification and Labeling of Products from Sustainably Managed Forests (Australia, May 26–31, 1996)

Sessions covered the following topics: principles of sustainable forest management; principles of certification of forest management systems and labeling of products from these forests; contribution of certification and labeling to sustainable forest management; current approaches to certification and labeling and lessons for the future; social issues related to certification and labeling; economic and trade aspects of certification and labeling; and implementation of certification and labeling. This meeting also had a special session on Challenges of Sustainable Forest Management, Certification and Labeling.

Intergovernmental Seminar on Criteria and Indicators for Sustainable Forest Management (Helsinki, Finland, August 19–22, 1996)

This seminar was a joint implementation effort supporting the work of CSD.

Forests for Life (San Francisco, May 8–10, 1997)

This conference was organized by World Wildlife Fund U.S. and World Wildlife Fund Canada. Its focus was on two main topics: landscape ecology — a means of implementing a North American forest vision and challenges to FSC products — and creating consumer awareness.

XI FAO World Forestry Congress (Antalya, Turkey, October 13–22, 1997)

Panels covered the following topics: forest and tree resources; forest biological diversity and the maintenance of the natural heritage; protective and environmental functions of forests; the economic contribution of forestry to sustainable development; social dimensions of forestry's contribution to sustainable development; policies, institutions, and means for sustainable forestry development; and an eco-regional review.

International Conference on Indicators for Sustainable Forest Management (Melbourne, Australia, August 24–28, 1998)

Promoted by IUFRO in collaboration with CIFOR and FAO for the purpose of developing scientifically based indicators. Panels covered the following issues: legal and institutional frameworks (legal planning; capacity building); productive capacity (inventories); economic functions and conditions (community needs, indigenous peoples issues, equity, labor safety); ecosystem health and vitality (key ecological processes and early warning systems, environmental stress factors, pests, and diseases); soil and water protection (physical, chemical and biological properties, water quality, yield, and biota); global carbon cycles (methodology, scale and use of forest products); biological diversity (ecosystems diversity, off-reserve management, species diversity, genetic diversity, fragmentation, and structural patterns).

New events and initiatives in forest certification are continuing to happen after UNCED. The newest addition to the debate on forest certification is the Sustainable Forestry & Certification Watch (SF&CW) (Box 2.5). The forest certification debate, which developed out of a need to provide a reliable outside mechanism to control forest practices, seems to raise enough suspicions that additional mechanisms are appearing for controlling its development. The SF&CW initiative intends to evaluate whether forest certification is really going to translate into better forest practices or is a "front" for continuing to deplete and harvest forests worldwide.

Box 2.5
Sustainable Forestry & Certification Watch

SF&CW is an independent, non-profit, nongovernmental, international organization aimed at enhancing the understanding of forest certification and its implications, particularly for sustainable forest management, international forest policy, trade in forest products, and consumer choice. In general, SF&CW encourages responsible stewardship of global forest resources by facilitating the identification and adoption of effective approaches to forest resource management and policy.

Why SF&CW?
Need for information on the various certification options. Forest certification is becoming an increasingly complex topic. There is a growing number of forest certification initiatives worldwide, and monitoring all these initiatives has become increasingly difficult for individual organizations. An independent source of information on the various certification options currently available or in development will assist those having to make decisions regarding forest certification (forest owners and industry, traders and retailers of forest products, governmental and intergovernmental agencies, etc.).

Uncertain impact of certification on forest management. There is uncertainty regarding the ability of certification to improve forest management practices. Certification largely emerged out of public concern for tropical deforestation. To date, the impact of certification on tropical deforestation has been, at best, negligible. Certification is mostly gaining ground in countries such as Sweden, where concerns related to forest-management practices did not exist. This raises questions about the relevance of the approach to achieve the core objective of sustainable forest management.

Public claims and consumer information issues. There is a need to monitor claims made in relation to forest certification and to assess whether they convey objective information to consumers and the public. Because of their close association with certification initiatives, many parties (including environmental organizations member of the Forest Stewardship Council) tend to issue contentious claims. For example, WWF maintains that FSC is "the only credible, independent guarantee that timber comes from well-managed forests."

Impact on national and international forest policies. Although forest certification is a market-based initiative, it has implications for national and international forest policy. At the national level, domestic forest policies may be undermined because of misconceived notions that if a forest is not certified, then it is not well-managed. At the international level, forest certification has been presented by several of its proponents as a reason for not pursuing other approaches, including a legally binding global forest convention.

Discussions on a role for governments. Many governments are concerned about the development of forest certification, including the legal acceptability of certain certification initiatives (i.e., against anti-trust [competition] legislation and international trade rules). In addition, there is much debate as to whether public forests should be inspected by private certification entities.

Source: www.sfcw.org

In general, most economic sectors that have used some kind of labeling have achieved some environmental improvements in practices (see 2.3). This desire by society to use labels and products' certification to ensure confidence that their values are being satisfied by producers is not only limited to the forest arena (Brockmann 1996). However, in no other sector has the issue of third-party certification resulted in so much intense debate and distrust. The challenges of establishing forest certification on the ground have been enormous, but undeniable successes have been achieved. However, despite all the efforts so far, some important elements have not been dealt with in both technical and political grounds (the subject of much of this book).

2.3   The certification of environmental claims in industrialized countries

Brett Furnas, Kristiina A. Vogt, Glenn Allen, Anna Fanzeres

To use a metaphor, the evolution of forest certification can be likened to the confluence of two rivers. One river represents the history of the debate about forest use and preservation. The other symbolizes the development of environmental certification as a policy tool. Consequently, it is helpful to understand how certification came to be used for verifying environmental claims in general, not just those pertaining to sustainable forest management. Industrialized countries, and the United States in particular, are good places to look at the different approaches being taken for using certification to communicate information about improved environmental performance of products and industrial processes.

The acceptance of certification as a means to promote environmental sustainability has differed on a global scale. For example, certification has been wholeheartedly embraced in Europe, while in the United States, a marketing-campaign approach by companies and organizations has been more common. In Germany, the public widely accepts labels issued by governmental agencies such as the German Institute for Quality Control and Labelling (RAL), but gives little credibility to self-declarations by companies (Broackman et al. 1996). In the United States, however, businesses seem to have a higher level of public approval. For example, Rainforest Crunch is a well-known product that was first put together as a project by Cultural Survival (Cambridge, Massachusetts) to satisfy their mission to aid indigenous groups. Ben & Jerry's, the ice-cream company marketing this product, pioneered the use of self-promotional labeling as a means of selling more of their product.

It may be that marketing in Europe is qualitatively different from that in the United States due to consumers' attitudes toward certification and labels. For example, in the United States, so many products are labeled "new," "light," "improved," or "environmentally safe" that many consumers do not pay attention to them. The introduction of another label in the United States, where so many labels already exist, will not have a dramatic impact unless the new label stands out from all the others. Despite consumer overexposure to many labels, consumer demand for healthier and more ethically produced products has fueled a boom in businesses attempting to satisfy these demands. Similar demands in Europe for healthy and ethically produced products have pushed fair trade labels into the mainstream (Durwael 1998). Based on surveys conducted in the United Kingdom, 86% of consumers were familiar with fair trade-labeled products and 68% said they were willing to pay more for these labeled products (Bowen 1997). In Sweden, Holland, and Belgium, about two-thirds of consumers were very aware of fair trade labels and what they symbolized (Bowen 1997). However, despite this strong consumer environmental awareness in Europe, it does necessarily reflect the real effects of these labels on consumer behavior. A survey conducted in German households demonstrated that two-thirds of the interviewees considered themselves environmentally conscious, but only one-third changed consumption patterns for purchasing cleaning chemicals, plastic bags, and utensils (Brockmann et al. 1996).

In the United States, perhaps the earliest examples of product certification date back to the Underwriters Laboratory and the Good Housekeeping seal. Active since the early part of this century, these programs were developed to assure consumers that products worked as advertised. Both of these programs are operated by private organizations and involve the verification of the factual information provided by manufacturers on product labeling or in supplemental literature. The cornerstone of these programs has been a reliance on independent laboratory testing that assesses product performance (Underwriters Laboratory 1998, Good Housekeeping Institute 1998).

In contrast, forest certification as a dominant approach to controlling forest management practices is quite recent (see 2.1). Certified tree farms have been around for decades, but the structured protocols being used in certification are a recent phenomenon. Forest certification is similar to the certification of manufactured goods in that it attempts to ensure production of environmentally sustainable products. The focus has shifted from verifying the functioning of a product to the functioning of the ecosystem (social and natural science portions) and the management system that created it. Forest certification is facing the same problems experienced in labeling non-forest-based manufactured products. In both cases, the public is unable to evaluate the manufacturing of products or the extraction of natural resources because they are too far removed from the process. This means that unbiased, independent verification is needed to show that fraudulent labeling is not occurring and that the environmental values of the public are being maintained.

Certification of organically produced food is an interesting issue to briefly mention, because of its potential relationship to forest certification (i.e., pursuing ecosystem health). There are many past issues that organizations interested in certifying organically produced foods have struggled with that are similar to those being experienced in forest certification. Despite the fact that organic farming initiatives were developed during the middle of the century (Conford 1988, Rodale 1999), certification of farms and their products only took off in the United States and Europe in the 1970s (Smith 1994, Soil Association 1999). Nevertheless, the process of developing criteria for organic certification seems to have been much slower than that which has occurred in forest certification. However, the issues and current developments in certifying organically grown food are very similar to the problems being experienced in forest certification. The increased interest of many segments of society in consuming healthy foods and/or knowing the origin of their products has captured the attention of large corporations. The practices used to grow and store food products by large corporations originally stimulated society to pursue alternative food products that were chemically and microbiologically safe. The large corporations have counteracted these societal-driven issues by developing their own interpretation of sustainability (Horsch 1999). Similar to forest certification, there has been an intensive lobbying effort by the industry to not allow for clear labeling schemes (such as "contains GMO," "irradiated," or "no rBGH") to be developed that would better inform the public. In the United States, these corporations have been sufficiently powerful to minimize government regulation that would require the use of labels.

The "third party" type of certification (see 2.4.1) can be linked to the environmental label — Blue Angel — issued since 1977 by the German RAL. It reflects the changes in environmental values that societies in developed nations were adopting at that time (see 2.1). This label introduced the approach of using an independent scheme for establishing the criteria for certification (involving industry, environmental organizations, and consumers' associations). This approach can be considered the origin of the third-party type of certification (see 2.4.1). The weakness of this particular label is its applicability to all consumer goods and its determination of a product's compatibility only in relation to similar products. However, some business segments (e.g., central heating systems, paints, and paper production) saw environmental improvements in their production after the development of criteria. This label has gained great popularity with industry. The use of this label has increased from having 80 certified products in the 1980s, to 3940 products in 77 different groups by 1993. Surveys have shown that it is a highly visible label to 43% of consumers when shopping, and that 62% of them believed that a neutral institution gives credibility to a label (Brockmann et al. 1996).

More recently, the state producer associations for timber and plastics in Switzerland and Germany have decided to issue declaration schemes for building materials and joinery products that consider their ecological aspects (Brockmann et al. 1996). These sorts of claims are often referred to as "first party" because of their origination from the same entity that manufactures the product (see 2.4.1). In the United States, the use of first-party certification has gained momentum.

Since the late 1980s, the American marketplace has witnessed a significant increase in the incidence of products with environmental claims made on the labels or in other forms of advertising (see 2.1). Many new products have been touted as "environmentally friendly" because of some improved aspect of the formulation, packaging, or manufacturing process that is believed to reduce negative environmental impacts, such as solid waste, energy consumption, or toxicity. Unfortunately, the basis for many of these first-party labeling messages to "second parties" (e.g., the consumers) has often been unclear. Some messages have been blatantly misleading or even fraudulent (Abt Associates 1993). As a hypothetical illustration of the confusion, consider the case of a spray-paint can labeled as "environmentally safe," without any further explanation. The manufacturer's implicit basis for this statement may be its removal of stratospheric ozone-depleting aerosol propellants from the formulation in accordance with laws applying to all such products. However, the claim may mislead consumers about the environmental impact of the product in relation to other aerosol paints, as well as non-aerosol paints. This is especially relevant if the particular product continues to contain high levels of toxic and smog-producing chemicals and the container cannot be easily recycled.

In the United States, four approaches have been taken in response to the problems with first-party environmental labeling claims:

1. Development of government rules for first-party environmental claims
2. Emergence of third-party certification programs run by private organizations
3. Tentative exploration of third-party certification by government agencies
4. Opposition by industry groups to certification, accompanied by the development of standardized, industrywide messages to consumers

The Federal Trade Commission's Guides for the Use of Environmental Marketing Claims, first released in 1992 and revised in 1996, provide government oversight over first-party claims. Although the guides do not have the strength of regulation, they represent the Federal Trade Commission's (FTC) official interpretation of the Federal Trade Commission Act which specifically prohibits unfair or deceptive advertising claims regarding environmental marketing. The FTC guides were designed to provide manufacturers with a "safe harbor" for the types of acceptable environmental claims they could put on their products without fear of legal challenge. The guides include general principles and specific examples of acceptable and unacceptable advertising for claims of general environmental benefit. They also include similar guidance for specific claims relating to source reduction and recyclable, biodegradable, and ozone-safe products (Federal Trade Commission 1996). In contrast, in Germany, such terms as "environmentally compatible" and "bio" are not defined by law, so legal margins for self declarations are possible (Brockmann et al. 1996).

Despite the late development of third-party programs in the United States, this effort has been aimed more strongly at making impartial scientific assessments about the relative environmental impacts of products. The two most notable third-party certification programs for consumer products in the United States did not emerge until the 1990s. They are run by private organizations: Scientific Certification Systems (SCS) and Green Seal (Abt Associates 1993). The former group has taken a route similar to Underwriters Laboratories by creating the SCS's Environmental Claims Certification Program and its Environmental Report Card. Both are intended to measure environmental performance, but not to set standards. On the other hand, Green Seal has entered the standard-setting business by using "life cycle assessment" to identify environmentally preferable products. In short, life-cycle assessment is a "cradle to grave" evaluation of all material and energy flows that occur during the manufacturing, distribution, use, and disposal of a product. Further information on both the Scientific Certification Systems and Green Seal are given in Boxes 2.6 and 2.7, respectively.

Box 2.6
Scientific Certification Systems (SCS)

Scientific Certification Systems, a for-profit organization based in Oakland, California has programs that: a) provide consumers with independent verification of environmental claims, and b) give information about the overall environmental performance of products. Through its Environmental Claims Certification Program, SCS takes an approach similar to that of Underwriters Laboratories and certifies only those claims that SCS engineers are able to scientifically verify. Certified products are authorized to bear labels featuring the SCS logo and a clearly stated environmental message. For example, SCS has certified cleaning products as "biodegradable" with a message explaining that the product "breaks down into carbon dioxide, minerals, and water." Although SCS does not set a standard recognizing superior environmental performance, the organization does set minimum performance thresholds below which it will not certify a claim. For example, SCS will not certify a claim that a product is more resource- or energy-efficient than required by law unless the product surpasses the legal standard in question by at least 10% (Abt Associates 1993, SCS 1998). The type of certification in which a statement is made pertaining to a single issue is often known as a "single criterion" program. SCS also issues environmental "report cards" by using "life cycle assessment" to measure the most significant environmental impacts for a particular product. SCS uses this methodology to create a graphically presented information label that quantifies the key impacts of a product in the same way that a nutritional label provides numbers about vitamins and calories (Abt Associates 1993, Scientific Certification Systems 1998).

Box 2.7
Green Seal

Green Seal, a not-for-profit organization based in Washington, D.C., goes beyond certifying the accuracy of information. The group awards a "seal of approval" to products it believes cause less harm to the environment than other products. Green Seal makes this determination by setting environmental standards for a product category. Using the principle of life-cycle assessment, Green Seal has developed numerous environmental standards for categories, including paints, newsprint, powdered laundry bleach, refrigerators, and alternative-fueled vehicles.

Before publishing a final standard, Green Seal circulates the draft standard for review and comment to manufacturers, trade associations, environmental organizations, and consumer groups. Nevertheless, seals of approval are likely to be criticized as inherently more subjective than single criterion programs because of the complex process of comparing different kinds of impacts. For example, how does one decide whether a pound of air pollution is worse than a pound of water pollution? (Abt Associates 1993, Green Seal 1998).

Although the German government runs the Blue Angel program in Europe and the European Union issues environmental labels to laundry and dishwashing machines (Brockmann et al. 1996), public agencies in the United States have been generally reluctant to enter the certification arena. Perhaps their reticence has been because the United States government has not wanted to appear to be throwing its weight around in the marketplace by endorsing one product at the expense of another. The exception to the lack of government involvement in certification programs has been in cases where there are single-issue claims tied closely to regulatory standards or legislative commitments. By means of an award label, the EPA's Energy Star program recognizes computers that are more energy efficient than the agency's standards (Abt Associates 1993). In 1993, the California Energy Commission similarly recognized refrigerators that went beyond federal energy-efficiency standards. In accordance with commitments in its Clean Air Act-mandated State Implementation Plan, the California Air Resources Board has begun to investigate the feasibility of developing special recognition labeling programs for consumer products that contain reduced levels of smog-forming, volatile organic compounds (California Air Resources Board 1994).

In recent years, industry trade associations have become more sophisticated in bringing their members together to address environmental issues. In opposition to third-party certification, they have argued for creating their own solutions to the problem of misleading claims in the marketplace. This is the same kind of attempt previously mentioned as taking place in Germany and Switzerland by producer associations for timber and plastics (Brockmann et al. 1996). In the United States, after the complete phaseout of stratospheric ozone-depleting substances in non-medical aerosols in the early 1990s, the consumer products industry launched a concerted effort of its own to provide a uniform message to consumers about the phaseout. The result was an industry-wide label bearing the message, "No CFCs — contains no CFCs which deplete the ozone layer" (Consumer Aerosol Products Council 1998).

The industry-led standardization of this first-party message also came in response to the FTC guides. Earlier claims made by individual companies implied that aerosols were suddenly "environmentally friendly" simply because they were no longer allowed by law to contain chlorofluorocarbons and other ozone-depleting chemicals. In addition to creating coordinated advertising messages, much of industry has come together in opposition to the third-party certification approach. Many in the private sector appear to philosophically disagree with the certification outfits about "who should be in the driver's seat" when it comes to providing information that can move the market.

In Europe, environmental claims are usually exposed to intensive scrutiny from organizations representing public interests (e.g., consumer associations, environmental NGOs). Especially important has been the role of large environmental NGOs. Despite the international approach to their work, they have also been "acting locally" and have been very effective in lobbying national governments and overarching organizations such as the European Union (Princen and Finger 1994). The role of the government in implementing regulations of practices (e.g., Blue Angel) has placed them in a different position in society than that of the federal government in the United States. This scenario has lead to a greater politicization of societal-environmental concerns (McCormick 1989). In many European countries, this approach materialized sooner than in the United States with the formation of Green political parties (McCormick 1989, Dobson 1990). The viewpoints of these political parties are concretely reflected in the design of institutional arrangements and regulations upon society (Eder 1996).

2.4   Analysis of forest certification approaches

2.4.1   Characterization of three approaches: Forest Stewardship Council, AF&PA Sustainable Forestry Initiative, and ISO

Brett Furnas, J. Scott Estey, Kristiina A. Vogt, Joyce K. Berry, Anna Fanzeres

Forest certification can take three basic forms (Barrett 1993, Ervin et al. 1996, Hansen 1997, Bruce 1998):

- *First-party certification:* an internal assessment by an organization of its own systems and practices
- *Second-party certification:* an assessment by a customer or outside trade organization
- *Third-party certification:* an analysis based on a set of accepted principles and standards (Barrett 1993, Ervin et al. 1996, Hansen 1997)

In addition to being differentiated based on the nature of the relationship between the auditor and the party being audited, the various certification initiatives can also be differentiated by the nature of the certification system and the products being certified.

Thus, two types of certification approaches are commonly discussed:

- *Systems-based:* certifies that the audited party has developed and adopted a management system which is conducive to environmental monitoring and improvements in environmental performance over time. This approach may also incorporate a set of improvement goals that the audited party will work towards.
- *Performance-based:* certifies that the audited party is meeting a specific set of previously adopted performance requirements.

The differences are often phrased in terms of "whether a criterion or indicator is prescriptive or descriptive" (Prabhu et al. 1996). However, the relevance of these dichotomies between systems is currently not as important, given the growing combination of approaches in the design and implementation of certification protocols.

Among the many forest certification initiatives being formulated and discussed around the world, most important in the United States are those proposed by the Forest Stewardship Council (FSC), the American Forest & Paper Association (AF&AP), and the International Organization for Standardization (ISO) (see 2.4.2). The relevance and comparisons of these different approaches have been discussed by Bruce (1998) for a specific European case and in a more general manner by Hauselmann (1997). In this and the following sections, we have chosen to focus our discussion on the United States given the availability of data and the relevance of its forest sector (both in volume and forestland areas).

A summary of the three dominant approaches being used in the United States follows:

Forest Stewardship Council Certification (FSC)

1. Third-party certification
2. Performance-based approach
3. Chain of custody (see 7.7 for definition) required to obtain product label. Focus on product labeling to communicate to consumers that products are being managed in an environmentally and socially sustainable manner

American Forest & Paper Association's Sustainable Forestry Initiative (SFI)

1. Second-party certification (recently, new initiatives for third-party certification have begun to be developed)
2. Systems-based and elements of performance-based approach
3. No chain of custody required

International Organization for Standardization's 14001 Initiative (ISO)

1. Third-party certification
2. Systems-based approach
3. No chain of custody required

2.4.1.1   Forest Stewardship Council

The first approach includes programs which have been accredited by the Forest Steward-ship Council (FSC) based in Oaxaca, Mexico. The FSC is an international, independent, non-profit, nongovernmental organization officially founded in 1993. Its origins can be traced back to 1990. It is a membership association with more than 170 members from 28 countries (Bruce 1998). In accordance with FSC statutes (FSC 1996), the representation within FSC (members and board representatives) is organized into two chambers — envi-ronmental/social and economic. In the statutes, the official representation for each cham-ber is 75% of the voting power for social, environmental, and indigenous organizations, with the remaining 25% for members with economic interests (industry, certifiers, con-sultants, etc.). This power distribution is a reflection of FSC's initial organizational strategy. However, the industry segment of the latter interest group has consistently complained about not having representation equal to the social and environmental-oriented organi-zations. The result of this pressure led the FSC during its last General Assembly in June 1996, to balance the voting power among the social, environmental, and economic interests by giving each 33% representation. Moreover, FSC places high importance on assuring a north-south balance (i.e., developed and less-developed nations) for its representation.

The Forest Stewardship Council is not a certifying organization, but an accrediting body that sets overarching standards for the certification organizations. FSC abides by *10 Prin-ciples* and respective *Criteria of Forest Management,* which focus on environmental, social, and economic values (see 1.2) (FSC 1999a).

In the United States, there are two FSC-accredited certifying organizations: Scientific Cer-tification Systems (with their Forest Conservation Program — FCP) and Rainforest Alliance (with the Smartwood Program). The FSC-approved certifiers determine whether a given operation meets the FSC standards by using their own assessment methods. The operation can then advertise their products as being "well managed" or "sustainable" according to the Smartwood Program, or just "well managed" according to SCS's Forest Conservation Pro-gram (SCS 1995, Smartwood 1998). Therefore, producers should be able to reach more dis-criminating markets where consumers are willing to pay more for the assurance of an "environmentally responsible, socially beneficial, and economically viable" product (FSC 1999a, see 3.1.2 and 3.1.3 for discussion of well-managed forest and sustainability).

The FSC approach to certification aims at product labeling as a means of communicating to the consuming public the authenticity of the claims regarding which products are produced from forests managed in accordance with worldwide-accepted standards (FSC 1996). To obtain this accreditation from the FSC, certifiers must present "a formal written methodology for verifying the chain of custody" (FSC 1996, see 7.7). Currently, as approved by the FSC board in September 1997, to carry an FSC label, wood products partially composed of recycled and uncertified raw materials are required to comply with the "Policy for Percentage-Based Claims." Solid-wood products must contain 100% wood from an FSC-certified forest. Assembled-chip and fiber products must contain at least 70% of their volume from an FSC-certified forest. These definitions, however, may still poten-tially be revised following further discussions among members and new recommendations to the board (FSC 1999b).

2.4.1.2   Sustainable Forestry Initiative

In the United States, the industrial approach to forest certification is represented by the American Forest & Paper Association's (AF&PA) Sustainable Forestry Initiative and the American Tree Farm System. Both initiatives have direct or close indirect links to the forest-products industry and do not have an umbrella organization accrediting their process.

In 1996, the American Forest & Paper Association established the Sustainable Forestry Initiative (SFI) program. This association represents individuals and companies which are actively involved in intensive management for industrial forestry. For example, in 1997, the AF&PA members accounted for approximately 84% of the paper production, 50% of the solid wood production, and 90% of the industrial woodland in the United States (*Tree Farmer* 1997).

The SFI initiative identifies 12 elements of an action plan which much be addressed as part of certification. These elements can be summarized under the following five topics:

1. Broadening the practice of sustainable forestry
2. Protecting water quality
3. Enhancing wildlife habitat
4. Minimizing the visual impact of harvesting
5. Providing opportunities for public outreach

Members of SFI are required to submit annual progress reports, but no outside certifier or "individual or narrow set of indicators are used to determine performance" (AF&PA 1995).

The first year after SFI was established, 17 of the 200 member companies initially committed to the principles of SFI were suspended because they failed to meet the requirements for membership, and a few members resigned. The lack of a standard set of indicators and sufficient data on how members were responding to the elements of SFI made it difficult to evaluate the success of this program. Initially, an expert panel of representatives from forestry, academia, and conservation groups was established to serve as outside observers on this initiative and to give it credibility. However, SFI has decided to actively pursue third-party certification. Therefore, its approach to certification will continue to evolve.

2.4.1.3 *International Standardization Organization (ISO)*

The systems-based environmental certification program of the International Standardization Organization (ISO) — the 14000 series — is the most significant process for the industry in general, and also for the segment of industrial forestry. In the United States, this approach gained some momentum, but did not achieve the same level of utilization as SFI (see 2.4.1.2). The ISO scheme of certification was launched in 1947 to develop internationally accepted standards for facilitating the exchange of goods and services worldwide in a postwar global economy (Hauselmann 1997).

It was not the original intent of ISO to certify forest-based products or sustainable forest management. The ISO 9000 series of standards for total quality management led to the development of the 14000 series. Not surprisingly, the latter set of standards resembles the former in that environmental performance is tied to the quality of an organization's operational and management practices. The ISO 14001 standard is the cornerstone of the 14000 series. It sets voluntary, industry-wide standards for adopting management practices that reduce environmental impacts and ensure regulatory compliance, while improving resource and energy-use efficiency of industrial operations. The standards have requirements related to corporate environmental policies, industrial process planning, use of best available technology, self-auditing, and a commitment to continual improvement. An independent third-party contractor accredited by the appropriate regional member of the ISO conducts the certification. In the United States, the accrediting ISO organization is the American National Standards Institute (Tibor 1996).

After UNCED, when the debate on sustainable forest management became part of the public and political agendas, ISO decided to establish a working group on forestry (TC 207) (see 2.2). The certification of forest operations using this approach has been very attractive to industrial enterprises because of their familiarity with the demands of the ISO 9000 label. However, the lack of support from NGOs for this initiative resulted in many industrial forest companies pursuing selective markets to use FSC type of certification (Elliot 1997). The pulp and paper industry has been the strongest supporter of the ISO initiative. To date, ISO 14001 has been used to certify forest management in five cases in Brazil, Finland, Sweden, and Indonesia (Ghazali and Simula 1997). Currently, Tukhill Economic Forestry Ltd., the largest private, forest-management company in the United Kingdom, is working working towards ISO 14001 certification (Shirley 1996).

In Canada, the ISO approach has been widely accepted (see 2.2). The ongoing development of Sustainable Forest Management System (SFM) standards is largely based on the ISO 14001 and 14004 standards (CSA 1996a, 1996b). Development of these standards is being led by the Canada Standards Association, a not-for-profit organization accredited by the ISO through the Standards Council of Canada. The standards include:

- Systems-based elements derived from the ISO 14000 series (i.e., formation of policy, public participation, plan development, information monitoring and assessment, review, and improvement)
- A set of suggested performance measures (i.e., increasing the number of standing dead-trees-per-acre to increase woodpecker habitat) which are to be selectively integrated into forest management plans through a process of public participation (CSA 1996ab)

One reason for Canada pursuing an ISO approach to forest certification is that the forest products industry, represented by the Canadian Sustainable Forestry Certification Coalition, favors standards which are internationally recognized and generic enough to be applicable worldwide. However, not everyone agrees with the approach Canada has taken in applying the ISO 14000 standards to forest management. The use of ISO 14001 in forest certification has been criticized as not being a substitute for the type of on-the-ground performance that is integral to the other performance-based forest certification approaches (GFPP 1996). Additionally, the ISO approach has not been considered relevant for forestry because of its product focus and because it currently has no field management or ecology requirement for its implementation (Northrup 1998). Furthermore, concern exists about the general effectiveness of the ISO 14000 series in prescribing a standardized system for global environmental management (Benchmark Environmental Consulting 1995, Hauselmann 1997).

Interestingly, the final draft report of the TC 207 working group (ISO 1997) does not make any reference to the ISO 14024 standards entitled *Environmental Labelling — Guiding Principles, Practices, and Criteria for Multiple Criteria-Based Practitioner Programmes — Guide for Certification Procedures*. In spite of its perplexing title and unofficial status as a draft standard, ISO 14024 could play a useful role in developing and improving forest-certification programs (e.g., FSC, SFI, SFM). Although the ISO 14024 does not provide guidance on what the performance measures should be, it does lay out a framework for developing, verifying, and certifying these measures. Much of the debate over forest certification has been about what *sustainable forestry* should be; ISO 14024 focuses attention on what *certification* should be.

Some of the key elements of ISO 14024 are:

1. Certification programs should identify the selectivity level of management for the program, and this selectivity level should be narrow enough to distinguish the labeled products as having less burden on the environment than alternatives.
2. Environmental standards should be set with the selectivity principle in mind.
3. All elements of the criteria should be backed up by a scientific basis of supporting studies that is "open for examination" in a transparent manner.
4. A documented protocol should be in place for assessing whether a product meets the standards.
5. The protocol for verifying initial and ongoing compliance should have sufficient rigor to maintain confidence in the program, and the conformity of this methodology should be able to produce repeatable and reproducible results (International Standardization Organization 1995).

None of the above elements is novel and all reiterate elements proposed by other approaches. However, the consolidation of these elements makes this approach useful when debating forest certification. Elements (1) and (2) can be correlated to the Blue Angel labeling system. The selectivity being proposed, however, must be developed in terms of parameters that the public (i.e., ultimate target of labels) understands the claims being made (for further discussions on selectivity, see 2.4.2). The current problem with general claims, such as sustainability (see 1.1), is that the public will not grasp what it really means. The third element discusses defining sustainability, which is still unresolved among scientists. Moreover, the ISO 14024 scheme demands a reliance on documentation (element 4) and harmonization of methods (element 5) which are also found in many other certification protocols.

2.4.2 Compatibility or dissimilarity of the Forest Stewardship, Sustainable Forestry Initiative, and International Standardization Organization approaches to forest certification

Brett Furnas, Kristiina A. Vogt, Anna Fanzeres

A comparison of the Forest Stewardship Council and International Standardization Organization approaches to forest certification has received enough attention (see 2.2). A World Wildlife Federation (WWF) report noted that there is "no inherent conflict between the FSC and ISO approaches: they are different, and can actually be complementary" (World Wildlife Fund 1997). This report also indicated that a number of forest products companies are interested in being certified by both programs. However, the WWF report pointed out that the FSC has already "drawn heavily on ISO documents in establishing guidelines for how certifiers should operate, and in designing the accreditation process" (World Wildlife Fund 1997). A study at the University of Edinburgh echoed the point that the FSC and ISO serve different purposes (Bruce 1998). This study also provided some evidence that the FSC may be more useful to small-scale forest enterprises that lack the resources to pursue both types of certification (Bruce 1998, however see 4.3.6, 7.4, 7.5).

There has been less discussion about the compatibility of the FSC and SFI styles of certification. The FSC and SFI approaches to certification differ fundamentally in their

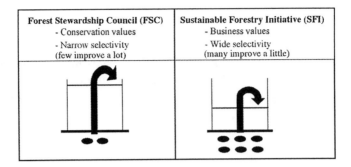

FIGURE 2.1
Two approaches to forest certification.

views of what constitutes sustainable forestry. They also disagree on the role that certification should play in the marketplace. However, both programs are similar in that they claim to aim towards the "sustainable" management of forests. Both cover a wide breadth of forest management issues, including ecological health, sustained timber yield, and the benefits to society.

The concept of "selectivity," as incorporated into the ISO 14024 standard, provides a useful mechanism for comparing the FSC and SFI approaches to forest certification. For this exercise, selectivity is the parameter for deciding what portion of the marketplace should be able to meet the criteria for certification (Tibor 1996). Put another way, how high should the bar be set? (see Figure 2.1). To illustrate this point further, a forest-certification program characterized by narrow selectivity may contain very restrictive criteria. Under the narrow-selectivity concept, only companies that represent less than 5% of industry-wide timber sales are able to meet these criteria without significant effort. On the other hand, certification characterized by wide selectivity may include less-restrictive criteria, so that more than 50% of the market can meet them by making less dramatic changes. A narrow selectivity program aims to recognize a few individuals who reach very high standards, whereas the advantage of a wide-selectivity program is that it can provide an incentive for a lot of individuals to change a little. This suggests that programs with opposite levels of selectivity are not necessarily incompatible.

The differences in selectivity between the two types of forest certification programs are linked to differences in values of what is meant by sustainable forestry, and what the role of certification should be. The FSC approach is one of narrow selectivity: it raises the bar high, for a few. It identifies and rewards the relatively small niche of forestland that is managed in an "environmentally preferable" manner, in relation to the larger realm of commercially producing forestland. This strategy is tied to the preservationist and restorative goals of FSC-accredited bodies which aim to check the destruction and degradation of natural forests and their indigenous communities (see 2.1). On the other hand, SFI takes a wide-selectivity approach. It raises the bar a little, for many, by setting self-imposed, minimum standards for the mainstream forest-products industry to meet.

The fact that the American Forest & Paper Association has made SFI certification a requirement for membership shows that SFI clearly raises the standards of forest management for a large portion of the industry in the United States. However, SFI does not raise the bar as high as the FSC program in terms of meeting many societal values for forestland (i.e., saving species or biodiversity, alleviating poverty, etc. — see 1.2, 3.1.2, and 3.2.2 for societal values relevant for sustainability). The SFI approach is geared more towards companies managing lands primarily for timber and/or game. Within the management site, SFI proposes for the protection of "special sites." In contrast, FSC emphasizes

TABLE 2.1

Amount of forestland in size classes by type of certification
(only includes FSC-accredited lands; FSC November 13, 1998).

Type of Forest Use or Owner	Certified Forestland Area (ha)					Total area by type, ha (% of all certified forests)
	<100 (# of sites) % within type	>100–1,000 (# of sites) % within type	>1,000–10,000 (# of sites) % within type	>10,000–100,000 (# of sites) % within type	>100,000 (# of sites) % within type	
Plantations	115 (2) 0.01%	4,035 (10) 0.4%	72,038 (22) 7.4%	467,697 (12) 47.8%	435,031 (2) 44.4%	978,916 (6.4%)
Natural — not industrial	171 (6) <0.01%	1,333 (5) 0.02%	35,611 (9) 0.7%	577,516 (11) 11.0%	4,620,103 (10) 88.2%	5,234,734 (34.3%)
Natural — industrial potential	85 (3) <0.01%	2,100 (5) 0.03%	21,423 (7) 0.4%	745,541 (12) 12.2%	5,320,959 (14) 87.4%	6,090,108 (39.9%)
Communal or Group-Certified	148 (3) 0.04%	1,586 (3) 0.4%	21,719 (6) 5.4%	381,262 (8) 94.2%	0	405,070 (2.6%)
Government-Owned	0	2,257 (5) 0.08%	3,348 (2) 0.13%	208,367 (5) 8.1%	2,357,063 (6) 91.7%	2,571,035 (16.8%)

the protection of "high conservation value forests aiming to maintain the overall biodiversity, productivity, and ecological processes." SFI contains no concrete criteria pertaining to biodiversity or chemical use (SFI 1998b). On the other hand, despite FSC's strong focus on the management of natural forests, the pressures and demands of the industrial-forest sector led to the development of specific standards for plantations.

In the United States, however, the development of specific standards for plantations was not sufficient to gain the support of the industrial-forest sector in FSC. The existence of the American Forest & Paper Association and the development of approaches such as SFI were more interesting to this economic sector. Because of the small need for North American forest industries, in contrast to Europe, to compete for more selective markets, SFI was a good approach to satisfy the local public. Therefore, in the United States, there are many more forest plantations certified under SFI than FSC.

In the United States, statistics of what types of forests have been certified are available for areas under the FSC approach (see 2.5). A small percentage of the total area certified is in plantations. Despite a great potential for these areas to be managed for industrial purposes, FSC has mainly pursued certifying forests which are managed for low-impact forestry (Table 2.1).

- Approximately 6% of the forests are being managed as plantations.
- About 40% of the natural forests have the potential for intensive management.
- About 34% are natural forests are not being certified for intensive management.

In the United States, another important approach is the system promoted by the American Tree Farm Association. This initiative is similar to SFI in that it attempts to raise the bar for sustainable management for a large group of forestland owners. It targets small forestland owners and requires them to write a management plan. This management plan moves these landowners towards documenting how they manage their lands. This is especially relevant since more than half of the privately held forestland is in the hands of these small-forestland owners, and only 5% currently have a management plan (Simpson 1998). This is not a minor point, since all the certification approaches require a management plan, and failure to become certified may be due to a lack of documentation (see 4.2.1).

Insofar as the FSC and SFI approaches to forest certification are driven by different values, each can play a complementary role in improving forest management on the ground. As long as each program clearly identifies the values on which certification is based, each serves a common purpose in providing information to consumers and the public at large. As such, the Federal Trade Commission's *Guides on the Use of Environmental Marketing Claims* and the draft ISO 14024 standards on environmental-labeling programs are potential tools for differentiating the messages of dueling forest certification programs. The FTC guides could provide safe-harbor language for advertising related to certification, whereas the use of ISO 14024 would entreat certifiers to clarify the selectivity level of their criteria.

2.5 Current statistics and characterization of certified forests by ownership type and land area

Kristiina A. Vogt, Anna Fanzeres, J. Scott Estey, Jennifer Heintz

Statistics on the ownership of forestlands around the world that have been certified are quite informative (Table 2.1–2.8). Much of the detailed documentation exists for forests certified under the Forest Stewardship Council approach (see 2.4.1.1). As of November 13, 1998 (quoted on January 29, 1999), a total forest area certified by FSC-accredited certification bodies was 12,334,552 hectares. Based on recent data, the total hectares must be decreased by 2860 ha, since the following two groups are no longer certified: Florestas Rio Doce, Brazil and The Solomon Islands Eco-Timber Trust, Solomon Islands. Presently, the FSC encompass six organizations:

- Skal, The Netherlands
- Institut für Marktökologie (IMO), Switzerland
- SGS's Forestry QUALIFOR Programme, United Kingdom
- Soil Association's Woodmark Program, United Kingdom
- Rainforest Alliance's Smartwood Program, United States
- Scientific Certification Systems's Forest Conservation Program, United States

The four latter organizations were the first to be accredited by the FSC and the numbers reported in this section are those that they have certified and are published in FSC documentation (Table 2.2). Some of these organizations predated the formation of the Forest Stewardship Council and have previous records of working as consultants and/or certifiers of industrial practices. In addition, several forests were certified prior to the development of the official FCS documentation on Principles and Criteria of Forest Management. The consequence of this has been that a considerable area of forestlands were previously classified as sustainably managed, or at least well-managed, even when they did not minimally comply with FSC standards (see Table 2.2).

The fact that forest certification protocols based on the FSC principles and criteria became an enormous laundry list of human-laden values is discussed in more detail in other sections of this book (see 1.1, 3.2). Besides the difficulties that this approach to forest certification will have to overcome technically, it is obvious that it will be necessary to have a serious discussion on the procedures used by certifiers. This self-evaluation process is fundamentally important if the FSC does not want to lose credibility and create an entangled mess in the minds of the public about what the correct message is, in terms of forest management.

The contributions of the FSC to the forest certification debate and forest-management practices in the industrial-forestry sector cannot be overemphasized. The response by the industrial forestry sector has been to develop parallel or competing approaches to forest certification. Despite the unfounded suspicion of environmental NGOs to these approaches, a positive consequence has been the promotion of management plans among many forestland owners who otherwise would not have pursued obtaining plans for their forests. In addition, it has resulted in the production of data on the qualitative state of forests.

The momentum to pursue non-FSC-type certification has been recent, but is rapidly increasing among the timber industry in the United States. Detailed statistics are just being produced on the numbers and the amount of land area by ownership that is certified using the non-FSC type certifications. Statistics for the non-FSC-type certification showed a considerable amount of land area certified under the SFI label. Under the SFI label, 21.1 million ha (52 million ac) of forestland had been certified as of January 1999. Since SFI does not individually certify forestlands, this number represents the total amount of forestland owned by SFI members. This statistic suggests that currently SFI has 1.7 times more forestland certified under their label compared to FSC-accredited bodies.

The process catalyzed by the FSC can lead to a significant change in the use and maintenance of forestlands in the United States. This is especially relevant if societal pressure continues to be exerted over organizations such as the American Forestry & Paper Association, and scientists move closer to concrete recommendations on how to conduct sustainable forest management.

The importance of another parallel approach can be illustrated by the initiative of the National Forestry Association. The Green Tag Forest certification program is a third-party certification system that was specifically developed for non-industrial woodland owners (Argow 1998). This certification follows the dogmas of the ISO Systems Management approach. The Green Tag program is currently available in only ten states, although there are plans for it to be made available in 36 states by the year 2000. Green Tag certification has certified nine forests in nine states in the United States as of November 30, 1998. The total acreage certified with this approach is a little over 405 ha (1,000 ac) (Argow 1998).

The influence of the promoters and supporters of the FSC-type of certification has also stimulated other types of developments in certification. In British Columbia, Canada, the debate on how to conduct and evaluate sustainable forest management involved significant public participation in the process, in addition to the input by industry and governmental agencies. The development of the British Columbia Forest Practices Code was a result of consensus building based on Canadian societal values of what were acceptable forest management practices to maintain ecosystem integrity and long-term socio-economic needs (Osberg and Murphy 1996). The forest industry in Canada, under much pressure, articulated a parallel answer that promoted the development of national standards for forest management.

The Canadian Sustainable Forest Management System became the first concrete conceptualization of an ISO type of approach to forestry (see 2.4.1.3). However, so far, no certified

TABLE 2.2

Pre-1998 forests that were certified, but no longer hold the certification label (as of November 13, 1998), and forests that have continued to maintain their certification. All forests had been certified through FSC-accredited organizations.

Ownership Certified (Location)	Certifying Organization	Year Certified	Length Certified (Years)	Forest Area Certified (ha)	Citation
Forests No Longer Certified					
Amacol Ltda (Pertel, Para, Brazil)	Rainforest Alliance	1991	5	59,000	1, 2
Bainings Community-Based Ecoforestry Project (Rabaul, Papua, New Guinea)	SGS Forestry	1994	5	12,500	1, 2
Chindwell Doors (Johor, Malaysia)	SGS Forestry	1994	1 initially	3,284	1
Demerara Timber Ltd (Guyana)	SGS Forestry	1994	1 initially	500,000	1
Perum Perhutani (Java, Indonesia)	Rainforest Alliance	1990	5	2,063,100; 2,831,500	1, 2
Plan Piloto Forestal (Mexico)	Scientific Certification Systems	1991	5	33,000	1
ISOROY, Gabon	SGS Forestry			210,000	2
African Charcoal (South Africa)	SGS Forestry			6,000	2, 5
Ulatawa Estates (Papua, New Guinea)	SGS Forestry			60,000	2, 5
Total forest area not maintaining certification as of November 13, 1998 = 2,946,884 ha (19.6% of previously certified)					
Forests Continuing to Maintain Their Certification					
Broadleaf Forest Development Project (Honduras)	Rainforest Alliance	1991	5	25,000	1, 2, 5
Collins Pine (Chester, California, U.S.)	Scientific Certification Systems	1993	5	38,300; 43,000; 37,600	1, 2, 5
Collins Pine/Kane Hardwood (Kane, Pennsylvania, U.S.)	Scientific Certification Systems	1994	5	48,300; 50,000; 46,400	1, 2, 5
Dartington Home Wood (Devon, United Kingdom)	Soil Association	1994	1	92	1, 5
Keneenaw Land Association Ltd. (U.S.)	Rainforest Alliance	1994	5	50,000; 62,726; 63,000	1, 2, 5
Menominee Tribal Enterprises (U.S.)	Scientific Certification Systems	1992	5	89,000; 97,500; 95,122	1, 2, 5

Pengelli Forest (Dyfed, Wales, United Kingdom)	Soil Association	1994	1	65	1, 5
Plan Piloto (Mexico)	Rainforest Alliance	1991	6	95,000; 86,215	1, 5
Portico S.A. (Costa Rica)	Scientific Certification Systems	1992	5	3,900; 10,000; 3,900	1, 2, 3, 5
Seven Islands Land Company (Maine, U.S.)	Scientific Certification Systems	1993	5	406,250; 390,000	1, 2, 5
Tropical American Tree Farms (Costa Rica)	Rainforest Alliance	1994	3	1,336; 1,380	1, 5
Ston Forestal (Costa Rica)	SGS Forestry			14,000; 14,929	2, 5
Big Creek Lumber (Santa Cruz, California, U.S.)	SCS			2,833; 2,720	2, 5
Flor y Fauna S.A. (Costa Rica)	Rainforest Alliance			3,500; 2,788	2, 5
CICOL (Santa Cruz, Bolivia)	Rainforest Alliance			40,000; 50,000	2, 5
Polish State Forest Service (Gdansk, Poland)	SGS Forestry			294,000	2, 5
Polish State Forest Service (Szczecinek, Poland)	SGS Forestry			622,563	2, 5
Polish State Forest Service (Katowice, Poland)	SGS Forestry			635,000	2, 5
Durawood Products (Zimbabwe)	SGS Forestry			24,850	2, 5
UZACHI (Mexico)	Rainforest Alliance			26,000; 24,996	2, 5
Stora, Ludvika Forest District (Sweden)	Scientific Certification Systems	1996		300,000	4, 5

Total forest area certified in November 13, 1998 = 12,124,552 ha (80.5% of previously certified)

1. Upton, C. and Bass, S., *The Forest Certification Handbook*., St. Lucie Press, Boca Raton, FL, 219, 1996.
2. de Callejon, D. P., Lent, T., Skelly, M. and Crossley, R., Marketing products from sustainably managed forests: An emerging opportunity, *The Business of Sustainable Forestry. Case Studies. A Project of the Sustainable Forestry Working Group*, The John D. and Catherine T. MacArthur Foundation, Chicago, IL, 3, 1998.
3. Diener, B. J. and Portico, S. A., Strategic Decisions 1982–1997. *The Business of Sustainable Forestry. Case Studies, A Project of the Sustainable Forestry Working Group*, The John D. and Catherine T. MacArthur Foundation, Chicago, IL, 12, 1998.
4. Fletcher, R. A., McAlexander, J. and Hansen, E., STORA: The road to certification, *The Business of Sustainable Forestry. Case Studies, A Project of the Sustainable Forestry Working Group*, The John D. and Catherine T. MacArthur Foundation, Chicago, IL, 14, 1998.
5. FSC, *Forests Certified by FSC-accredited Certification Bodies*, Forest Stewardship Council, Oaxaca, Mexico, 7 pages, 13th November, 1998c.

TABLE 2.3

Natural forests not certified for industrial uses by ownership, location, certifying organization, and land area; only forests certified by FSC-accredited organizations are included (FSC-November 13, 1998).

Ownership (Location)	Certifying Organization	Type of Ownership	Forest Area (ha)
CICOL (Bolivia)	Soil Association	Communal	52,000
Haliburton Forest and Wildlife Reserve (Canada)	Rainforest Alliance	Private reserve	19,184
Proyecto Desarrollo Bosque Latifoliado,	Rainforest Alliance	Communal	25,000
Cooperativa Agroforestal Colon (Honduras)	Rainforest Alliance	Communal	7,970
Magnifica Communitá di Fiemme ASL (Italy)	SGS Forestry	Communal	11,000
Deramakot Forest Reserve (Malaysia)	SGS Forestry	Government	55,000
Sociedad de Productores Forestales Ejidales de Quintana Roo;	Rainforest Alliance	Communal	86,215
UZACHI;	Rainforest Alliance	Communal	24,996
San Juan de Aguinaldos;	Rainforest Alliance	Communal	3,000
Echeverria de la Sierra (Mexico)	Rainforest Alliance	Communal	3,000
European Union-IRECDP Sulu, Kilu, Mareka Regions (Papua New Guinea)	SGS Forestry	Communal	4,310
Gdansk;	SGS Forestry	Government	294,000
Katowice;	SGS Forestry	Government	645,000
Szczecinek;	SGS Forestry	Government	622,563
Krakow (Poland)	SGS Forestry	Government	173,166
Stora Soderhamn Forest District;	SCS	Private	222,000
Assidomän Skog & Trä (Sweden)	SGS Forestry	Private	256,000
Soluthurn;	SGS Forestry	Government	2,095
Foundation Rollirain (Switzerland)	Insitut für Marktokölogie (Switzerland)	Private	17
Camphill Village Trust;	Soil Association	Communal	13
Coed Cymru, Wales;	SGS Forestry	Group	40
The Woodland Trust (United Kingdom)	SGS Forestry	Trust	9,850
Darcie Mahoney, California;	Rainforest Alliance	Resource Manager	353
Menominee Tribal Enterprises, Wisconsin;	Rainforest Alliance	Communal	95,122
Metropolitan District Commission's Quabbin Reservoir Lands, Massachusetts;	Rainforest Alliance	Government	23,577
Pennsylvania Department of Conservation and Natural Resources;	SCS	Government	485,615
Minnesota Aitkin County Land Department;	Rainforest Alliance	Government	89,876
Minnesota Department of Natural Resources, Aitkin County;	Rainforest Alliance	Government	146,719
The Kan Property, California;	Rainforest Alliance	Private	17
The Krantz-Kahan Property, California;	Rainforest Alliance	Private	24
Arcata City Forest, California;	Rainforest Alliance	Government	486
Still Waters Farm, Washington;	Rainforest Alliance	Private farm	60
Ecoforestry Management Associates, Oregon;	Rainforest Alliance	Resource Manager	142
Ecoforestry Institute, Oregon;	Rainforest Alliance	Private Institute	170
Forest, Soil and Water, California;	Rainforest Alliance	Resource Manager	182
Vermont Family Farms;	Rainforest Alliance	Resource Manager	2,666
Robert Whittaker Family Property, California;	Rainforest Alliance	Private family farm	1,175
Randy Jacobszoon, California;			
Menominee Tribal Enterprises, Wisconsin	Rainforest Alliance	Resource Manager	1,545
(United States)	SCS	Communal	95,546
Muzama Crafts Ltd. (Zambia)	Soil Association	Communal	1,273,700

TABLE 2.4

Resource managers certified by location, certifying organization and land area. This only includes those certified by an FSC-accredited organization (FSC November 13, 1998).

Ownership (Location)	Certifying Organization	Forest Area (ha)	Plantation = P; Natural = N
Edward Tunheim, Santa Cruz, California	Rainforest Alliance	8,094	P
Darcie Mahoney, California	Rainforest Alliance	353	N
Randy Jacobszoon, California	Rainforest Alliance	1,545	N
Ecoforestry Management Associates, Oregon	Rainforest Alliance	142	N
Forest Soil and Water, California	Rainforest Alliance	182	N
Vermont Family Forests, Vermont	Rainforest Alliance	2,666	N
Eric Huff, California	SCS	1,497	?
Mike Jani, California	SCS	1,497	?
Steve Staub Forestry, California (United States)	SCS	90	?

forest products are available and no forests have been certified using the Canadian Standard Association (CSA) set of standards. This lack of certification of forestlands is logical considering that the Canadian standards development for this approach was only concluded in 1996 (Abusow 1997, de Callejon et al. 1998, LaPointe 1998). This statistic will probably change in the near future. For example, at least 15 forest products companies and 12,000 private woodlot owners were implementing the Canadian Sustainable Forest Management system standards on their lands in 1997 with the idea of obtaining the certification label in the future (Asubow 1997).

The ISO Management System approach is also the preferred tool of the Australian and New Zealand forest industries for assessing their forestlands (Mankin 1998). However, future statistics will show how effectively this system is implemented in these countries in relation to the FSC-type approach. In addition, in these two countries the intensity of the debate on sustainable forest management led to the development of codes of practice (McCormack 1996, Visser 1996). Probably one of the most interesting aspects of these initiatives has been the development of dialogue between previously opposing segments of society. Environmental groups, government officials, industry, and labor representatives had to work jointly to transform confrontations into solutions. This formula is far from perfect, but has significantly contributed to advancing forest management practices. For example, these initiatives are developing low-impact logging techniques to maintain forest structure, promoting better planning for minimizing land disturbance (i.e., fragmentation and introduction of exotic tree species, opening of roads), and refining the methodology to evaluate natural systems. These new proposed practices might provide forest management with unprecedented changes in their tools (see 2.1, 3.1.3, Chapter 6).

Tropical countries also are responding by developing improved forest management techniques. Research institutions in countries such as Costa Rica and Brazil have been working to produce field manuals for forest management (Beek 1996, Amaral 1998). Furthermore, guidelines for logging/harvesting of natural forests have been developed for countries such as Fiji, South Africa, and Malaysia, (Jusoff and Mustafa 1996, Strehlke 1996, Warkotsch et al. 1996). Moreover, certification initiatives by FSC-accredited certifiers in the less-developed nations are helping to create dialogue with governmental agencies where the logging industry plays a strong economic role. Countries such as Malaysia and Indonesia are working to develop national protocols (see 2.2) due to their viewpoint that international initiatives, such as the FSC, are a threat to their sovereignty. In Indonesia, FSC-type certification continues to evolve despite problems with past operations that were originally certified (i.e., part of the area certified by Rainforest Alliance has been decertified; see Table 2.2). The weak legitimacy of governmental initiatives in the eyes of consumers from developed nations (see 2.1) is recognized by the Indonesian government. Therefore, the Indonesian government submitted their protocol for FSC approval (issued in 1998). The question then becomes whether or not all these new efforts are necessary to develop a protocol. Would it be better to direct more energy and resources into developing a set of regional standards as proposed by FSC (FSC 1994)?

Presently, clear statistics that show the amount of forestlands certified under the different labels are not organized sufficiently for an adequate evaluation of their contribution to sustainable forest management around the world. The little existing data on the amount of forestland certified by the different organizations can be summarized as follows:

- SFI (U.S.)– **21.1 million** hectares
- FSC (Worldwide)– **12.3 million** hectares
- Green Tag Forest (U.S.)– **405** hectares

The amount of forestland that will be certified under the different labels is expected to increase. These numbers have to be interpreted in context of the sharp growth of certified areas in a very short time period (Bruce 1998). For example, the amount of area certified by FSC-type certification was around 30,000 ha in 1994 (Bruce 1998) and the area has grown dramatically since that time (Tables 2.1, 2.5). Even the Sustainable Forestry Initiative of the AF&PA was just developed in 1996, so the addition of new members to the organization should result in an increase in the amount of forestland under its program.

This increase in certified areas, however, cannot be evaluated simplistically as meaning that the concept of sustainable forest management is spreading. One very important aspect to be considered is the type of forestland being certified. The global statistics on ownership and land area certified under the FSC label showed that most individual tracts of land areas were greater than 10,000 ha in size. When summing up the number of individual tracts of certified land by type, almost half are greater than 10,000 ha in size, a quarter are >1000–10,000 ha in size, and 18.9% are >100–1000 ha in size. The following statistics summarize the predominance of large tracts of forestland to be certified under the FSC label (Table 2.1):

- Most (**91.7%**) of the certified government-owned forests had land areas that individually were **greater than 100,000 ha** in size.
- Most (**87–88%**) of the certified natural forests with industrial or non-industrial potentials had land areas that individually were **greater than 100,000 ha** in size.
- Most (**92.2%**) of the certified plantations had land areas that individually were **greater than 10,000 ha** in size.
- Most (**94.2%**) of the communal/group certifications were on land areas **greater than 10,000 ha** in size.
- When summing up all the certified forestland areas are **less than 100 ha** in size, these small landholdings covered less than **0.1%** of the total certified forestland area.

This low percentage of land area in small landholding sizes (<100 ha), accounted for 8.3% of the total number of individual tracts of land (Table 2.1). The importance of small

TABLE 2.5

Natural forests certified for industrial or private uses by ownership, location, certifying organization, and land area; only forests certified by FSC-accredited organizations are included (FSC November 13, 1998).

Ownership (Location)	Certifying Organization	Type of Ownership	Forest Area (ha)
Programme for Belize (Belize)	Rainforest Alliance/ Soil Association	NGO	95,800
CIMAL;	Rainforest Alliance	Wood Company	87,562
Empresa Maderera Taruma Ltda;	Rainforest Alliance	Wood Company	83,467
Empresa Agroindustrial La Chonta Ltda (Bolivia)	Rainforest Alliance	Wood Company	120,000
Mil Madeireira Itacoatiara Ltd. (Brazil)	Rainforest Alliance	Wood Company	80,571
Tembec Inc.-Huntsville Division;	SCS	Industrial	939
J.D. Irving Limited (Canada)	SCS	Industrial	190,969
Portico S.A.;	SCS	Wood Company	3,900
Fundacion TUVA (Costa Rica)	Rainforest Alliance	Industrial	500
Masaryk Forest (Czech Republic)	Soil Association	Private/Industrial	10,441
Yaguarete Forest L.P. (Paraguay)	SCS	Private/Industrial	16,000
AssiDomän Skog & Trä, Värmland & Karlskoga;	SGS Forestry	Private/Industrial	60,700
Korsnäs Aktiebolag Skog;	SGS Forestry	Private/Industrial	664,200
Stora, Ludvika;	SCS	Private/Industrial	300,000
AssiDomän Skog & Trä, Älbsvyn;	SGS Forestry	Private/Industrial	676,000
AssiDomän Skog & Trä, Kalix;	SGS Forestry	Private/Industrial	778,000
Stora, Mora;	SCS	Private/Industrial	300,000
Stora, Sveg;	SCS	Private/Industrial	285,000
Stora, Strömsberg;	SCS	Private/Industrial	92,000
AssiDomän Lycksele;	SGS Forestry	Private/Industrial	727,000
AssiDomän Skog & Trä, Värnamo;	SGS Forestry	Private/Industrial	141,000
AssiDomän Ostersund;	SGS Forestry	Private/Industrial	180,000
Skogssällskapet AB;	Soil Association	Private/Industrial	72,000
AssiDomän Skog & Trä AB Växjö;	SGS Forestry	Private/Industrial	171,000
Hedemora Skogsförvaltning (Sweden)	SGS Forestry	Private/Industrial	397,790
Dyfed Wildlife Trust, Wales (United Kingdom)	Soil Association	Private/Industrial	65
Big Creek Lumber Company, California;	SCS	Timber Company	2,720
Collins Pine Company, California;	SCS	Timber Company	37,600
Blencowe Managed Forest Lands, California;	Rainforest Alliance	Private/Industrial	4,878
Kane Hardwood/Collins Pine Company, Pennsylvania;	SCS	Timber Company	46,400
Keweenaw Land Association, Michigan;	Rainforest Alliance	Land Association	63,000
McClellan Mountain Ranch, California;	Rainforest Alliance	Private Ranch	133
Seven Islands Land Company, Maine;	SCS	Timber Company	390,000
Tree Shepherd Woods, Washington;	Rainforest Alliance	Private	8
Two Trees Forestry, Maine;	Rainforest Alliance	Industrial	3,572
Northeast Ecologically Sustainable Timber, New Hampshire;	Rainforest Alliance	Private/Industrial	1,366
Wylatti Timber Management Company Ltd., California;	Rainforest Alliance	Private/Industrial	365
Our Lady of the Redwoods Abby, Inc., California;	Rainforest Alliance	Private	163
Surface Road Associates, Washington;	Rainforest Alliance	Private	12
Red Hills Lumber, Massachusetts;	Rainforest Alliance	Lumber Company	2,945
Individual Tree Selection Management, Inc., Oregon (United States)	Rainforest Alliance	Private/Industrial	2,042
Durawood Products (Zimbabwe)	SGS Forestry	Private/Industrial	24,850

TABLE 2.6

Listing of ownership, locations, certifying organization, and amount of land area of forests certified as plantations under FSC-accredited certifiers (FSC November 13, 1998). The forest area in plantations is an overestimate, since forest areas are not listed separately as to how much is in natural and plantation.

Ownership (Location)	Certifying Organization	Forest Area (ha)	Plantation = P Natural = N
Zoniënwoud;	Soil Association	2,452	P and N
Heverleebos;	Soil Association	4,684	P and N
Meerdaalwoud (Belgium)	Soil Association	3,146	P and N
Floresteca Agroflorestal Ltda (Brazil)	SGS Forestry	3,000	P
Eucatex S.A.;	SCS	29,340	P
Duratex S.A. Botucatu;	SCS	47,904	P
Plantar S.A. Fazenda de Almas (Brazil)	SCS	3,118	P
Klabin Fabricadora de Papel e Celulose S.A. Paraná (Brazil)	Rainforest Alliance	221,000	P
Flor y Fauna;	Rainforest Alliance	2,788	P
Tropical American Tree Farms;	Rainforest Alliance	1,380	P
FUNDECOR (Costa Rica)	SGS Forestry	14,929	N and P
Umweltbehörde Hamburg (Germany)	Skal	4,938	P and N
Koninklijke Houtvesterij Het Loo;	Skal	8,058	P and N
Gemeentebossen Ede;	Skal	2,044	P and N
Gemeentebossen Arnhem;	Skal	1,304	P and N
Gemeente Apeldoorn;	Skal	571	P and N
Gemeente Renkum;	Skal	283	P and N
Gemeente Lelystad;	Skal	465	P and N
Staatsbosbeheer Regio Flevoland-Overijssel;	Skal	18,200	P and N
Staatsbosbeheer Regio Groningen-Drenthe (The Netherlands)	Skal	21,714	P and N
Craigpine Timber Ltd;	SGS Forestry	2,305	P
Rayonier New Zealand Ltd. (New Zealand)	SGS Forestry	42,720	P
Futuro Forestal (Panama)	Rainforest Alliance	23	P
Bracken Timbers;	SGS Forestry	4,500	P
SAFCOL Kwa-Zulu Natal Province;	SGS Forestry	51,922	P
SAFCOL Mpumalanga South;	SGS Forestry	51,196	P
Soetmelk Boerdery;	SGS Forestry	300	P
Kanhym Landgoed;	SGS Forestry	602	P
Majuba Hill Farms;	SGS Forestry	1,458	P
SAFCOL Eastern Cape;	SGS Forestry	42,714	P
Mondi Forests North;	SGS Forestry	214,031	P
SAFCOL West Cape;	SGS Forestry	53,530	P
SAFCOL Mpumalanga North (South Africa)	SGS Forestry	64,378	P
Dipped Products Ltd.;	SGS Forestry	4,582	P
RPK Management Services;	SGS Forestry	5,476	P
Horana Plantations Ltd. (Sri Lanka)	SGS Forestry	2,668	P
Wildlife Trust West Wales;	Soil Association	92	N and P
Duchy of Cornwall;	Soil Association	1,004	N and P
National Trust;	Soil Association	240	N and P
Bioregional Charcoal Company Limited;	SGS Forestry	2,037	N and P
Big K Charcoal Merchants Limited;	Soil Association	862	N and P
Elveden Farms Ltd.;	Soil Association	1,600	N and P
Plashett Wood;	Soil Association	150	N and P
Wrekin District Council (United Kingdom)	Soil Association	452	N and P
Tunheim (Resource Manager, California);	Rainforest Alliance	8,094	P
Collins Lakeview Forest, Oregon;	SCS	29,150	N and P
Maine Forestry;	SCS	1,402	P
The Brown Tract, Oregon (United States)	Rainforest Alliance	110	N and P

TABLE 2.7

Communal or group certifications of forestlands by ownership, location, and area; only those certified by FSC-accredited organizations (FSC November 13, 1998).

Ownership (Location)	Certifying Organization	Forest Area (ha)	Plantation = P Natural = N
CICOL (Bolivia)	Rainforest Alliance	52,000	N
FUNDECOR (Costa Rica)	SGS Forestry	14,929	N and P
Proyecto Desarrollo Bosque Latifoliado;	Rainforest Alliance	25,000	N
Cooperativa Agroforestal Colon (Honduras)	Rainforest Alliance	7,970	N
Magnifica Communitá di Fiemme ASL (Italy)	SGS Forestry	11,000	N
Sociedad de Productores Forestales Ejidales de Quintana Roo;	Rainforest Alliance	86,215	N
UZACHI;	Rainforest Alliance	24,996	N
San Juan de Aguinaldos;	Rainforest Alliance	3,000	N
Echeverria de la Sierra (Mexico)	Rainforest Alliance	3,000	N
European Union-IRECDP Sulu, Kilu, Mareka Regions (Papua, New Guinea)	SGS Forestry	4,310	N
Kanhym Landgoed (7 farms) (South Africa)	SGS Forestry	602	P
Bioregional Charcoal Company Limited (21 woodlands);	SGS Forestry	2,037	N and P
Big K Charcoal Merchants Limited;	Soil Association	862	N and P
Coed Cymru, Wales;	SGS Forestry	40	N
Camphill Village Trust (United Kingdom)	Soil Association	13	N
Menominee Tribal Enterprises, Wisconsin;	Rainforest Alliance	95,122	N
Maine Forestry (United States)	SCS	1,402	P

TABLE 2.8

Data on government-owned (includes local, county, state, and federal) certified forests; only those certified by FSC-accredited organizations (FSC November 13, 1998).

Ownership (Location)	Certifying Organization	Forest Area (ha)	Plantation = P Natural = N
Deramakot Forest Reserve, Sabah (Malaysia)	SGS Forestry	55,000	N
Gemeentebossen Ede;	Skal	2,044	N and P
Gemeentebossen Arnhem;	Skal	1,304	N and P
Gemeente Apeldoorn;	Skal	571	N and P
Gemeente Renkum;	Skal	283	N and P
Gemeente Lelystad;	Skal	465	N and P
Staatsbosbeheer Regio Flevoland-Overijssel;	Skal	18,200	N and P
Staatsbosbeheer Regio Groningen-Drenthe (The Netherlands)	Skal	21,714	N and P
Gdansk;	SGS Forestry	294,000	N
Katowice;	SGS Forestry	645,000	N
Szcezecinek;	SGS Forestry	622,563	N
Krakow (Poland)	SGS Forestry	173,166	N
Wrekin District Council (United Kingdom)	Soil Association	452	N and P
Quabbin Reservoir Lands, Massachusetts;	Rainforest Alliance	23,577	N
Pennsylvania Department of Conservation & Natural Resources;	SCS	485,615	N
Minnesota Aitkin County Land Department;	Rainforest Alliance	89,876	N
Minnesota Department of Natural Resources, Aitkin County;	Rainforest Alliance	146,719	N
Arcata City Forest, California (United States)	Rainforest Alliance	486	N

forestland holdings for the maintenance of healthy forest ecosystems can be illustrated in many ways. In highly developed nations such as United States, the fragmentation and encroachment on forestlands create a challenge for the design of sustainable practices. It also demands coordination among neighbors due to edge effects and other variables (see 5.2.1.2). In regions where forestlands still occupy vast and continuous tracts of land (i.e., Amazonia), small landowners/settlers contribute an apparently insignificant amount of timber extraction, but isolated individuals lead to highgrading of the forest (Otavio and Mendes 1998). If the promotion of forest certification aims to improve forest practices worldwide (see 1.2), then this type of landholding will have to be more seriously targeted.

Another important aspect of this discussion is whether forest certification is promoting an alternative for deforestation worldwide. Statistics on the type of forest cover, its ownership, and the purpose for it certification do not support that this is happening. The numbers show that for the total of 175 forestlands that were certified in 1998 by FSC-accredited bodies, no single type of forest use or owner category predominated (Table 2.1).

TOTAL NUMBER OF CERTIFIED FORESTLANDS BY CATEGORY:

- 24.4% were in natural forest not managed for industrial purposes
- 28.6% were in plantations
- 24.4% were in natural forests managed with industrial potential
- 11.9% were in communal or group certifications
- 10.7% were in government owned lands

Despite the appearance when counting the number of owners that there is no bias towards any particular type of forestland ownership pursuing certification, close to a third (34.3%) of the total area (as of November 13, 1998) is certified as natural forests, rather than for intensive management for timber (Table 2.1). Currently, a small proportion of forestland areas certified under the FSC label have been communal or group certifications — with 2.6% of the total forestland certified under this type of ownership (Table 2.1).

Communal land area never exceeded 100,000 ha in size. Most of the communal or group-certification (94.2%) land area was found in the >10,000 to 100,000 ha size class. This large land size represented 40% of the total number of individual communal tracts of land (of course, each communal tract of land has many owners). When counting the total number of communal lands certified, 70% of the total were comprised of large tracts of land that were in the > 1000 to <100,000 ha size categories. Although less than 1% of the communal land area was in the smaller size classes (<1000 ha), it accounted for 30% of the total number of individual tracts of land (Table 2.1). Most of the communal lands were certified as natural forest with no industrial forestry (Table 2.7). These lands are generally not managed for timber export. They are managed to satisfy local or regional consumption of timber and non-timber forest products (see 7.8). However, two of the group certifications are in plantation forests, so forest management for timber export does exist in this category.

> For natural forests not managed for industrial purposes, 88.2% of the land area is in tracts greater than 100,000 ha in size, even though this represents only 24.4% of the number of individual tracts of land in this category.

These statistics for natural forests mainly reflect four land tracts in Poland that represent a third of the total certified forestland area. Small landholding sizes do not exist in this category since less than 1% of the land area is found in tracts less than 10,000 ha in size

(Table 2.1). Although little area existed in this category, 48.7% of the individual tracts of land were recorded in this category, suggesting that the holdings are smaller per owner. The classification of the type of ownership in this grouping showed that much of this land is communal, government-owned, or held in private reserves (Table 2.3). In the tropical locations, indigenous communities harvest timber and non-timber forest products from these natural forests, but do not actively manage them for export of forest products. This grouping also included the nine certified resource managers who are presently found only in the United States (Table 2.4).

Similar to the natural forests not managed for industrial purposes, natural lands with industrial potential have a similar proportion of land area in large land tracts and little in smaller tracts. For example:

> For natural forests managed for industrial purposes, 87.4% of the land area is found in tracts greater than 100,000 ha in size and less than 1% of the land area is found in tracts less than 10,000 ha in size (Table 2.1).

When counting the individual number of certified forests within a category, 63.4% of the total tracts of forestland were found in land areas >10,000 ha in these natural forests with industrial potentials (Table 2.1). Ownership in the >10,000 — 100,000 ha size class (12.2% of the total land area ownership in this category) is mainly dominated by mid-size companies associated with the timber industry (Table 2.5). In the United States, most of the companies in this size class are family-owned companies that have managed the land for several generations and pursued certification under the Rainforest Alliance or SCS label (see 7.6). In the >100,000 ha land area grouping, 79.6% of the total natural forests managed for industrial purposes are found in Sweden and consist of 14 different tracts of land (Table 2.1, 2.5).

> Out of the total area of FSC-certified forests globally, plantations accounted for a small part of the total (6.4%) land area (Table 2.1).

On an area basis, both the >10,000–100,000 and >100,000 ha groupings had similar land area certified in plantations and accounted for 92.2% of the total area, but for only 29.2% of the number of individual tracts in that category (Table 2.1). The greatest number of individual tracts (45.8% of the total number) were found in the >1000–10,000 land area class even though this accounted for only 7.4% of the total area. The total land area in plantations is probably an overestimate of the actual certified area in plantations since many areas are classified as natural and plantation, with the area in plantations not specified (Table 2.6). The greatest number of total tracts in plantations occurred only in South Africa, where ten individual areas have been certified as plantations (Table 2.6). Brazil and Sri Lanka have the next highest number of tracts of land certified as plantations — with Brazil having four, certified plantations and Sri Lanka having three.

> Government-owned certified forestlands accounted for less than a fifth (16.8%) of the total certified forestlands globally in 1998 (Table 2.1).

The government category consists of city, county, state, and federally owned lands. Government-owned lands that have been certified are located in Malaysia, The Netherlands, Poland, the United Kingdom, and the United States (Table 2.8). More than half of the government-owned lands were classified as natural with no management; less than half were classified as lands with the potential for industrial forestry (Table 2.8). Government-owned lands were found predominantly in the largest land area classes (91.7% of the total found in the >100,000 ha size category) when comparing within the same category (Table 2.1). This category represented 33.3% of the total number of individual tracts of forestland — suggesting a few individual tracts of land with large areas. Over half of the individual land tracts were quite large for government owned lands — 61.1% of the total number of individual tracts were greater than 10,000 ha in size. No land areas less than 100 ha were recorded for government-owned lands (Table 2.1).

3

Definitions and Current Values Integrated into Certification Protocols

Kristiina A. Vogt, Anna Fanzeres

To understand the advantages and disadvantages of the current certification protocols, it is important to understand some of the terminology that has driven its development. These terms commonly discussed in certification have had a strong influence on the approaches and tools used to assess ecosystems. The two most common terms heard in certification are *criteria* and *indicators*. Much time and energy has been placed into explicitly defining criteria and indicators since these are the tools that express the values inherent to assessment protocols. Because these terms are so central to certification, they will be discussed below.

Other terms that are important to understand are *sustainability* and *good forest management*. In attempting to define the tools for certification, sustainability of the social and natural system is one of the primary goals of pursuing certification. Because of the difficulty in defining sustainability, and therefore the indicators to assess it, there has been a movement away from using sustainability by some certifying groups (see 2.1). This movement has been to define the forest as well managed. This raises the question of what it means to be well managed and what the indicators are that one would look for to measure it. These terms are important here because, in order to become certified, they define the condition which forestland owners are expected to move towards, or to satisfy, when managing their forestlands.

Since values continue to crop up as terms that drive assessment protocols, it is important to understand which values are embedded in certification. In the current certification protocols (especially those under the umbrella of the Forest Stewardship Council), human values for species play a significant role in defining the structure of protocols (see 2.1). Since humans value species and no extinction of species is acceptable, species are frequently used as the indicators of system health. Because of this, it is important to understand the relevance of using species as indicators of system condition. It is also important to identify in which systems species and ecosystem functions are strongly linked so that species become good indicators of system health. In order to familiarize the reader with this subject, the scientific basis for the role of species in ecosystems is presented in Section 3.2.1.

Certification also requires the integration of the natural and social systems in a manner similar to what is being attempted as part of ecosystem management (Vogt et al. 1997a). There is an acceptance that people cannot be isolated from the natural system and that the functioning of the natural system cannot be understood without determining how the social system affects it (see 2.1). Understanding this fact has led researchers to combine in the same protocol indicators for both the social and natural system. The question that needs to be asked is: how well do they integrate in the protocols? Since the integration of the social and natural sciences is recent, it is important to understand how past attempts at integration have worked and why they haven't. The importance of values in defining how each discipline is interpreted is relevant if frameworks for the assessment of forest practices are to be developed that concretely evaluate the driving variables of the system.

3.1   Defining terminology

3.1.1   Criteria and indicators

Kristiina A. Vogt, Anna Fanzeres, Daniel J. Vogt

Principles, criteria, and indicators are topics that have received much attention by the certifying organizations, NGOs, the forest industry, and the scientific community. Principles are the overarching ethical/philosophical guidelines that define sustainable forest management. Criteria and indicators are tools used in defining whether sustainable management is occurring. The development of certification protocols under this structure must be understood in light of the changes within society (especially in developed nations) and the growing influence of NGOs (see 2.1, 2.2, 3.1.4). This framework has become the ultimate unquestionable mechanism on how sustainable forest management should be conducted and evaluated (see 1.2). The consequences of such inflexible approaches and alternatives are discussed in several chapters of this book. However, to set the stage for this discussion we would like to present some of the most-quoted definitions of principles, criteria, and indicators being used in the forest certification debate.

Bueren and Blom (1997) conducted an analytical exercise that summarized the development of principles, criteria, and indicators. They proposed that three processes, or types of actors, have been at the forefront of these initiatives:

1. Political process, involving international organizations (e.g., United Nations agencies, ITTO, NGOs)
2. Operational process, involving certifying organizations (e.g., FSC, ISO, national governments, NGOs)
3. Scientific process, where the Center for International Forestry Research (CIFOR) has been a focal point. However, others have also played an important role in this process (i.e., the World Resources Institute [WRI], the International Institute for Environment and Development [IIED], and the Tropenbos Foundation) (see Box 3.1)

Box 3.1
Research organizations working on forest certification issues

Center for International Forestry Research (CIFOR)
CIFOR is headquartered in Indonesia and was established as part of the Consultative Group on International Agricultural Research (CGIAR). Its objectives are: improving the understanding in developing countries of the physical, biological, economic, and social aspects of forest

systems; increasing sustainable production from forests through management, species selection/improvement, utilization, marketing, and policy; providing information and advice to support policy decisions on forestland use; and, increasing national forest research capacities. The first major research project undertaken by CIFOR was on "Testing Criteria and Indicators for the Sustainable Management of Forests" at the forest-management unit level. This initiative has been divided into two testing phases to be completed by 1998/1999 when a "tool box" approach to sustainability assessment at the management unit level will be released (Prabhu et al. 1996).

World Resources Institute (WRI)

This organization is based in Washington, D.C., and is known worldwide. It conducts policy studies that revise indicators and certification initiatives.

International Institute for Environment and Development (IIED)

This organization is based in London and was deeply involved in the groundwork of the TFAP. It maintains close ties with ITTO. Presently, IIED is conducting policy studies of incentives for certification, the economics of certification, and forest resource accounting.

Tropenbos Foundation

This Netherlands-based institution released a study in early 1997 entitled *Hierarchical Framework, Forest Management Research*. This was a product of CIFOR researchers and has been widely used by all organizations working on principles and criteria.

Bueren and Blom (1997) also pointed out problems with the tools being developed for certification, such as "a lack of consistency in the use of terms and concepts, based possibly on an incomplete understanding of their meanings." However, their suggestions for improvements were limited because they focused exclusively on a better hierarchical arrangement of these categories.

The search for more clear and operational definitions of principles, criteria, and indicators has also been pursued by CIFOR (Prabhu et al. 1996). For this institution, nine attributes must be present in a working set:

1. Relevance of the criteria and indicators
2. Unambiguously related to the assessment goal
3. Precisely defined
4. Diagnostically specific
5. Easy to detect, record, and interpret
6. Reliability, especially as indicated by replicability of results
7. Sensitive to stress on the forest management, ecological, or social systems
8. Provides a summary or integrative measure over space and/or time
9. Appealing to users

This list of parameters is well thought out and is based on a good scientific background and common sense. During the last five years, CIFOR has placed considerable resources into promoting numerous field tests with existing protocols and has promoted meetings to refine existing parameters for the evaluation of sustainable forest management. However, most of the efforts of this organization have been in examining and testing pre-existing frameworks used to identify principles, criteria, and indicators. It has not proposed very strong new tools that might be more appropriate to evaluate the responses of forest ecosystems subjected to human interference and influence.

The definition of a principle extracted from the *Concise Oxford Dictionary* and presented by Prabhu et al. (1996), is "a fundamental truth or law as the basis for reasoning or action." Criteria are the components of this structure that would operationalize sustainability by providing endpoints (Prabhu et al. 1996). Also, Prabhu et al. (1996) stated that "criteria are the intermediate points to which the information provided by indicators can be integrated." These societal desirable endpoints became the driving force when analyzing a natural system. The major consequence of this approach is that the legacies driving the systems' responses are not being taken into consideration (Vogt et al. 1999b, see 5.2).

Criteria were also defined by Wijewardana et al. (1997) as follows:

> Vital functions and attributes of forest ecosystems (such as biological diversity and forest health); the multiple socio-economic benefits of forests (such as timber, recreation and cultural values); and in most processes, the policy framework (laws, regulations, and economic measures) needed to facilitate sustainable forest management.

This definition incorporates socio-economic sustainability as a factor in the decision-making process as to whether a forest is deemed certifiable. Most of these social and economic factors are an externality to the Forest Management Unit (FMU) itself, although they will drive how the ecosystem functions as a unit (Vogt et al. 1999ab). Ecosystem management explicitly links the social and natural sciences as an integral part of managing systems — it states that the biological portion of the system cannot be studied in isolation from the social system (Vogt et al. 1997a).

> Simplistically, *criteria* can be considered the main components of ecosystems that have been measured by researchers attempting to understand their functioning. Criteria encompass all the components or variables that help to produce an ecosystem, and as such, could be termed as an ecosystem "catalogue."

Depending on the scale of examination, an ecosystem's criteria may comprise all abiotic and biotic components of that system. These may include such measurements as those made on plants (woody and non-woody); soils; mammals; reptiles; amphibians; insects; microbes (symbiotic and pathogenic); insects; nutrients and trace elements (gaseous and mineral); atmospheric and soil gases; and water in its many forms. As part of the repertoire of ecosystems' variables, it is also important to consider the social and economic structures overlapping physically within the natural system or imposing effects (despite distance) on the natural components. Institutional arrangements and mechanisms (see 2.2) and societal values (see 3.2.2) are fundamental elements, given the fact that natural systems are embedded in them, such as the egg yolk in the egg white (Vogt et al. 1997a)

Each of these criteria can be further broken down into smaller components that begin to identify the diversity of structures and functions within each group (called indicators, and more specifically at the local level, verifiers).

> *Indicators* are defined as variables that express the functioning of the criterion and are therefore the tools that are used to assess whether the criterion requirements have been satisfied (see Chapter 4, 5.1).

The identification of indicators alone has produced an enormous literature base. Most of this is a result of attempts to identify and catalogue all the variables that could be considered relevant in assessing ecosystems in general and societal structures (UNDP 1990, Constanza 1992, McKenzie et al. 1992, Schneider 1992, Vogt et al. 1993, Hammond 1995, Jordan 1995, Markham 1995). This focus on indicators is reflected in the significant data requirements of most certification protocols (see Chapter 4).

In accordance with Prabhu et al. (1996), these are "primary tools of evaluation" which are involved in "information sharing." The international or more generally applicable exercises on the development of indicators have chosen this hierarchical unity for the description of thresholds. The implication of this choice is that the large-scale harmonization (national, international, or even global) being sought leads to an extremely subjective language when describing these parameters.

The lowest hierarchical level present in certification protocols is known as "verifiers" (Bueren and Blom 1997). The considerations to these parameters, which would really allow for on-the-ground verification, have been minimal. Bueren and Blom (1997) recognized this when they stated that there "*may be a need* to describe the way indicators are measured in the field" (italics are ours). On the other hand, CIFOR states that these parameters are where "thresholds can be defined" (Prabhu et al. 1996). However, because they are "extremely site-specific" (Prabhu et al. 1996), they have been ignored or have received little attention in the field tests conducted by this organization.

It is important to understand that "criteria and indicators are not, in-and-of themselves, performance standards or measures" (Wijewardana et al. 1997). When criteria and indicators are assessed over different periods, it is really performance which is assessed. Performance-based assessments can occur only if the same types of data have been collected at several different time intervals (i.e., forest inventory or growth data collected annually for several years). If more attention had been given to the development of verifiers, the working limitations of this framework, based on principles, criteria, and indicators, would probably have become clear and alternatives would have been sought.

3.1.2 Sustainability

Kristiina A. Vogt, Brian Rod, Toral Patel-Weynand, Anna Fanzeres, Bruce C. Larson, Indah Kusuma, Christie Potts, Allyson Brownlee, Jennifer Heintz, Heidi Kretser, Andrew Hiegel

There is an assumption that certification will result in sustainably managed forests (Granholm et al. 1996, Upton and Bass 1996, Viana et al. 1996, FAO 1997). Because of this, the definition of what sustainability entails becomes important. There is, however, no universally accepted definition for sustainability. Defining sustainability poses many problems since the terminology is vague, not easily quantified, and no "sustainable" project exists to be used as a standard. The concept is too young, and the time-scale is too long to adequately assess whether or not any management system is sustainable. In fact, the inability to define sustainability has driven some of the current interest to move from using the term, to talking about "good forest management" or "forest stewardship" (Mankin 1998). Perhaps sustainability has been perceived to be unattainable because it has been viewed as a term that must be defined, rather than as a concept which expresses the persistence of forests through time. How sustainability has been defined and interpreted since its identification as a useful term for management and sustainable development follows.

3.1.2.1   Definitions of sustainability

In 1987, the World Commission on Environment and Development developed the first widely accepted definition of sustainability (see 2.1). Their definition of *sustainability* was the "ability to meet the needs of the present without compromising the ability of future generations to meet their own needs" (WCED 1987).

This introduced the idea of using sustainability as a mechanism for determining the thresholds for permissible development to achieve the present and future needs of communities.

Sustainable development has been defined more holistically, but without addressing ecosystem health as "meeting human needs without depleting natural resources or irrevocably damaging the system which produces those resources, while establishing equitable and viable patterns of living throughout the world" (Crump 1991).

This definition did not expand on the interdisciplinary nature of this concept. It also failed to address the complexities of sustainability and the links between each of the disciplines.

The need for an interdisciplinary approach to the definition of sustainability has been perceived by a number of scholars and policymakers, despite the difficulties and low practical successes that have resulted from attempts to operationalize this concept (see 2.1, 2.2). An example of these efforts is the "International Conference on the Definition and Measurement of Sustainability: the Biological and Physical Foundations" sponsored in June 1992 by the World Bank. The conclusion of this meeting can be summarized as follows (Munasinghe and Shearer 1995):

1. "Sustainability involves complex interactions — biogeophysical, economic, social, cultural, and political — and that sustaining the global life support system is a prerequisite for sustaining human societies."
2. "Maintaining biological productivity is the key to sustainability."
3. "The factors necessary for the maintenance of biological production include fertility or nutrient availability, energy, adequate moisture, proper substrates, subcritical levels of toxic substances, and an adequate and genetically varied stocks of biological organism. Such factors form the biogeophysical foundation of sustainability."

The recognition of interactions between natural and social components has faltered because of the inability of most scholars to grasp or identify the existing mechanistic links. Recognition must to be given to the work of Merino (1996) and Rocha et al. (1997) for beginning to describe these links. Moreover, one of the field tests conducted by CIFOR in the Brazilian Amazon beginning in 1998 and a meeting held in December of the same year, resulted in innovative perspectives that are beginning to identify those parameters that would expose these links (Murrieta 1998, unpublished manuscript and Anna Fanzeres, personal communication).

In this book, we define sustainability in a manner similar to how it is defined in ecosystem management (Vogt et al. 1997a):

> Social- and natural-science resistance and resilience to disturbances (could be human or not) that allows the management unit to function within its natural range of variability.

This definition of sustainability results in the identification of measurable variables, because the recovery rates or lack of change in an ecosystem component can be tracked in response to disturbance. In a similar vein, Grove (1997) defined sustainability as a measurement of the resistance and resilience of the social sciences. For the natural system, determining sustainability is much more tractable if one defines sustainability as the resistance and resilience of systems and, ultimately, their integrity (Vogt et al. 1997a). Kay and Schneider (1994) defined the following three conditions as being integral to ecosystem integrity:

1. Knowing the current state of the system compared to normal environmental conditions
2. The ability to deal with stress in a changing environment (resistance to stress and the recovery rate after a stress)
3. The ability to continue to self-organize (change from within) in a changing environment

In certification, definitions of sustainability have been expanded from a primary timber-production focus to include sustainability of all of the following (Goodland et al. 1991):

1. Forest-dwelling people and those dependent on the forest for their livelihood
2. Extraction of non-timber forest products such as latex, oils, and nuts
3. The environmental services from the existence of the forests
4. Biodiversity

Other definitions of sustainability are quite similar to the four given by Goodland et al. (1991), although they are worded differently. For example, Tietenberg (1996) defined sustainability as:

1. Non-declining well-being so that future generations are not prevented from achieving a level of well-being similar to the current generation
2. Non-declining value of the natural capital so that the remaining stock of natural capital does not decrease
3. Non-declining physical service flows from selected resources

3.1.2.2 Integrating social and natural sciences in sustainability

Others have stated that for any human endeavor to be sustainable it must satisfy three criteria concurrently — the system must be economically, ecologically, and socio-politically sustainable (Zonneveld 1990). For assessing the sustainability of forestlands for certification, we would add silvicultural sustainability to the three criteria mentioned by Zonneveld (1990). Therefore, to be successful, sustainability assessment of forest resources has to incorporate each of the following elements:

- Economic sustainability
- Ecological sustainability
- Socio-political sustainability
- Silvicultural sustainability

Of these four elements required for assessing sustainability, economic and silvicultural sustainability are the easiest to measure. Frequently, economic sustainability is the primary tool used in assessments because it produces more concrete, quantitative answers for natural-resource uses. In general, both economics and silviculture have tested and adapted tools over a long time period, so more confidence exists in their abilities to assess system characteristics (Constanza et al. 1991, Smith et al. 1997). For example, silvicultural sustainability measurements have not varied significantly within the last century (Smith et al. 1997). Despite the existence of well-developed tools, silvicultural sustainability is not really assessed as part of many certification protocols. This happens because few indicators directly take into account the silviculture required to produce a particular forest condition on a given piece of land over the stand developmental stages required for that forest. The reason is the "snapshot" approach used by certifiers when evaluating a forest management unit (FMU). The value-laden parameters used do not consider many relevant variables (i.e., evaluation of regeneration patterns; presence or absence of key components of the system; and impacts on and responses of the soil substrate, etc.).

Since the economic measurement of sustainability could be more easily performed quantitatively, there has been an attempt to place a monetary value on the scarcity of resources and their efficient allocation. When considering forest sustainability from the economic standpoint, the real question is: "Is the scheme efficient, profitable, and economically feasible?" — all factors that can presently be assessed with existing tools. Moreover, it must be kept in perspective that in the neoclassical economic viewpoint, sustainability means being replaceable (Clark 1991, Adamowicz et al. 1996, Pendleton 1994, unpublished manuscript). Most theorists in the field of economics will state that markets have a tendency to allocate a resource to its most efficient use in a most efficient way by means of the profit incentive. This movement of goods to the highest-valued use by society occurs without much planning and is a beneficial aspect of this type of analysis since there are relatively low planning costs. Since the New Deal years of President Franklin Roosevelt, cost benefit analysis has created great interest due to its possibility of utilizing magic numbers to justify or reject projects (Larry Lohmann 1997, unpublished manuscript). Because of the ease in using economic models (even when there exists great disagreement among different economists), assessment in the past of sustainability or sustainable development mainly relied on economic evaluations (Constanza et al. 1991).

In contrast to economic sustainability, the tools for assessing ecological and socio-political sustainability are perceived to be in the intermediate stages of development. Much of this is due the complexity of human-dominated ecosystems and the long-term temporal and spatial scales that drive their functioning (Vogt et al. 1997a).

> Currently, ecological and socio-political sustainability become enveloped in individual or ethnocentric values (Tuan 1974) that make their assessment based more on values and not non-valued-based assessments that incorporate human values (Vogt et al. 1999b).

Assessments of all four elements of sustainability are further complicated by the fact that each does not function in isolation from the others. Feedback loops exist so that a change in one element causes a change in the expression of one of the other elements. This suggests the need to develop tools that are able to integrate the assessment of these four elements instead of dealing with each one separately.

The difficulty in implementing ecosystem-based management is a result of the short time period in which theory and tools for the integration of the social and natural sciences (Vogt et al. 1997a, Aley et al. 1998) have been developing. One of the basic tenets of ecosystem management is the integration of the social and natural systems. Most people recognize the strong interactions that are integral to ecosystem management. However, they are struggling with the development of the tools for implementing management which effectively integrate this interaction.

Certification has attempted to deal with this integration by listing information needs specifically within each field (see 4.2.1). However, in striving for sustainability, existing protocols present a compartmentalized list of parameters under the major categories of societal interest (ecological, social, and economic). But this list is not balanced; given the fact that the destruction of forests and associated values was one of the strongest value-laden reasons for the promotion of certification, the amount of information required in the social sciences is much less than in the natural sciences (see 4.2.3). This disproportionate emphasis on ecological parameters generates a subliminal message that natural-science information is much more important for assessing a site.

> This problem can ultimately be overcome if successful weighting of the information needs
> of any certification occurs, and there is an ability to identify when the social science
> information is more important than the natural science, and vice versa. There is a need to
> identify the minimum information required for any assessment protocol (see 5.1.2).

The most important constraint on using the word sustainability is that there must be a mechanism or approach for actually measuring it. If the definition is broad, it may be difficult to operationalize it in practice. Because of the difficulty of defining sustainability in more concrete terms there has been a push to use the term "good forest management" (see 3.1.3) as the mode of operation in assessing systems in certification. Good forest management can be defined clearly, has broad acceptability in the scientific and non-scientific communities and can be measured (see 3.1.3, 4.3.5.3). Certification is faced with the same problems which have occurred with ecosystem management where the drive to define terms dominates the discussions (Vogt et al. 1997a). When there is difficulty in clearly conceptualizing the words being used, implementation at the ground level is even more difficult.

Sustainability can be almost entirely defined as a social value. This raises all sorts of conflicts on whose values this definition should be based. How much should the local/traditional, national, or international values determine the assessment of sustainability? This is an important consideration, because the values at these different levels often cause conflict in how resource utilization should occur. Embedded in the certification protocol are the values particular to many national and international organizations which desire to maintain systems in particular states. For example, the deforestation of tropical moist forests might benefit few people or communities at the local level (even if in a short-term perspective). However, on a larger scale, forest devastation might impair global environmental services (e.g., climate, availability of products, tourism, etc.). Who, then, should decide how these systems should be managed and whose values should be overridden — is it to benefit the larger constituency?

Most discussions of sustainability directly incorporate benefiting the indigenous communities dependent on the forests for their livelihood (see 3.1.4, Goodland et al. 1991). This makes it important to have a framework that allows the conflicting values to be assessed, and to determine, if particular values are allowed to dominate, what the implications are for the natural system (see Vogt et al. 1999b). It is important that the values which are allowed to dominate, be considered in the context of the entire system. It is clear that societal factors determine what should be sustained, how it should be sustained, and who should benefit from what is being sustained. The natural resources will probably place constraints on how sustainability will be expressed as a social value. Natural-resource information should be used to determine the limits or thresholds toward which a particular system can be pushed, without changing its resistance and resilience characteristics.

Merino (1996) wrote that forest certification is different from other types of certification because of its focus on sustainability and the use of sustainability to develop its analytical framework. Incorporated into social sustainability is an ability to alleviate poverty (see 3.1.4); identification of the importance of the local populations and stakeholders to be involved in determining how resources are used; and having a long-term commitment by inhabitants for sustainable management.

Merino (1996) identified the following as the social elements of sustainability:

- Alleviation of poverty conditions
- Sustainable population pressures
- Social participation
- Resource tenure systems
- Mechanisms of conflicts-of-interest resolution and consensus building

According to Merino (1996), because of the need to manage for the future, sustainable management requires long-term commitment by the inhabitants of an area. The analogy to the *Tragedy of the Commons* has been mentioned previously (see 2.2), and seems to be a dogma among pursuers and defenders of forest certification. However, when we recognize that patterns of stability in land tenure are measured in decades (see 7.5), this definition means that certification will not be possible, since the basic requirement for long-term habitation on a given piece of land cannot be achieved. Interestingly, in the United States, the few timber companies which have been successful in being certified (e.g., Seven Islands in Maine, Kane Hardwoods in Pennsylvania) are those in which one family has owned the land for a long time. These families have a management philosophy for good stewardship. For example, one family has owned and managed Seven Islands in Maine for 150 years, while a family in Kane, Pennsylvania has owned Kane Hardwoods/Collins Pine for four generations.

If social- and natural-systems' sustainability were important tenets of certification, the five social elements of sustainability listed by Merino (1996) would have to be satisfied for a forest to be certified. The five social elements are also factors which make it extremely difficult for small landowners to pursue certification, because of the time and financial impact of satisfying all these elements of sustainability (see 7.6).

These issues raise important questions on whether alleviation of poverty should be a part of all certification assessments and whether this would be an unrealistic demand for a forestland owner, who would provide the services of a governmental agency (see 3.1.4). However, it must be pointed out that this is a debate bigger than the forest certification arena. This is the role played by many NGOs which, especially in developing countries, are attempting to fulfill the role of governmental agencies. Moreover, this matter also raises the question of who the stakeholders should be who are making the decisions on how land is managed (see 7.3, 7.4, 7.7). The long-term commitment expected from forestland owners/dwellers may also be very difficult to satisfy even in developed countries, which are considered more stable. For example, in the United States, statistics for the states of New York and New Hampshire show land-tenure turnover to be 12+ years, on average.

To overcome the latter problems, an alternative proposition was proposed the by FSC when they introduced the concept of group certification (FSC 1998a). Some pilot initiatives have been established by FSC-accredited certifiers in Costa Rica, the United States, the United Kingdom, and the Solomon Islands. However, group certification does not exempt individuals from their commitment to FSC principles despite the fact that the "group level" will ultimately ensure the adherence to the overall commitment. The implications and consequences of this approach to forest certification must still be evaluated in terms of the effects on the ground (see 7.8).

The interpretation of this approach to certification has been to certify an individual as a "resource manager." The potential problems for this initiative can be illustrated by a case presented at our forest certification class by a resource manager who did not, himself, interpret this situation as being problematic. This resource manager had been hired to cruise and process the selling of the timber of a private forestland that was quite large in area. Given the fact that the manager was certified, the timber sold from this area was marketed as certified wood. However, the problem that could have arisen is that the landowner could have decided to build condominiums on the property. The deal that had been signed by the resource manager did not preclude the owner from using the land for other purposes, because he was not directly committed to the principles of certification. The owner would have, therefore, profited from a supposedly premium price for this timber, and later would have received profits that could have negatively impacted the maintenance of the forest. The resource manager would not have been affected either, because he was not the owner of the land. Since property rights are very strongly protected by North American legislation, the forestland owner has the right to implement any activities on the land. This scenario highlights how difficult it is to verify to consumers that their purchase of certified wood ensures the maintenance of that forest's ecosystem.

3.1.2.3 *Issues related to achieving sustainability*

The growth of forest certification has spawned several competing approaches which are discussed in detail in Chapter 4. These approaches make various claims of what it means to be certified, including claims of "sustainability." The unfortunate result is that consumers may become confused by the different labels and approaches, and in some cases, may be misguided by vague and even misleading claims about "sustainability." This has increasingly become a problem for certified forest products, including those based on legitimate claims (see 2.1).

There are many problems in certifying forests which claim that "sustainability" or "sustainable forest management" is being assessed. Four major problems in certifying forests for sustainability can be summarized as follows:

1. Definitions of sustainability (see 3.1.2)
2. Insufficient tools, techniques, and indicators to assess sustainability (see Chapters 4, 5.1, 5.2)
3. Issues of scale (see 5.2.1.2)
4. Excessive costs and data requirements associated with assessing "sustainability" (see 7.6, 7.7)

This is not to suggest that the objectives of "sustainability" should not be a long-term goal of certification. However, most certification approaches do not actually evaluate sustainability, so those existing claims are unjustified.

A great difficulty lies in our inability to define sustainability because of the lack of measurable criteria and indicators to assess it. For instance, the Society of American Foresters defines sustainability as the ability of forests to maintain their health, diversity and overall integrity into the future, while being impacted by human activities and uses. Yet the terms and indicators used in this definition such as, "forest health" and "integrity" are controversial, poorly defined, and difficult to measure and model, particularly over the long term and across different spatial scales (Vogt et al. 1997a). Sustainability has proven to be hard to define because it refers to ecological and social systems. It is also difficult to develop good measures or indicators for assessing sustainability that can deal with the problems of scale, impermanence, uncertainty, and the complexity of these dynamic systems (see 5.2.1.2). In almost all cases, there is no baseline (e.g., pre-colonial or natural forest) to which the current condition can be compared. Even the pre-colonial forest is difficult to define in the northeastern United States. For example, a major hurricane impacted this region prior to the European settlers moving there. Some evidence exists that hurricanes can have major impact on the species composition, especially when the hurricanes are followed by hot fires. Henry and Swan (1974) documented how a hot fire followed the hurricane of 1785 that destroyed the hardwood regeneration, and led to changed forest composition and structure.

Unfortunately, sustainability has become a word that encompasses misleading connotations and different meanings and values (see 3.2.2) that people attach to it. All the ideas and values associated with this vague term lead to confusion for the consumer. Moreover, a definition of sustainability — that requires the protection of environmental quality while meeting human needs over the long-term — sets up unavoidable trade-offs and paradoxes. Some loss of ecological integrity and species abundance and/or diversity is inevitable in harvested stands. Therefore, defining sustainability requires the determination of who should or should not benefit from management activities (see 3.2.2). Finally, claims of sustainability have to face the challenge of addressing cumulative impacts (Beanlands 1995).

Problems with assessing forests as being sustainable are frequently due to scale issues. Problems of scale can be grouped into three categories (see 5.2.1.2, 7.5):

1. Differences in certifying small- and large-scale forests
2. Assessing management factors that cover much larger spatial and temporal scales than a single-forest ownership
3. The recognition of conditions and factors beyond the control of a forest manager

Ecological and social processes are critical to an ecosystem's functioning. These processes occur across different spatial scales (e.g., stands and watersheds) and over different periods of time (e.g., annual cycles and rotations). Many ecological processes and driving variables operate beyond the scale of individual forests or forest managers, so what happened one hundred years ago, or what currently happens on neighbor lands may affect a candidate forest. Only poorly developed techniques exist to control or even measure such large spatial- and temporal-scale phenomena. In addition, our understanding of landscape processes (e.g., disturbance cycles, population viability) are just developing (Levin 1995).

> Some of these processes should be assessed on a time scale of 500–1000 years (Holdren et al. 1995), far longer than most certification approaches address.

The data needed to analyze such large- and multi-scale processes is often unavailable and is exceedingly expensive and difficult to obtain. Without the necessary data and tools, it is difficult for certifiers to justify making claims that require data from longer time scales.

Finally, some of the principles and criteria for "sustainability" may be unrealistic, because they are beyond a forest manager's control. For instance, maintaining biodiversity is often cited as a criterion for certification (see 3.2.1). However, population viability and species diversity are usually influenced by large spatial- and temporal-scale processes and contexts. It would be ridiculous, for example, to expect a city park in the middle of Denver, Colorado to sustain a viable grizzly bear population, even though the park itself may provide suitable habitat for grizzly bears. Similarly, one cannot expect a forest to maintain biodiversity if it is surrounded by a fragmented landscape that has many landowners and different land uses. Therefore, it would be very difficult to certify many forests.

In addition to the problems of defining "sustainability," there exists a need for measuring whether a forest or forest product meets established criteria. However, good indicators, or adequate or practical tools and techniques to measure or model the functioning of a forest, are just evolving. As part of this, there is a need for the ability to determine what information is relevant to collect for different sites, instead of assuming there is uniformity of data needs (see 4.3.5, 5.1).

In many cases, the tools and techniques to assess "sustainability" are insufficient, inaccurate, or overly complex. For example, practical assessment techniques to measure ecological impacts of a forest edge are unavailable (see 5.2.1.2), yet this is an important criterion for several approaches. As noted earlier, there are many difficulties in assessing large-scale processes. Existing indicators and methods generally focus only on the short-term; even traditional economic projections are made only several years into the future. Other indicators are only suggestive of "sustainability" and cannot make adequate assessments or predictions. It is also difficult to generalize across regions. There have been attempts to develop regional standards, but no regional criteria have yet been established. Regional criteria also need to deal with weighting the importance of different information, instead of assuming similar data needs in a region (see 5.1). Moreover, it is often impractical for the assessors themselves to do the time-consuming and expensive long-term research. It is generally unreasonable to perform studies, such as population viability analysis, for even one species. Thus, it may be necessary to base judgments on insufficient or incomplete data, if any data at all is available. This leads to significant questions about the reliability of certification decisions as they are presently approached.

The cost of auditing a forest is a major obstacle to greater enrollment in forest certification, particularly for non-industrial private forests (see 7.4–7.7). Excessive data requirements to assess sustainability usually worsen time and cost constraints — some certification approaches have over one-hundred criteria (see 4.2.1). In testing existing certification approaches, it quickly becomes apparent that much of the required data may be unavailable or incomplete. Some data cannot practically be obtained or is prohibitively expensive to collect; other indicators are not necessarily relevant or necessary (see 4.3.5). Assessors are then forced to decide whether to certify a forest or forest product in the absence of this data, leading to questions of subjectivity and the legitimacy of the assessment. Finally, some forest managers feel that the costs of satisfying criteria to be "sustainable" are too restrictive, such as requiring public involvement in management decisions. Although the ideals of truly sustainable forest management are good, they may not be realistic or may conflict with norms of property rights. In the United States, many certification requirements can be satisfied when a landowner follows the Best Forest Management practices of their state, or other federal, state, and local regulations (see 5.3). However, certification requires only that the Best Management Practices for that state are followed, but not how these practices could replace other criteria or indicators that need to be satisfied when being certified.

3.1.3   Good forest management

Alex Finkral, Bruce C. Larson, Anna Fanzeres, Kristiina A. Vogt, Christie Potts, Heidi Kretser, Jennifer Heintz, Allyson Brownlee

The concept of sustainable forest management is not a new idea. As matter of fact, the term and its basic principles were outlined in Germany and France in the Middle Ages (circa 500 A.D. to about 1500) (Winters 1974, Linnard 1982, McCormick 1989, Neeson 1991, Elliott, 1996). The scientific minds of that period developed the concept of tallying timber trees and the practice of dividing forested areas into plots, according to the number of years they had been exploited (Winters 1974, Seip 1996). There were also concerns articulated about regeneration through sparing seed trees and allowing resprouting of stumps (Winters 1974, Linnard 1982).

The traditional definition of sustainable forest management is oriented towards ensuring continuous resource yields. More recently, this concept has begun to change (see 2.1), but more on theoretical grounds than in the methodological parameters being used. As described in Section 2.5.1, there has been a recent move for the development of codes of practices and field guidelines for timber harvesting. These new parameters are being developed for large-scale operations and also small/community-level initiatives (Dykstra and Heinrich 1996, Amaral 1998, IUFRO 1997).

FAO has placed a great deal of emphasis on mechanization of harvesting, better planning of roads, and for the felling of trees (see Chapter 6). These tools are considered to be the major assets for foresters in tropical countries, working for governmental agencies or private enterprises, who are attempting to balance a forest's development with its conservation (Dykstra and Heinrich 1996). The "technological fix" approach to the development of these recommendations is obviously in accord with the United Nations' paradigm of sustainable development (see 2.1). This is also the approach used in countries such as Brazil and Costa Rica, where field manuals stress the development of forest-management plans (Beek 1996, Amaral 1998).

If the introduction of planning is successful, especially in tropical countries, dramatic improvements in logging operations should occur. The present model of forest use in most tropical countries can be characterized as "mining of the forest." This practice involves not only highgrading (i.e., extraction of a few commercially valued timbers and extraction of the best specimens), but also random building of roads with no commitment to maintaining forest cover. As previously stated, the manner in which logging occurred in the tropics was an important catalyst for the development of forest-certification initiatives (see 2.1). However, the technological solutions being proposed exemplify the dichotomy that exists between existing scientific knowledge of tropical forests, and the practices being incorporated into guidelines for sustainable forest management. This does not mean that the existing protocols have to be made even more data-collection intensive. Instead, emphasis should be given to identifying those variables that really dictate how the system is responding to human activities within the natural constraints of the site (e.g., soil characteristics, vegetation composition/structure, key fauna elements, natural disasters — floods, blowdowns, fires, etc.) (see Chapters 5 and 6).

At the beginning of the forest certification debate, the goal was to achieve sustainability. Recently, however, there has been a push away from using sustainability as a goal of management, because it is difficult to define (see 3.1.2). Instead, the current goal of certification is to obtain good forest management (see 4.3.5.3). According to Drescher (1998), the rationale for sustainability is to get what we want out of our land. He stated that the aspiration for sustainability is a noble endeavor that assumes impermanence of the resources. Therefore, the pursuit of sustainability is not relevant, and we should simply be striving to do good forestry on the ground. Doing good forestry should first strive to satisfy the need for maintaining ecosystem health. It should also be followed by a discussion of its benefits (so that the arguments for balancing the social- and natural-science needs disappear) (Drescher 1998). Many of the certification protocols reflect this philosophy and have recently moved away from sustainability to evaluating whether a forest is well managed.

Good stewardship can be defined as landowners obtaining higher value from their land by utilizing the knowledge and skills of resource professionals to address good management using better forest practices on their lands. Forest management can be defined as the application of forestry principles and practices to retain the health of the forest while also satisfying landowners' objectives. The goals for each landowner are intimately tied to their value system, how they want to interface with their land and the economic stability of the owner.

The management of forest ecosystems is a pursuit that merges, in different combinations and to variable degrees, biophysical sciences with economics, politics, and human values. With few exceptions, the best forest management is a result of contextual consideration; disciplinary integration; and creative, appropriate application. Although the goals of forest management can be as disparate as individual personalities, good management strategies share the same components:

- Integration of the tenets of multiple disciplines
- Knowledge and creativity of the people involved
- A planning stage
- Explicit understanding of the pressures and motivations that underlie the goals for a forest
- Thorough consideration of constraints, sensitive species, and unique situations
- Public acceptance, or in some cases, tolerance, of a management strategy
- Established and well-maintained paths of communication among administrators, managers, landowners, and laborers
- Forest-management plans and silvicultural goals based on a sound foundation of science — one which is geographically and temporally appropriate
- Maintenance of site productivity through soil conservation
- Maintenance of a vegetative cover on the land

As with most complex endeavors, forest management always benefits from a planning stage. Before any silvicultural prescriptions or financial analyses are written, it is of paramount importance for a manager or a management team to understand explicitly the pressures and motivations that underlie the goals for a forest. Understanding who owns a forest and what management's responsibilities to them are helps define the limits of management. What is at stake? What can be gained or lost? What benefits are expected from the forest and are they sustainable?

A thorough consideration of constraints, sensitive species, and unique situations is also a critical component of the planning process. Realizing from the outset that management possibilities can, or should, be narrowed in certain places eliminates "surprise" conflicts in the future. A comprehensive forest inventory and background research usually provides the kind of data needed to make such considerations. Furthermore, it is critical to examine a forest within the context of the surrounding landscape. If the proposed goals and objectives of a forest are completely incongruous with those of the surrounding region, it could be an indication that those goals are biologically impossible or socially inappropriate.

Although in temperate and boreal regions some biology and ecology have been incorporated into the design of forest management practices, social factors have only recently been recognized as potentially relevant. Public acceptance, or in some cases, tolerance of, a management strategy is vital for its success (see 3.2.2). Research reveals that the public holds a stake in how both publicly and privately owned forests are managed. The role of forest managers is changing to include more communication with stakeholders and more sensitivity to the public's tolerance.

Communication within a forest-management operation is also essential. Established and well-maintained paths of communication among administrators, managers, landowners, and laborers provide a means for fostering respect, trust, and rectitude at all levels.

Forest management plans and silvicultural goals should be based on a sound foundation of science — one that is geographically and temporally appropriate. What is ecologically justifiable in some forests around the world is simply impossible or unreasonable in others. Similarly, some management practices and methods that seemed feasible in the past are outdated and inappropriate today. By keeping abreast of the forest science in the surrounding region, a forest manager can apply progressive ideas and techniques that are supported by the most current information.

Foremost among these concerns is the maintenance of site productivity through soil conservation (see 5.2.1.1). Management practices that reduce the fertility of a site leave fewer options for the future. Management practices aimed at conserving soil, however, offer a safety net; in these cases, even failures do not cause irreversible damage. Regardless of specific management practices, managers of forested ecosystems have a responsibility to maintain a vegetative cover on the land. In addition, the protection of water and wildlife should always share space at the center of forest management ideals.

Forest management should not be based strictly on quantitative formulae or static ideas. At its best, forest management is an art that requires the integration of sciences based upon judgment and context. Cautious planning and clearly defined goals and objectives provide an initial direction. The best forest management then aims to maintain the delicate balance among the desires and needs of people, ecosystem health, and the yield of benefits long into the future.

The numerous individuals and organizations working on forest certification have placed considerable energy into defining good forest management and in determining what indicators will reflect this forest condition. However, good forest management is not defined by certifiers as presented above, but as indicators for achieving value-laden desired endpoints that maintain forest ecosystems (or restore them) into patterns believed to be correct. Moreover, certifying organizations under the FSC umbrella have had great difficulty in providing a satisfactory working definition of sustainable forest management. The lack of regional standards for concretely guiding certification operations on the ground (see 1.2) and the insistence on using the same protocol language creates subjectivity when utilizing the protocol.

The development of a working definition of the concept of good forest management is important because of the difficulties of small landowners in being certified (see 4.3.6, 7.5). This parameter would allow this important segment of forestland owners to show that their forests are well managed.

> Achieving good forest management is a goal that might provide the solution for small landowners to be able to successfully pursue certification (see 3.1.3).

Small landowners are unable to satisfy all the requirements of achieving sustainability as it is currently defined because of the many values that are embedded in the protocols. Current protocols are structured towards large landowners who are better able to achieve more values of their lands. Certification appears to require large land holdings, so that the cost of paying for the certification can be recovered; additionally, the problem of chain of custody is easier to implement on large land holdings (see 7.7). On small-scale land holdings, it is probably better to promote improved stewardship or good forest management as the label to pursue.

3.1.4 Certification and poverty eradication

Toral Patel-Weynand, Kristiina A. Vogt

The timber certification debate has listed "poverty alleviation" as one of the benefits of becoming an environmentally sensitive "green" producer and owner in the forestry sector (see 3.1.2.2). To examine the role of poverty alleviation in timber certification, there is a need to agree on the definition of poverty. How certification intends to impact the impoverished to alleviate their status must also be addressed. There is a need to have a clear understanding of how the poor use forests and where certification can be a useful tool in eradicating poverty. Furthermore, it is important to realize that increased income potential through the certification of timber and wood products is a very different issue in developed and developing nations. In addition, if the certification assessment protocol is to be robust, then indicators exploring the links between poverty and forest use must to be considered. When addressing the question of poverty eradication on a global level, the problems of developing countries become much more urgent. Addressing the needs of the poor in developing countries is much more basic than that which is relevant for small landowners in developed countries.

3.1.4.1 Sustainable development, forests, and poverty

Given that currently some 1.3 billion people are living in abject poverty, sustainable development approaches are desperately needed that will incorporate their many benefits into society today and into the future. The need for sustainable approaches is especially critical for the world's forests and the people whose livelihood is dependent on them. Dealing with forestlandscapes is important since forests once covered 40% of the world's land surface but have been reduced by one-third, mostly during the last fifty years (Myers 1995). Forests can supply an exceptional array of goods and services, but they are often overexploited and underutilized, since only a few of their many products are harvested.

The extent of world forest cover demonstrates the importance of forests to sustainable development and to the reduction of poverty. FAO estimates that world forest cover in 1990 was 3.4 billion ha or 26% of the earth's land area [Note: The FAO defines forests as "ecological systems with a minimum of 10% crown coverage of trees."] The FAO also estimated that another 1.6 billion ha of land (another 12% of the earth's land surface) contain woody vegetation and other woodlands consisting of shrubs and scrub that often have forest characteristics (FAO 1994).

Temperate and boreal forests occupy 1.64 billion ha or 48% of the world's forest cover (FAO 1994). More than 70% of these are located in the Russian Federation (45%), Canada (15%), and the United States (13%) (FAO 1994). The temperate zone includes two basic ecological formations: mixed temperate forest and boreal forest. The mixed temperate forests consist of coniferous, broadleaf, deciduous, evergreen, and other forest types found in the non-tropical zone and on mountain ranges in subtropical and tropical countries. The boreal forests extend between the Arctic tundra and the temperate zones in a circumpolar belt of mainly coniferous trees. The boreal forests cover 920 million ha and make up 27% of the earth's forest area. Temperate zone forests are widely recognized for their enormous contribution to global industrial timber supplies as well as for their non-wood products, recreation, and environmental services. In addition to their economic importance, temperate and boreal forests play an important role in biodiversity conservation and as sites of carbon sequestration (FAO 1994).

The tropical forest zone contains 1.76 billion ha (52% of the world's forest cover) and is divided into six ecofloristic zones: (1) the tropical rain forests; (2) the moist deciduous forests; (3) the dry zone; (4) the very dry zone; (5) the desert zone; and (6) the hill and mountain forests. Of these six types, 96% of tropical forests occur in formations (1), (2), (3), and (4). Nearly 500 million people, most of them poor, live in or near forests and depend on them for food, fuel, fodder, timber, and income (World Bank 1991). More than 200 million people live in tropical forest clearings. These people have been categorized as: groups who have lived in the same forest for generations, often referred to as indigenous or tribal peoples, as well as non-indigenous forest dwellers; people who have more recently moved into the area, often described as settlers or squatters; and people who live part-time in the forest working as small-scale harvesters of forest products. Tropical forests are recognized for their contributions to food security by providing sources of food, income, jobs, fuelwood, medicine, and construction materials. Forest wildlife is a significant source of protein for forest dwellers and the rural poor in many countries. Tropical forests are also recognized for the richness of their plant and animal life. In addition, they are recognized in the existing literature for providing environmental services such as the prevention of soil erosion and the regulation of runoff from storms (FAO 1994). Moreover, FAO and UNDP have deemed the management of tropical forests important for the production of wood and other products and services, for conservation, and for improving the quality of life of the millions of people who live near or in these forests (FAO 1993, Patel-Weynand 1997a).

The conventional economic contributions of forests are well documented, particularly the value of wood energy, solid wood, and fiber products. Worldwide consumption of wood totals nearly 3.4 billion cubic meters a year, of which about 50% is consumed as fuelwood, building poles, beams, and other round timber for the construction of houses in developing countries (World Bank 1991). An estimated 3 billion people depend on fuelwood as their main source of household energy (FAO 1994). The annual value of fuelwood and wood-based forest products to the global economy is over $400 billion, or about 2% of global gross domestic product (FAO 1995a). This value is calculated to have grown by 2.7% annually from 1961–1991 (FAO 1994). However, forests contribute almost three times as much to the economies of developing countries in comparison to the economies of developed countries. The opposite is true regarding the contribution of exports of wood-based products, which reach about $98 billion per year and involve about one-quarter of all global timber production (FAO 1995a).

While there is continuing debate regarding how temperate and boreal forests should be managed, the aggregate size of temperate zone forests in industrial countries is generally stable or even increasing slightly through reforestation efforts (FAO 1995a). However, tropical zone forests are undergoing rapid deforestation, especially in developing countries. The FAO estimated that 15.4 million ha of tropical forests were lost each year during the 1980s and that the area of severe forest degradation is perhaps even larger. In looking at the management and conservation of closed tropical forests, the FAO showed that tropical American countries are unable to satisfy their own wood needs and to balance their trade deficit (i.e., totaling $242 million). The tropical American countries comprise 651.6 million ha of closed tropical forests or 56.0% of the global total and have a population density of only 335.8 million inhabitants. In most of these countries, deforestation is a subject of concern from an ecological, political, and economical standpoint (Patel-Weynand 1997a).

The rates, causes, and effects of deforestation differ greatly from one country or region to another. These differences are due to population density and growth rates; the extent and quality of forest resources; levels and rates of development; the structure of property rights; and cultural systems. Recent estimates suggest that nearly two-thirds of tropical deforestation worldwide is due to farmers clearing land for agriculture. However, many other economic activities are also directly or indirectly linked to deforestation (for example, mining, access roads for logging, etc.). The largest losses of forestlands are occurring in the tropical moist deciduous forests, which are also the best-suited for human settlement. In the 1980s, an estimated 61 million ha were deforested — that comprises more than 10% of the moist deciduous area (FAO 1994). The Consultative Group on International Agricultural Research (CGIAR) recently released a study stating that subsistence farmers using slash-and-burn practices could cause the loss of almost half of the world's remaining tropical forests. In addition, logging conglomerates aggressively pursuing timber rights could eliminate the remaining half not threatened by poor farmers. The loss of these forests undermines the ability of countries to achieve sustainable economic growth to alleviate poverty.

3.1.4.2 Certification and poverty alleviation

To play a role in poverty eradication, assessment protocols for certification must include a clear understanding of forest-use patterns of the poor. This can include a need to understand complex social, economic, and ecological issues related to forest conservation and use.

Rural populations depend not only on the environmental services provided by forests, but also on forest products. Forest products contribute significantly to food security during seasonal or periodic famines, or shortages of crop-based foods (FAO 1978). In some areas, forests are the primary source of protein, energy, oils, medicines, and even staple foods. At least three-quarters of the world's people depend largely on folk medicine, a significant amount of which is derived from forests (FAO 1995a). Rural populations depend on tree products to sustain soil fertility and structure; to feed livestock; and to maintain desired moisture regimes and water flows. This intense use of and dependence on forests by communities within or in close contact with these ecosystems leads to changes in its composition and structure. Careful examination of the anthropological literature demonstrates that there has never been any case of maintenance of the original biodiversity of forests when populations grow and participate in the market economy (Rui Murrieta, personal communication).

3.1.4.3 Fuelwood and timber

Wood is the dominant domestic fuel for rural people in developing countries and for many of the urban poor, as well. In many parts of the developing world, wood is also the principal structural material for constructing shelter and housing. Wood is the preferred fuel because, in most cases, it can be gathered, distributed, and used at a very low cost. For the poor, the option of using wood often is driven by the lack of alternatives to wood fuel or locally available organic materials. Commercial fuels, even where they are available, require cash outlays for stoves and related equipment, which are generally beyond the reach of the rural poor (FAO 1978).

One consequence of growing rural populations is an increase in the pressures on locally available forest resources and other sources of woody material. Under these pressures, the source of wood fuel changes progressively, from collecting deadwood to the lopping of live trees, the felling of trees, the total destruction of tree cover, the loss of organic matter to the soil, and eventually to the uprooting of stumps, and removal of shrubs. Subsequent to this, agricultural residues and animal dung are diverted to fuel use which is detrimental to soil structure and soil fertility (FAO 1978).

The steady disappearance of wood in the vicinity of the community leads to increased social hardship, as well. Progressively, more of the time of household members must be devoted to gathering fuel. As the situation deteriorates further and the household is forced to purchase its wood fuel, a heavy burden is placed on the household budget. Eventually, the shortage of wood fuel can affect the nutritional well-being of the people. In parts of West Africa, people's food intake has been reduced to one cooked meal a day. In the uplands of Nepal, only vegetables that can be eaten raw are grown. In Haiti, the introduction of new food crops with better nutritional value into the wood-poor hills is impeded by the fact that these new crops require more cooking (FAO 1978).

3.1.4.4   Food and the environment

In order to obtain their daily food, an estimated 200 million people living in tropical forest areas practice "slash and burn" farming (shifting agriculture) on perhaps 300 million ha of forestlands. Traditional systems of shifting agriculture, which employed a lengthy fallow period under forests to restore the fertility of soils which are only capable of supporting agricultural crops for a limited number of years, have largely broken down. Growing population pressures and migration into forests areas by landless people from elsewhere have forced a shortening of the fallow period to the point where it is insufficient to restore soil fertility or to recreate a usable forest crop. Similar trends are discernible in the more open savannah woodlands (FAO 1978).

In addition to crop production, there are many other ways in which rural communities in developing countries can draw upon the forests for food. Bush meat and honey provide supplementary food sources, as do a wide variety of tubers, fruits, and leaves. Fish production in swamps or mangrove forests can also be an important protein source, since mangroves and swamp forests offer a most valuable protective and productive habitat for fish (FAO 1978).

In many areas, trees are a source of fodder. In Nepal, leaves make up about 40% of the annual feed of a buffalo and about 25% for a cow. In dry forest areas, livestock often cannot survive without forest grazing. In the Sahel, leaf fodder is the principal source of feed in the dry season. Excessive grazing of trees during the recent, lengthy drought in the Sahel contributed significantly to the large-scale destruction of the vital tree cover (FAO 1978).

In addition to pressures on the forest from shifting cultivators, expanding rural populations require more land on which to grow food. In many countries, forests are the largest remaining land bank — the one land cover whose decrease will release major areas for increased crop production. To the extent that this process releases land that can sustain the growth of crops to food production, such a development is logical and should be planned. But over large areas, the pressures of growing populations force landless farmers onto soils which cannot sustain crop production and onto slopes which cannot be safely cultivated, at least with the techniques and resources available to them. The consequences of these practices on wind and soil erosion; silting; flooding; and drought are well known (Patel-Weynand 1997b). Because some 10% of the world's population lives in mountainous areas and another 40% in adjacent lowlands, half of the world's population is directly affected by unsustainable extension of agriculture onto forestlands (FAO 1978).

The process of environmental degradation following the destruction of the tree cover is often accelerated by the pressures of fuelwood harvesting (Patel-Weynand 1997b). These pressures tend to be most severe in the neighborhoods of large towns and cities. As mentioned above, wood is the preferred fuel not only of the rural poor, but also of the many urban poor, who use it principally in the form of charcoal. The large concentrated demands that ensue from urban populations have led to treeless wastes in periurban areas in many parts of Africa, Asia, and Latin America (FAO 1978).

3.1.4.5   Income and employment

In addressing the poverty alleviation issue, certification assessment standards can help build the asset portfolios of the poor through a number of initiatives that enable access to common areas and sustainable management through multiple use. For example, forests and trees can give rise to cash crops such as mushrooms, chestnuts, walnuts, and pine kernels. Bamboo can be cultivated for shoot production. In many countries, trees are grown at the small land-holder level to provide fuelwood for sale to urban and semi-urban areas. In India, the income from gathering and selling fuelwood is an important part of the economy of villagers living near forested areas, especially for the poor and landless in these villages. Tree farming can also provide profitable industrial wood crops, such as the pulpwood grown by farmers in the Philippines (FAO 1978). The export of rattan by Southeast Asian countries is valued at approximately $300 million per year. Among non-wood products, the gum arabic produced from trees as a farmer crop in Sudan accounts for exports worth $70 million per year (FAO 1995a).

Non-timber forest products (NTFPs) are critical to rural inhabitants and provide important sources of food, income, medicine, and shelter (Pierce 1999). Hundreds of millions of people use NTFPs on a regular basis and the economic value of these products is estimated to be in the tens of billions of dollars annually (CIDA 1992, Pimentel 1997, Pierce 1999). Yet, the subject of NTFPs has only recently been better addressed within forest certification protocols and there has been no endorsement of NTFPs in any certification scheme. However, it is important to keep in mind that certification protocols involving NTFPs present challenges that may require the modification of current methods of assessment. For a holistic assessment, NTFPs must be recognized and the ecological, social, and economic issues surrounding their management must be dealt with within the context of the forest management unit and its surrounding landscape.

One of the main reasons it is difficult to address the role played by NTFPs in the certification process is the informal nature of the NTFP markets because they exist in what is referred to by researchers as the "invisible economy." Informal markets and subsistence utilization of natural resources and the social structures and networks that drive these economies are not compatible with traditional, neoclassical economic models and theories, including the market-based certification process. For example, in Alaska, subsistence involves complicated patterns and networks of production, processing, and distribution, and exchange and consumption which help to maintain a complex web of institutional relationships involving authority, respect of wealth, obligation, status, power, and other components of social structure (Pierce 1999). Grass and Muth (1989) and Pierce (1999) noted that the introduction of traditional market behaviors can stress the structure of subsistence cultures.

Certification by itself may not be adequate to address economic, ecological, and social complexities of NTFP management and trade. However, it may be a tool to promote better management of these critical forest resources through harvester education programs, biological and market research and rural-development assistance programs (Pierce 1999). If certification of NTFPs is widely pursued, care must be taken to ensure that policies are in place to include NTFPs from small stakeholders so that they do not suffer the negative and financial consequences of competing with certified products.

When assessing the economic value of non-timber products (which include fuelwood and non-wood products), a three-tiered utilization at the local, national, and international levels must be taken into account. Some products are marketed, processed, and utilized far from forests by industries, urban dwellers, or distant international economies. Others enter only into the national or regional economy, while yet others are collected and used only locally by forest dwellers and adjacent rural populations. Considerable overlap exists between these categories of products. For example, even products collected in large quantities for export can provide local income and employment through harvesting. In addition, products consumed and processed locally may also have local or national significance. Many non-timber goods, however, are not traded in markets and those that are traded are rarely reported in national and international statistics of forest-product trade. This lack of reporting strongly implies that the overall value of non-timber products is seriously underestimated in the statistics currently available. Recent studies in various tropical countries have shown that non-timber products can add significantly to the value of forests (ITTO 1990).

Forests provide timber and other raw materials for local craftsmen and small-scale artisan and processing activities in addition to the income and employment generated by their industrial exploitation. For example, it has been estimated that in 1985, approximately 59,000 people were employed in cutting and transporting cane in Indonesia, while an additional 17,000 were employed in processing and manufacturing (ITTO 1990). Throughout the developing world, doors and other woodwork for builders furniture, tools, and agricultural inputs, such as fence posts, are made locally within the community. These products, together with wooden handicrafts and other products of non-wood raw materials, can also be marketed outside the community (FAO 1978).

Small-scale rural enterprises taken as a whole are a major source of rural livelihood in developing countries. Within the small-enterprise total, small-scale processing and manufacturing enterprises account for the largest component. Within these processing and manufacturing enterprises, a survey of six countries revealed that small-scale, forest-based, rural enterprises constitute an important part of the rural sector. They often represent from one-sixth to one-third of all processing enterprises by both number and employment (FAO 1987a).

Forest-based processing enterprises are important for more than just employment. For example, they:

- Provide above-average incomes to entrepreneurs, their families, and their employees
- Transfer skills through informal training
- Generally contribute to the local and national economies, and sometimes to exports

An important consideration is that earnings from these enterprises improve farmers' income security, and therefore reduce pressures that can lead to over-exploitation of the agricultural land-base (Patel-Weynand 1997a). In addition, the landless, women, and other agriculturally disadvantaged groups are well-represented among those getting income and employment from forest-based processing enterprises (FAO 1987a).

Forest-based processing enterprises are characterized by: (1) technologically simple operations which demand limited skills and low capital; (2) predominantly rural location; (3) reliance on the entrepreneur and his or her family for much of the labor input; and (4) very small size (average employment ranges from two to four workers per enterprise). The greater part of these enterprises produce either: (a) furniture, builders' materials for woodwork, agricultural implements, vehicle parts, and other products of wood; or (b) baskets, mats, and other products of canes, reeds, vines, grasses, and other similar materials. Both of these product groups serve predominantly rural household and agricultural markets, and are often the principal sources of supply to the latter. The close integration with agriculture is reflected in the seasonal pattern of operations and in the dependence on agricultural incomes to generate much of the demand for their product. Most entrepreneurs operate their forest-based activity jointly with other processing, service, or agricultural activities, so that they seldom occur as separate enterprises (FAO 1987a).

Forestry can also contribute to rural incomes in less-direct ways. If other alternatives for raising the income of the rural poor are not promising, the establishment of fuelwood lots may provide a means to raise their incomes. This could also allow dung and agricultural residues to be added back to the soil to increase or maintain crop yields (FAO 1978). In this way, certified forests may also contribute to a more equitable distribution of income.

3.1.4.6   Poverty alleviation and environmental protection

Areas of land under forest cover that are owned by the poor are quite small in developing countries. The small land-ownership sizes make it difficult for these forest owners to pursue certification (see 7.4 and 7.5 for a discussion of the constraints that land-ownership size places on certification). Because of the small area, the low amount of wood products and NTFPs that can be harvested may not be able to compete with similar products from large certified enterprises. To address the question of equity at the market level, consideration must be given to making certification standards more inclusive and better adapted to different sizes of forestlands (see 2.5, 7.2).

Development lessons learned in the past 25 years strongly suggest that simply protecting the forests and natural landbases from degradation will not ensure sustainability in the long term. In fact, these approaches will be short-term solutions unless the social part of the equation is given proper attention in the poverty-eradication debate. Equity in resource use, empowerment, institutional and human-capacity building are necessary for investment in resource conservation. With regard to certification, there is a need to find the social and natural science indicators and bridges that foster compatibility in actions and growth.

There is a need to establish a clear understanding that the poor and indigenous are knowledgeable on issues related to environmental protection. Certification is potentially a tool that would allow a realistic framework to be developed for small communities that would allow for the generation of concrete social benefits and environmental protection at the same time.

Certification can help address environmental concerns by integrating them into economic-growth strategies. A potential model would be created through the incorporation of two sets of incentives and information systems:

1. Those which reduce uncertainty about the future (institutional mechanisms that deal with conflict resolution in the access and use of resources)
2. Those which send out the appropriate signals about price and quantity in the marketplace (fiscal policies, such as taxes and subsidies, "social" pricing, etc.)

Other necessary elements include (Pearce and Warford 1993; see 2.1 for more discussions on these macro-level measures):

- Modifying the presentation of environmental and economic statistics so that the environmental impact of economic change can be discerned and the services of the environment highlighted
- Revising systems for appraising investments and policies to ensure that they adequately reflect and integrate the environment

3.1.4.7   Ecology of forest use by the poor

The people found in forest areas can be grouped into three broad categories:

1. People who have lived in the area for generations (indigenous and non-indigenous forest dwellers)
2. People who have recently moved into the area (settlers or squatters)
3. Nonresident groups who enter periodically to extract selected resources

Despite the fact that the first two groups actually reside in the forest, the last group often plays a significant role in the deterioration or shrinkage of the forest (World Bank 1991).

Three definitions of sustainability have been applied to the use of forest areas of all types:

1. Continuous flow of timber products or other specific goods and services, some of which may be essential for sustaining the livelihood of indigenous forest dwellers
2. Continued existence of the current ecosystem
3. Long-term viability of alternative uses that might replace the original ecosystem.

The extent to which any particular land area can satisfy any or all of the criteria suggested by these approaches is highly site specific. This depends on soil quality, topography, markets, and the availability of skills and technology (World Bank 1991).

The conversion of forest to other land uses or to wasteland takes a variety of paths and seldom involves the actions of only one group of decision makers. Common forms of deforestation include deliberate clearance and conversion; degradation and subsequent clearance following logging; and gradual conversion as a result of the shortening of fallow cycles in shifting cultivation (World Bank 1991).

In general, the rural poor are not involved in the direct conversion of large-scale forest areas to other uses or in the industrial logging of forests. However, they are affected by and occasionally increase the environmental changes caused by these activities. A discussion of forest changes caused by the activities of the rural poor follows.

3.1.4.7.1 *Ecological impacts of fuelwood and timber use by the poor*

Careful logging of natural forests which encourages the continuance of forest cover, seldom brings about total deforestation. In natural tropical forests, only a relatively small proportion of the standing volume consists of marketable trees worth the costs of felling and extraction. This is especially true for veneer and export sawlog operations, but also holds for lower-valued products, such as construction lumber. Clear felling is not usually practiced in moist tropical forest logging. Even so, the residual stand can be significantly damaged as the construction of logging roads and trails, clearance for yarding, and other operations entail considerable additional felling and soil disturbance (see Chapter 6, World Bank 1991).

Even more, when loggers ignore any limits on small-diameter trees, concentrate their harvests on a few more valuable commercial species or accelerate their entry to previously logged sites, the forests do become degraded. When the logged forests are close to towns and cities or become more easily accessible, the urban and rural poor can be involved in the degradation process. Many marginal economic activities such as the conversion of logging residue to charcoal become economically viable. As mentioned above, the demand for fuelwood and timber for housing materials by rural people and for charcoal by urban dwellers can also lead to deforestation in cases where the demand exceeds the regenerative capacity of the forest. This problem is especially severe in tropical dry forests (World Bank 1991). When the regenerative capacity of the forest is sufficient to meet the demand of the local population for fuelwood and housing materials, use of the forest by the local poor meets the first two definitions of sustainability listed above.

If sustainability is defined as the continued provision of specific products or services, then the purposes behind industrial forest management are compatible with three of the widely desired functions of forest ecosystems: timber production, carbon fixation, and protection of water resources. However, the compatibility of industrial management with biodiversity and the welfare of forest dwellers are the subjects of vigorous debates. The selective preference of commercial systems for high-value or rapidly growing species modifies the original structure and composition of the forest, which affects the availability of niches for certain species and may diminish biodiversity (Shiva et al. 1994). Moreover, the concentration of land for the implementation of most of these projects in tropical regions creates social instability (Carrere and Lohmann 1996).

3.1.4.7.2   *Environmental impacts of forest use for food production by the poor*

Shifting agriculture can be a sustainable system of food production by small rural populations, provided the land base is large enough to allow for the regeneration of soil fertility. As population pressures increase, fallow periods become shorter, and the landscape becomes more fragmented — the ability of the forest cover to reestablish itself is diminished. Increased farming pressure on the forest creates the same conditions as poor logging practices, and eventually the forest ecosystem may degrade (World Bank 1991).

Mechanical or manual techniques are used to sometimes completely remove the existing forest vegetation and to prepare the site for subsequent use (see Chapter 6). The regenerative capacity and sustainability of various alternative uses of forestland varies widely. Some of the world's most productive and robust agricultural lands were once forested. Conversely, some agricultural settlement projects in forested areas have been serious failures, especially in the tropics. Forest soils, despite their ability to support vibrant forest ecosystems, are often rendered extremely poor in nutrients when cleared, and may be subject to excessive erosion and other forms of soil degradation (see 5.2.1.1). Site selection, technology, and the institutional setting are key determinants of the rate of conversion of forests to other uses. In view of the diminishing area of forestlands, especially moist tropical forests, highly demanding environmental and policy analyses should precede any significant new development or utilization efforts. These analyses should include assessments of vegetation; soils; hydrology; the institutional and incentive framework; and the value of conservation for all concerned, particularly forest dwellers.

3.1.4.7.3   *Ecological impacts of income generation and employment of the poor on forestlands*

When forest resources grow scarce, shortages of wood and other raw materials serve as a constraint on the continued activity of forest-based enterprises. Their small size and very limited income surpluses mean that the forest-based processing entrepreneurs do not have the resources to invest in long-term forest projects. Their importance in generating non-farm employment and income is critical to reducing pressures on forest and agricultural resources, thereby indirectly preventing environmental degradation.

Since timber demands less total nutrient resources than many other forest goods (although nutrient loss at harvesting can be substantial), community woodlots will be most easily sustainable on marginal lands which should not compete with forests set aside for non-timber goods. Theoretically, non-timber products can usually be harvested with little or no disturbance to soil or timber tree regeneration. However, current research is showing that the development of a demand for non-timber forest products can cause the loss of that species from the ecosystem and may have more long-term impacts on ecosystem level carbon and nutrient cycles (Jennifer O'Hara, unpublished). Therefore, harvesting of non-timber forest products should be pursued with caution until it can be clearly shown that these effects do not exist.

If managed properly, non-timber goods harvested by the rural poor for direct export or to provide inputs to forest-based processing enterprises can play a neutral or positive role in forest conservation and development. Their collection from the forests, if done on a sustainable and nondestructive basis, poses a negligible threat to the maintenance of a continuous forest and results in minimal changes in the functioning of the undisturbed primary forest. This is especially true if the accumulated knowledge of traditional forest dwellers is tapped to avoid over-exploitation and the forest dwellers themselves benefit from the harvesting. Furthermore, incorporating the growth of plant species, which yield non-timber products into reforestation programs for wasteland or logged-over forests, can reduce the discount rate on the investment in a timber crop (ITTO 1990). However, scientists have recently begun to question the economics and sustainability of harvesting non-timber products by other than low-population densities of human beings (Godoy and Bawa 1993).

Non-timber products can also be obtained from secondary forests, taking pressure off the primary forests. Secondary successional forests often grow up on land converted to pasture and lying near settled areas. In Brazil, sales of products from the babassu palm, which grows in secondary forests, provide up to 40% of the income for 450,000 households based principally on the labor of women and children (May 1991).

3.1.4.8   Role of gender and forest use in poverty eradication
Toral Patel-Weynand

Gender equality is not only a matter of social justice, but is also beneficial to society and the economy. Although the gender gap is narrowing, women still tend to be less educated than men, to work more hours, and to be paid less. Eliminating gender inequalities: (1) leads to significant productivity gains; (2) provides large societal benefits; and (3) enhances poverty reduction efforts. There is strong evidence that investing in women will generate benefits for society in the form of lower child mortality, higher educational attainment, better nutrition, and slower population growth — all of which will ensure the quality and sustainability of economic growth. Improving women's access to credit is linked to increased holdings of non-land assets, to improvements in the health of female children, and to increased probability that girls will enroll in school. Independent access to land is associated with higher productivity and, in some cases, with greater investments by women in land conservation. In summary, by directing public resources towards policies and projects that reduce gender inequality, policymakers not only promote equality but also lay the groundwork for slower population growth, greater labor productivity, a higher rate of human capital formation, and stronger economic growth (World Bank 1995).

Because of the traditional gender division of labor, women have specific needs and interests in forestry that are different from those of men. In most rural areas, trees are important in rural economies, largely as a result of the uses to which women put them. In many societies, it is women who find and transport the household's fuelwood for cooking and heating; who gather wild fruit and nuts; who find fodder for the domestic stock; and who make medicines and other products from woody materials. Women also earn what little cash income they have from activities that relate, directly, or indirectly, to trees and forests. Because women are aware of the utility of trees on the homestead, they take good care to plant and maintain them. Women are repositories of knowledge that men may not have about forest products, plant attributes, and traditional methods of tree management. Thus, in many rural societies, a special relationship exists between women, the family, and trees (FAO 1987b, World Bank 1991).

Women collect fuelwood for more reasons than simply cooking. The homestead fire provides many other benefits: (1) drinking water is boiled and washing water is warmed on the fire, while fish and meat are smoked above it; (2) the fire provides light at night, and heat to dry a wet harvest; (3) it may also be used to cure tobacco, boil water to extract natural medicines from leaves and bark, and make dyes; (4) the smoke from fires is used to keep insects away; and (5) in some countries, household fires are used to keep livestock warm on chilly nights. Fires also have many social and ritual uses, particularly as the focal center for evening conversations. Thus, the fuelwood collected and transported by women has many functions. Therefore, when supplies diminish, much more than the family meal is threatened, because the basis of village life is altered (FAO 1987b).

Small-scale, forest-based enterprises, such as the collection and processing of raw materials into useful products, are major sources of income for the poor, especially for rural women, including those from landless families (Patel-Weynand 1997b). Based on a study in Uttar Pradesh, India, nearly 50% of the income for poor women came from forest and common land compared to only 13% of men's income (FAO 1987b).

Women and men make very different uses of forest resources — even of a single species of tree. For example, in Pananao in the central mountains of the Dominican Republic, the control of, responsibility for, and labor involved in exploiting the palm tree vary according to the parts of the tree being used. Men use the wood for construction; women use the fiber in handicrafts; and both men and women use the tree's products for animal fodder. The division of control, responsibility, and labor also shifts with place and activity. Near the homestead, only women are involved, but on pastureland, men exercise all three functions exclusively, except for the collection of fuelwood. The divisions are mixed on cropland, where women contribute labor; and on forest remnants — areas of previously over-exploited forest — where women are given responsibility for the trees and provide the labor, but men have control of the resource (FAO 1987b). The ways in which women traditionally use forest resources are becoming increasingly unviable. There are four main reasons for this, each of which has a cumulative, negative impact on the lifestyle of rural women (FAO 1987b):

1. In many areas, traditionally useful multipurpose tree species are becoming increasingly scarce as desertification and deforestation take their toll. Women therefore have longer to walk to collect fuelwood and other forest products, and this adds hours to their already long working days.

2. As more and more men find employment in the towns and cities, women are forced to carry out jobs previously done by men. This leaves little time for the lengthy business of collecting and processing forest products, however important they may be to the family economy.

3. New technologies are changing land use, reducing the availability of minor forest products that women have traditionally used as a source of additional income. These technologies are frequently introduced without providing women with other income-earning alternatives.

4. Development projects often improve conditions for men, leaving women with as much, or even more, to do than before.

These four factors need to be considered by forest certification protocols if they are to help "restore the balance between women's needs, and the forest and tree resources available to them" (FAO 1987b). Each includes issues related to both poverty and gender.

3.1.4.9 *Policies and institutions affecting forest use by the poor*

Poverty must be viewed from the perspectives of both poor individuals and poor communities, for two reasons. First, although strategies to earn incomes are largely microeconomic decisions made by individuals, group norms and conventions play important roles in both defining access by individuals to natural resources and determining the attitudes of individuals toward common property resources. Second, a separate emphasis on poor communities highlights the fact that poverty in most developing countries is associated with low productivity and low capital formation, aspects which cannot be alleviated by redistributive policies alone (Jagannathan 1989). The linkages between poverty at the individual and community levels within the environment can be summarized as follows (Jagannathan 1989):

State of Being Poor	State of the Environment
Individual with low consumption and low expectations of future income	Habitat covering living conditions and access to amenities
Society or human organizations of poor communities	Wider concept of habitat-covering land, water, and atmospheric resources of the area or sub-region

The linkages shown above interact with each other to form the following relationships (Jagannathan 1989):

- A poor individual interacting with his or her immediate environment—poverty is manifested by low income and consumption levels and with suboptimal choices made in consumption of basic needs (shelter, safe drinking water, education, and health care). These suboptimal consumption decisions lower a person's physical quality of life, and become visible as degraded physical habitats and preventable environmental health risks.

- A poor individual interacting with the land and water resources around his or her living area—poverty is indicative of a lack of access to capital, technology, and alternative sources of livelihood, and can also be associated with suboptimal individual production decisions, because of high private discount rates. For example, farmlands and pasturelands could be over-exploited, investments in soil conservation neglected, and crops be grown on steep mountain slopes.

- Communities of poor individuals interacting with their immediate environment—social norms and conventions determine how communal open-access resources (e.g., village wells, common lands, modern service facilities provided by the state) are utilized. These socially enforced rules determine attitudes toward conservation and waste disposal. For example, in most poor societies, the recycling of agricultural and domestic wastes is widely institutionalized. Socially determined conventions and norms usually change slowly, and are often unable to adapt to changed circumstances caused by demographic and economic pressures.

- Communities also determine access to productive resources, such as land and water—although most countries utilize a formalized legal system to define access to land and water resources, high transaction costs in utilizing the formal system result in the continued importance of traditional rights to usage. The interface between the formal and informal system often determines the incentives for sustained use of productive assets such as land.

Public policies may affect one or more of the above-mentioned poverty-environment relationships:

- Some policies have been targeted at the individual poor producer or consumer (e.g., special credit programs, guaranteed producer support programs, food subsidies), which result in reducing the dependence of the poor on their immediate environment.

- A second group of more general policies has attempted to lift whole communities out of the poverty trap by seeking to augment the human capital of poor communities by spreading literacy, skill formation, and preventive health care.

- A third group of policies has addressed spatial deficiencies by improving the physical quality of life in communities and by integrating rural areas with market centers through physical infrastructure investments in roads, housing, schools, health centers, water supply, electricity, etc. Investments of this nature increase the mobility of labor and capital between locations.

- A fourth set of policies has attempted to remove or reduce institutional constraints by specifically addressing the problem of access by the poor to productive resources (e.g., land reforms, land redistribution, input delivery systems).

Taken together, all of these poverty alleviation measures are designed to improve access by poor individuals and communities to income earning opportunities by integrating their skills and work capabilities with the dynamic sectors of the economy. The policies, in theory, should reduce the direct dependence of the poor on the local natural resource base for subsistence. In practice, public investments and policies are affected by the rent-seeking behavior of the elite, and filter down unevenly to poor individuals and communities, undermining the maintenance of spatial and distributional equity (Jagannathan 1989).

With respect to forests, the incentives for sustainable or non-sustainable use by the poor and non-poor are the result of the interaction of a broad range of sector policies affecting forests. The impact of intersectoral policies on the forest sector were uncovered during the 1980s, when policy analysts found that the roots of deforestation and forest degradation often lay outside the forestry sector. The classic example is the situation where a country's property laws and land reform legislation require settlers to clear land in order to secure title to forested areas — large areas that could have been exploited on a sustainable basis for commercial timber and non-wood products have been lost in this way. Repetto has suggested that one useful way of conceptualizing how intersectoral policy linkages affect forests is to visualize a set of concentric circles moving outward from the forest (FAO 1994):

- At the hub are policies that directly affect forest management: forest revenue structures; tenurial institutions governing the privatization of forestland and enforcement of traditional use rights; reforestation incentives; and administration of timber-harvesting concessions.

- In the next circle are policies directly influencing the demand for forest products: trade and investment incentives to promote wood-using industries; and energy pricing to encourage fuelwood substitutes.

- In the third circle are policies directly affecting extensions of the agricultural frontier and the rate of conversion of forestland: agricultural credit, tax, and pricing incentives for frontier agriculture, including policies affecting the price of new forestland; incentives for cultivation at the intensive — as opposed to the extensive — margin; and the concentration of landholdings, as well as public investments that indirectly spur frontier expansion in the form of road building and public services, such as agricultural research and extension.

- In the outer circle are macroeconomic policies that indirectly affect deforestation: exchange-rate policies affecting tropical forest export products; policies affecting capital markets which influence investors' time horizons; demographic policies; trade and investment policies affecting labor absorption; and rural-urban migration.

How these policy linkages are defined and interpreted depends on whether forest issues are: (1) viewed from a national (macro) or a forest unit (micro) perspective; (2) evaluated using development-oriented or resource-oriented concepts of capital, space, and location; and (3) analyzed with macroeconomic or microeconomic methods, therefore establishing macro or micro sets of policy priorities (FAO 1994). Note that trade liberalization raises important questions regarding the social distribution of wealth, resources, and income. More open markets tend to concentrate wealth and redistribute it to economically efficient groups at the expense of less-advantaged and less-efficient segments of society. These shifts in wealth often require public intervention to adjust for imperfect competition and market failures. However, recent empirical studies contradict the view that logging for the international timber trade is a major cause of deforestation and environmental degradation. The evidence suggests that, in many countries, a large portion of logging is for domestic consumption and not for trade. For example, because the majority of tropical forests are cleared for agriculture and the majority of wood consumed for energy, only about 6% of the total amount of wood cut in the tropics enters the international timber trade (FAO 1994).

While forest conditions affect the opportunities for national development, the development process also shapes what these forest conditions are, what they do, and what they will become. In general, as development proceeds, population pressure on land increases and then declines; urban demand and prices for wood products and energy increase; urban incomes and savings rise; non-farm employment opportunities grow; road systems and water-resource developments expand; governmental capacities to protect forests, subsidize forest growth in agricultural areas, and cooperate with local populations in forest management increase; and the strength of environmental interests relative to extractive interests and the extent of their international integration increase.

3.2   Value-laden issues of certification

Two topics in particular are strongly expressed in existing certification protocols and appear to drive the structure and the design of most protocols:

- Saving species and biodiversity
- Values and desires of certifying organizations, society, and landowners

It is important to understand these two topics and their scientific basis. A lack of understanding of their ramifications and predictive ability has resulted in the identification of incorrect criteria and indicators (see 1.2). The result of this is the development of certification protocols that do not appear to be able to assess an ecosystem as to its "sustainability" or whether "good forest management" is being practiced (see 3.1.2, 3.1.3).

This discussion will be followed by a presentation of non-value-based topics that need to be considered in any certification assessment (see 5.2). These non-value-laden topics can be separated into social- and natural-science based legacies that need to be assessed as to their relevance for assessing each ecosystem (Vogt et al. 1999b).

3.2.1 Species and ecosystem relationships

3.2.1.1 Human value for species and their link to ecosystems
Kristiina A. Vogt, Karen H. Beard

Because species are frequently the level at which systems are examined to determine if they are degrading, it becomes important to understand whether or not it is an appropriate indicator of ecosystem health and sustainability. This is an important topic of discussion since most certification protocols use the number of species as a central vehicle to identify indicators. However, using species as an indicator of the ecosystem is probably not valid in most cases and actually reflects the human value for species. The question that arises is how effectively species can be used to assess a site. In other words, is it sufficient to emphasize species during an assessment of an ecosystem, and do species reflect how that system functions?

> In most examples, species appear to be poor indicators as to whether a system is in a trajectory to being degraded, or whether more species will be lost from that system. However, in other cases, species can be very useful indicators to monitor, especially when keystone species are present, or a species is part of a dominant functional group controlling system function.

Certain ecosystem types (e.g., grasslands) have been identified where strong positive relationships exist between species diversity and the stability or productivity of that system (Johnson et al. 1996). However, these relationships have not been as common in woody perennial species, where the diversity and response of a species to its environment appears to occur at a level other than species. Only in those systems where consistent relationships have been shown between species and the system function should the species level be the focus of monitoring. This subject will be further discussed in the next section to illuminate where it might be relevant to monitor species. Discussion will also present the scientific evidence and the controversies surrounding the species and system functional relationships. Most of this discussion will demonstrate why a species focus is probably not appropriate as one of the primary analytical tools for certification assessments.

Historically, there has been interest in maintaining and collecting data on diversity at the species level. For example, species have been the focus of concern over the loss of biodiversity and the "conceptual, biological, and legal focus in conservation " (Meffe and Carroll 1994). This has occurred even though ecologists generally agree that biodiversity at all levels of biological organization is being lost globally at an unprecedented rate (Myers 1984, Wilson 1988, Pimm et al. 1988).

Species are often the focus of policy decisions because of their different values in society. Some of these values are:

- Species are valued intrinsically by the sociocultural system (Ehrenfeld 1976, Norton 1987, Shiva et al. 1991)
- Species have great economic values (Farnsworth 1988)

- Species are important in community dynamics and have synergistic effects on other species (Paine 1966)
- Species important in ecosystem dynamics, altering energy flows and nutrient cycles (Naiman 1988), performing ecosystem functions such as the production of oxygen (Norton 1986), and playing key roles in ecosystem resistance and resilience (Margalef 1963, Wilson 1988)

Therefore, the measurement of the number of species in a system, or species richness, is often measured to represent ecological communities and ecosystem functions. However, the ecological significance of species richness as a measure of community and ecosystem functions is questionable. There is little empirical data showing that species richness is the best measure of biodiversity. First, it has been recognized that there may be other more important levels of diversity, such as genetic diversity, population diversity, species diversity, biotic assemblage diversity, functional diversity, and whole-system diversity (Vogt et al. 1997a). These other levels of diversity may be more important than that played by species diversity in particular systems; the relative importance of the roles played by these various levels of biological organization are system dependent. Second, the relationship between all levels of biological organization and ecosystem processes has also come into question as particular levels of biological organization do not appear to be related to those factors humans value in the system. Therefore, although there is no question that biodiversity loss at the species level is an important concern (Soulé 1986), the appropriateness of using species richness as a measure of biological diversity has not been generally verified.

Discussion on species and ecosystem relationships needs to include society's value for species maintenance in the environment. United States federal laws are also based on maintaining species requiring forestland owners to manage their lands so that species on their lands are not lost or diminished in numbers (see 5.3.2). As one of their primary goals, all assessment protocols have to explicitly deal with how to maintain species. However, it is important to recognize that species may not be the indicator to monitor if there really is an interest in sustaining them and their habitat. If this is not recognized, we may hasten the path of degradation and species loss under the false premise that we are effectively monitoring a system. Again, it is important to know when the species approach works and when it does not. The next section will discuss our scientific understanding of species within an ecosystem framework. This discussion will also give clues as to where species should be used as the level of analysis and assessment.

3.2.1.2 Definitions and measurement techniques for species diversity
Karen H. Beard

There are many definitions for the term biological diversity (Machlis and Forester 1994). Most definitions treat diversity as a qualitative state that can be identified to exist at the genetic, species, ecosystem, or landscape levels (Wilson 1988). There is not much agreement about how to measure diversity (Hulbert 1971). The measurements of the potential indicators of biodiversity range from number of species or species richness (i.e., measured as alpha, beta, or gamma diversity) (Scott et al. 1987), to abundance and distribution of populations (Krebs 1972) to the degree of genetic variability in and among species (Soulé 1986). There are also more conservation-oriented measures that are discussed, such as the number of endangered species and centers of species richness with high endemism (Myers 1984).

According to Machlis and Forester (1994), others measures of biological diversity focus on:

- Ecosystem functions (Ray 1988)
- Community interactions (Janzen 1988)
- Natural communities (The Nature Conservancy 1975)
- Successional stages (Franklin 1988)
- Ecological redundancy (Walker 1992)

Therefore, the term biological diversity needs to be well defined in each context since the definitions have very different meanings. The use of the term would result in the identification of a suite of different indicators.

In most cases, indicators of ecosystem functioning will need to be identified. Although, species are often used to represent ecosystem functioning, the relationship between species and their functional roles is often unclear. In fact, functionality is not a characteristic used to differentiate or confirm taxonomic relationships. In many situations, it is likely that biological levels of organization other than species would be more important to understanding ecosystem stability and functioning. The question for each new system then becomes: where in the biological hierarchy are functionally distinct phenotypes found that contribute to ecosystem functioning? (Jonsson et al. 1997).

There are some clear examples for plant systems where the plant species are not the level in the biological hierarchy that should be used to assess ecosystem health. For example, for woody plant communities where the diversity of mycorrhizal species tends to be high, the diversity of the mycorrhizal species colonizing plant root systems is often more important than plant species diversity in assessing ecosystem health (Vogt et al 1997). Mycorrhizal species have different capacities of responding to the heterogeneous environment of the soil or to disturbances that are not uniformly expressed in the soil environment of the individual tree. Tree species may not have the ability to respond to these environments either; thus, these associations make plant species more plastic in their response to disturbance. Therefore, the important measure of how well the ecosystem will respond to disturbance may be more dependent on the number of mycorrhizal species which enable the plant to adapt to the disturbance rather than on the number of plant species.

In other circumstances, it is likely that the intraspecific level of diversity is the most important level of biological organization, since that is what is responding to a disturbance at the system level. For example, trees have tremendous phenotypic and genotypic plasticity at the species level, so that one species will demonstrate a multiple of responses to any given disturbance. Therefore, for woody plants, the diversity of adaptive responses (i.e., changing growth rates and how they respond to disturbances) may be linked to the genetic level, as opposed to the species level (Sultan 1987).

Studies show that woody plants' variable responses did not occur at the species level (NAPAP 1993). For example, using seedling studies, a wide range of sensitivities to pollutants was found between strains of a given species as well as among species (Barnard et al. 1990). Genetic diversity has been found to be greatest between families within species in other studies (Jonsson et al 1997). The results suggest that a high diversity of perennial woody species will not be necessary for a tree to successfully occupy a site and to respond to disturbances. Therefore, for this system, the diversity of mycorrhizal species or of genetic responses at the individual species level should be determined, as opposed to the number of species present as an indicator of ecosystem health.

Depending at what level the system has been affected, as well as the characteristics of the system, changes in the system to preserve the functioning of the system will be observed. As was observed in grassland ecosystems, and will be discussed below, the species composition changed as a result of the disturbance; therefore, there was a change in the modes of interspecific interactions to maintain a particular system output (Tilman 1996). However, if the change is in the genetic composition of the community or within a species, then there will be different phenotypic compositions manifested that may still maintain a particular system output. If genetic diversity is lost from these systems, then different phenotypes need to arise to maintain a particular system output if an ecosystem function is to be maintained. These different levels may be viewed as the levels at which adaptation must occur for the functioning of the ecosystem to be maintained and clearly will not always be at the species level. The level in the biological hierarchy at which changes to preserve ecosystem functioning is observed indicate the level in the system that should be monitored to assess ecosystem health.

3.2.1.3   *The development of the relationship between species diversity and ecosystem stability*
Karen H. Beard

The idea that species diversity influences ecosystem function can be traced back to Odum's (1959) textbook in which he based his approach to ecology around Tansley's (1935) concept of the ecosystem (McNaughton 1993). This textbook likely motivated MacArthur's (1955) suggestion that the more alternative pathways for energy flow that exist in ecosystems, the less likely that disturbance would be able to destabilize the system (McNaughton 1993). Although MacArthur's argument described interactions among faunal species and, therefore, was mostly related to community stability as opposed to ecosystem stability, it had tremendous implications for ecosystem ecology. The next significant contribution came from Elton (1958), who tried to explain how more "complex" systems were more stable than "simple" systems. The early arguments from these two theorists laid the foundations for the idea that diversity was a stabilizing force in communities and ecosystems — an idea that prevailed from the mid-1950s through the mid-1970s.

Although the early arguments presented by MacArthur (1955) and Elton (1958) seemed intuitive, ecological theorists began to advocate counter-arguments to these ideas in the mid-1970s. Ecologists began building mathematical models of simple and complex systems which suggested that more complex systems were actually more susceptible to instability and species loss (e.g., May 1973, DeAngelis 1975, Pimm 1979). Thus, theoretical studies on the relationship between complexity and stability concluded that systems with more species (more interspecific interactions per species or connectance, and stronger interactions) were not as likely to be as stable as systems with fewer species (May 1973). The diversity-stability controversy is a result of the different conclusions obtained between these intuitive and theoretical arguments about the relationship between diversity and stability.

There are two major problems with the intuitive and theoretical suggestions described above. First, although the observations and model systems were suggestive, neither had been tested empirically. Second, neither explicitly addressed the relationship between species diversity and stability in terms of ecosystem function, but rather as community stability (this idea is explicitly tested elsewhere and will not be discussed here; see Smedes and Hurd 1981, Dodd et al. 1995, Frank and McNaughton 1991). The controversy has not ended about whether species diversity stabilizes systems, and contradictory arguments are still being presented for both sides. Therefore, according to the literature, the relationship between diversity and ecosystem stability has not yet been resolved (Johnson et al.

TABLE 3.1

This table summarizes the experimental tests of the diversity-stability hypothesis. These experiments measured the response of the ecosystem as a change in a primary productivity measurement to some type of disturbance.

Ecosystem	Disturbance	Productivity Measure	Relationship Between Diversity and Stability
Grasslands, California (McNaughton 1968)	NA	Changed Biomass/Productivity	No Relationship
Grasslands, India (Singh and Misra 1968)	"Disturbed"	Changed Biomass/Productivity	Negative
Grasslands, New York (Hurd et al. 1971)	Nutrient pulse	Aboveground Biomass	Positive
Grasslands, New York (Mellinger and McNaughton 1975)	Nutrient pulse	Above- and Belowground Biomass	Positive
Grasslands, Serengeti (McNaughton 1977)	African buffalo grazing	Aboveground Biomass	Positive
	Change in precipitation	Aboveground Biomass	Positive
Grasslands, Serengeti (McNaughton 1985)	Four ungulates grazing	Aboveground Biomass	Positive
	Change in precipitation	Aboveground Biomass	Positive
	Several ungulates grazing	Aboveground Biomass	No Relationship
Grasslands, Minnesota (Tilman and Downing 1994)	Drought	Aboveground Biomass	
Grasslands, England (Silvertown et al. 1994)	Change in precipitation	Aboveground Biomass	Positive
			Positive
Grasslands, Minnesota (Tilman 1996)	Drought	Above- and Belowground Biomass	Positive
Grassland plant spp. (Wardle and Nicholson 1996)	NA	Belowground Biomass	Positive
		Belowground Biomass	No Relationship

1996, Chapin et al. 1998). However, experimental research is uncovering how species diversity affects the functioning (photosynthesis, decomposition, and nutrient cycling) of the systems they inhabit although these generalities being developed from the experimental tests of the theory are particular to certain types of systems, such as grasslands (see Table 3.1).

3.2.1.4   Keystone species, ecosystem engineers: where species are good indicators of the ecosystem
Karen H. Beard

Ecological concepts have been developed to describe the relationship between single species and ecosystems. For example, there is the indicator species concept, which is used to describe species that represent other species, environmental conditions or ecosystem processes of interest (Zonneveld 1983). The implication with the indicator species concept is that abundance and biomass of species indicate the presence of certain environmental attributes, but that the species itself is not necessarily important in the ecosystem. The concept can be a useful management tool because one species is easier to measure than species diversity and ecosystem processes, although its effectiveness greatly depends on the selection of an appropriate species (Landres et al. 1988). It has been suggested that functional groups may serve as better indicators of ecosystem processes than single species (Vogt et al. 1997a). Functional groups may eliminate the problem of finding an appropriate indicator species and would not be as subject to natural population fluctuations that might complicate the outcomes. The importance of species in ecosystems and the pivotal role that they might play is exemplified by keystone species. When keystone species are lost from ecosystems, their loss will result in a disproportionately larger effect on some property of the system (Paine 1966, Bond 1993, Jones et al.1994, Power et al. 1996, Jones et al. 1997). Based on the keystone concept, keystone species play an essential role in ecosystem structure (Jones et al. 1994, Jones et al. 1997), ecosystem function (Bond 1993) and/or the maintenance of community diversity (Paine 1966). Keystone species are often found in functional groups that contain few species. Keystone species may not be obvious because some are keystone only at particular times and in particular places. The keystone species concept is important because, despite its complexity, it can help managers identify species important to the ecosystem and deserving of protection (Vogt et al. 1997a).

Many studies have been conducted to determine the role of individual species in ecosystems. Some of the most notable studies illustrating keystone qualities of species are listed in Table 3.2. This list illustrates the range of ecosystems and species that have been found to serve keystone functions. From this list, it is clear that keystone species can play a variety of essential roles in the community and ecosystem (i.e., altering community dynamics, ecosystem structure, and ecosystem functioning). Keystone species appear to play many different roles in the community or ecosystem and many ecosystem types appear susceptible to keystone species.

Other examples of species playing important roles in controlling ecosystem function can be found with species that have been called "ecosystem engineers" (Jones et al. 1994). These species are not necessarily keystone because they do not necessarily have a disproportionately large effect on some property of the system. Jones et al. (1997) summarized examples of ecosystems and species where physical engineering is most likely to be important. He identified the following characteristics to ecosystems where species function as engineers:

- Dominant organisms are massive, persistent structures (e.g., forests and coral reefs)
- Plant cover is extensive (e.g., forests)
- Animals build or destroy massive, persistent, abiotic structures (e.g., beavers, gophers, termites, coral)
- Large animals are abundant (e.g., elephants, bison, hippopotami)
- Abiotic substrates are amenable to biogeomorphic action (e.g., digging and burrowing animals, lichen, and plant root chemical exudates
- Ecosystem has persistent structure (e.g., tree, mound, carapaces)
- Many abiotic resources are integrated (e.g., animals affecting rivers, streams, soils, or sediments)
- Environments are extreme (i.e., positive interactions may be more important in harsh environments)

Although these generalizations describe situations where ecosystem structure is likely to be changed by single species, these characteristics can also be used to generalize systems where species are likely to change ecosystem functioning, since structure and function are related.

It is clear that species play important roles in the structuring and functioning of ecosystems, maintenance of communities, and representing the attributes of a system, however, the relationship between species and ecosystems remains relatively unpredictable. It most likely remains unpredictable because identifying which species represent which functional groups for particular ecosystem processes and the relative importance of different species for different ecosystem functions is often difficult. Undoubtedly, the importance of species in ecosystems will vary on a case-by-case basis. However, generalizations such as those above about which systems are likely to contain ecosystem engineers and thus, indicator species of ecosystem health, are useful in determining where species could be monitored as surrogates for ecosystem health.

3.2.1.5   *Main limitations of past species/ecosystem studies*
Karen H. Beard

Although there is general consensus that there is a positive relationship between species diversity and ecosystem stability based on the studies described above (Table 3.1), the evidence that tests the relationship between species diversity and ecosystem stability is severely limited. There are very few empirical studies and the studies that exist have flaws in their experimental designs that make the generalized use of the results difficult. Therefore, the diversity-stability relationship has not been resolved, despite the fact that the results of these experiments might appear convincing (i.e., most results indicated that the relationship between species diversity and the degree to which an ecosystem is stable is positive—except McNaughton 1968, Singh and Misra 1968, McNaughton 1985). The main limitations of these studies fall into four categories: lack of appropriate controls, insufficient surrogates for productivity measurements, lack of appropriate disturbances, and limitations of the study system. These four points will be discussed below.

It is important to mention that several other studies have investigated the relationships between species diversity and ecosystem stability (Ewel 1986, Berish and Ewel 1988, Ewel et al. 1991). However, these studies did not look at primary productivity; rather, they investigated soil fertility changes with diversity. Although these papers support the hypothesis that diversity begets stability, they will not be further considered here. Another recent study investigated the effects of grassland species on soil microbial biomass, the microbial respiration:biomass ratio and plant litter decomposition (Wardle and Nicholson 1996). This study included only two species mixtures and monocultures. They found that monoculture versus two species mixtures influenced soil processes negatively and positively in a non-additive manner. Even though this study measured changes in microbial respiration:root biomass, a potential measure for the stability of the system (Wardle 1993), it is not considered further here because it included combinations of multiple species in its treatments.

Other lines of evidence for testing the relationship between species diversity and the stability of the system come from theoretical studies. King and Pimm (1983) and Chapin et al. (1998) modeled grassland communities, but have not empirically tested model results — they compared their results to empirical data. The results of both of these studies indicated that species diversity is positively correlated with ecosystem stability. A more recent study on this relationship was conducted by Doak et al. (1998) who suggested that the positive diversity-stability relationship often found in grassland communities could be a statistical artifact. They argued that stability measures, such as biomass and productivity, in these studies showed that species diversity leads to greater stability because of statistical averaging of fluctuations in species abundance when more species are present.

More experiments are being conducted in grassland communities that are also producing positive relationships between diversity and ecosystem stability (Hooper, in press; Hooper and Vitousek, in press). Although there are few studies that explicitly addressed the

TABLE 3.2

Summary list of classical studies on keystone species.

Cause of Species Loss	Species	System	Ecosystem/Community Measurement	Effect of Species Loss on Measurement	Reference
NA	Mammalian herbivores (Lemmings)	Arctic	Soil nutrients (N and P), community composition, nutrient content of plant tissue	(–) soil nutrients (N and P), (0) community composition, (–) nutrient content of plant tissue	Batzli et al. (1980); McKendrick et al. 1980
NA	Fiddler crab	Salt marsh	Soil drainage, soil oxidation-reduction potential, decomposition of belowground plant debris, primary productivity	(–) soil drainage, (–) soil oxidation-reduction potential, (–) decomposition of belowground plant debris, (–) primary productivity	Bertness (1985)
NA	Snow geese	Arctic	Primary productivity, N cycling	(–) primary productivity, (–) N cycling	Cargill and Jefferies (1984a,b)
Experimental removal	Fish	Lake	Primary productivity, nutrient cycling, algal biomass, community structure	(+,–) primary productivity, (+,–) nutrient cycling, (+,–) algal biomass, (+,–) different species	Carpenter et al. (1993); Carpenter (1988); Carpenter and Kitchell (1985), (1987)
NA natural distribution	Earthworms Sea otters	Forests Marine	Ecosystem structure Community composition, primary production	(–) ecosystem structure (+) sea urchin populations, (–) primary production	Darwin 1881 Estes and Palmisano (1974)
NA— Species outbreaks	Invertebrate herbivores	Forests	Primary productivity, water yield, nutrient cycling, streamwater chemistry	(–) primary productivity, (–) water yield, (–) nutrient cycling, (0) streamwater chemistry	Dyer (1986)
Experiment removals and additions	Detritivous fish	Tropical stream	Sediment deposition, composition of algal and invertebrate assemblages	(+) sediment accrual, (–) total invertebrate density, (+,–,0) density of different taxa	Flecker (1996)
Acid deposition	Snails	Forested	Passerine egg shell success, clutch care, nests occupied	(–) egg shell success, (–) clutch care, (–) nests occupied	Graveland et al. (1994), Graveland and van der Wal (1996)

NA	Pocket Gophers	Grasslands	Ecosystem structure, plant/herbivore community, soil nutrients	(–) structure, (–) species diversity, (+, –) soil nutrients	Huntly and Inouye (1988)
NA	Snails	Desert	Amount of nitrogen, productivity, soil formation	(–) nitrogen addition, (–) productivity, (–) soil formation	Jones 1990
NA	Hippopotami	Wetlands	Ecosystem structure	(–) ecosystem structure	McCarthy et al. (1998)
NA	Beavers	Forested wetlands	Stream morphology, hydrology, carbon and nutrient cycling, decomposition rates, community composition	(–) ecosystem structure and dynamics	Naiman et al. (1986), Naiman (1988)
Experiment removal	Starfish	Marine intertidal	Community composition	(–) species diversity	Paine (1966)
Experiment removal	Moose	Forests	Plant community composition, litter production, soil C, N, CEC, field N availability, potentially mineralizable N, microbial respiration rates, potential mineralizable N, microbial biomass	(+) plant diversity, (+) litter production, (+) soil C, (+) N, (+)CEC, (+) field N availability, (+)potentially mineralizable N, (+)microbial respiration rates, (0)potential mineralizable N, (0) microbial biomass	Pastor et al. (1988)
NA— Species pulse	Salmon	River	Community composition, primary production, nutrient cycling	(–) community composition, (–) primary production, (–) nutrient cycling	Richey (1975)
Over fishing	Fish	Coral reefs	Community composition, species diversity, reef complexity, productivity, and resilience	(–) community composition, (–) species diversity, (–) reef complexity, productivity, and resilience	Roberts (1995)
Hunting	Ungulates	Savanna	Plant community composition, ecosystem structure, fire regime	plant community composition, ecosystem structure, fire regime	Ruess (1990)
Metamorphosis	Tadpoles	Freshwater ponds	Standing crop of phytoplankton, N state shift dissolved to particulate, primary production, phytoplankton community	(+) standing crop of phytoplankton, (+) N state shift dissolved to particulate, (+) primary production, (0) phytoplankton community	Seale (1982)
NA	Prairie dogs	Prairie	Plant community composition, net primary production, nitrogen yield	(–) plant diversity, (–) net primary production, (–) nitrogen yield	Whicker (1988)
Experiment removal	Salamanders	Forests	Invertebrate community, decomposition rates	(+) invertebrate numbers, (–) larger prey items, (+) decomposition rates	Wyman (1988)

diversity/stability hypothesis, the general conclusion from those that have attempted it is that diversity increases ecosystem resistance and resilience to disturbance (as was found in Johnson et al. 1996). The limitations for these studies are now discussed.

3.2.1.5.1 *Lack of appropriate controls for diversity*

One of the most obvious problems with studies testing the relationships between species and ecosystem stability is their lack of appropriate controls for diversity. To conduct an experiment that uses diversity as a contrasting variable, there needs to be a direct control of diversity (Naeem et al. 1994). Otherwise, it is not possible to attribute the observed response to diversity, as opposed to some other factor. The underlying problem with these experiments is that diversity and other potentially confounding variables are inter-correlated and better manipulation of species diversity and the environment is required to overcome these problems (Huston 1997).

In every one of the studies mentioned above (Table 3.1, except Tilman et al. 1994, Tilman 1996; Silvertown et al. 1994; and possibly Singh and Misra 1968), naturally species-rich and species-poor communities were used to test the hypothesis. The species-rich communities were typically older and the species-poor communities were typically younger. However, in each of these studies, only species diversity was considered relevant and the ages and species compositions of the communities were considered unimportant. The assumption that older and younger communities would not have different compositions of species is unfounded and it is likely to have greatly affected the results. As an example, older fields may be dominated by plants species that have higher root:shoot ratios than those in younger fields because of increased competition, and this would greatly affect the response of the community to disturbance (Tilman 1988, Wedin and Tilman 1990). Thus, the results of these studies cannot be used to address the diversity-stability hypothesis as described above because the species composition in the two plant communities likely affected the diversity-stability relationship (Givinish 1994). The results from these studies are questionable because species composition or, more specifically, the functional characteristics of the species in the older and young communities were not considered.

Tilman (1988) and Silvertown et al. (1994) created more- and less-diverse plots using long-term addition of nitrogen fertilization, as opposed to using different-aged fields. However, there are problems with this method, as well (Givinish 1994, Chapin et al. 1998, Huston 1997). For example, (N) nitrogen-rich sites are known to favor plants with lower allocation to roots versus leaves, higher stomatal conductance and greater photosynthetic capacity (Tilman 1988). These attributes are also known to make species more susceptible to drought. Therefore, although species diversity was lowered in N-enriched plots, the treatment likely differentially eliminated the drought-resistant species (Sala et al. 1995, Chapin et al. 1998). Again, in these studies, the composition of the communities was not considered, even though species composition likely affected the diversity-stability relationship. Tilman (1994) later attempted to account for this factor by determining the number of C_4 and C_3 plants in the different plots; however, C_4 plants only comprised a low number of the species in Tilman's study system (Givinish 1994). Therefore, considering their abundance does not effectively account for the expected shift to drought-sensitive plants with N fertilization. The results from his study are also questionable because species composition and/or functional differences in the different plant communities were not considered.

3.2.1.5.2 *Insufficient surrogates for productivity measurements*

Another problem with these experiments is the use of surrogates for primary productivity, as opposed to measuring it directly (Johnson et al. 1996). The measurement of total primary productivity is difficult. However, the drawbacks of the surrogates used in place of measuring primary productivity directly is not recognized or addressed in any of these studies, and unfortunately, the choice of state variable greatly affects the conclusions. Not one of the studies effectively measured total primary productivity of the system (although it was measured in Naeem et al. 1994) and most used only changes in aboveground biomass as a surrogate for productivity. The measurement of aboveground biomass as a surrogate for total primary productivity is clearly a huge oversight. Belowground productivity must be considered since it can be a major portion of net primary productivity in some ecosystems (Vogt et al. 1982, Vogt et al. 1996). In grassland ecosystems, belowground biomass is at least an order of magnitude greater than aboveground biomass (Laurenroth et al. 1997), and belowground production accounts for 60–70% of the total plant community energy (Mellinger and McNaughton 1975). Belowground productivity is also known to be a particularly sensitive indicator of environmental stress (Vogt et al. 1993). Therefore, belowground productivity would be a particularly appropriate measurement to include in these studies to determine the ability of a system to return to its ground state productivity level following a disturbance.

Only two of the studies included a measurement of belowground biomass (Mellinger and McNaughton 1975, Tilman 1996). In both cases, belowground biomass was measured using root cores. There are tremendous shortcomings in using this method to measure belowground productivity, especially of grassland ecosystems, although it is commonly used (Vogt et al. 1998). However, the results from these two studies are probably the most complete because at least some measurement of belowground productivity was made.

Tilman (1996) used root:shoot ratios to analyze his belowground data. Root:shoot ratios are important because plant growth is balanced by plant carbohydrate assimilation by the shoot and nutrient assimilation by the root. Thus, plant growth is not only dependent on the availability of these compounds, but by a complex function of the root:shoot biomass ratio (McNaughton et al. 1982, Tilman 1988). Since nitrogen and diversity were strongly connected in his study plots, this suggested that root:shoot ratios were important to consider in the study (Givinish 1994). However, Tilman (1996) only presented root:shoot ratio data for the first and final year of the study. Unfortunately, root:shoot measurements taken only at the beginning and end of the study are not sufficient to estimate productivity, since how the ratio is developed and maintained is of primary concern in understanding the effects of the disturbance on the primary productivity of the system. Belowground measurements have to be taken more completely in future studies.

3.2.1.5.3 *Lack of appropriate disturbances*

Another potential problem with these studies is the application and appropriateness of the disturbances to the systems. To appropriately test whether diversity influences the stability of a system, a perturbation that moves the ecosystem away from its original state should be introduced. In addition, perturbations should only emulate natural disturbances. For example, adding nutrients to a nutrient-rich site would not likely represent an appropriate or natural disturbance (McNaughton 1993). The best types of disturbances to use in these studies are natural disturbances that do not differentially impact any part of the plant community.

In the first two studies, McNaughton (1968) and Singh and Misra (1968) did not measure ecosystem stability by taking measurements before and after a disturbance. The measurement of stability used in these studies, derived from Margalef (1963), was average biomass/net production. This measurement of "stability" measures the consistency of the rate of productivity and is not a true measure of stability as defined above. Stability is defined as the change in productivity with respect to a movement away from and recovery to some ground-level productivity. Therefore, neither of these studies successfully tested the diversity-stability hypothesis because neither study introduced a disturbance to the system.

The nutrient pulse disturbance used in Hurd et al. (1971) and Mellinger and McNaughton (1975) is also not likely to be an appropriate disturbance because it is likely to affect different-aged communities differently and this was not taken into consideration. In the studies conducted by McNaughton (1977, 1985) in the Serengeti, grazing was used as a disturbance to measure the resistance and resilience of more and less-diverse natural communities. However, selective consumption is widely observed in herbivores (Chew 1974, McNaughton 1978, Jones and Hanson 1985, Kavanagh and Lambert 1990, Majer et al. 1992, Price 1992, White 1993, Wilson and Jefferies 1996). The grazers more heavily grazed in the more diverse plots (McNaughton 1977, McNaughton 1985) and the grazing likely positively influenced these communities (Chew 1974, Mattson and Addy 1975, McNaughton 1977, Batzli et al. 1980, McNaughton 1985). Therefore, the conclusion that older, more diverse plots had greater resistance and resilience to grazing than less diverse plots, when they are selectively grazed upon and positively affected by grazing, is unfounded considering that the results are complicated by a number of factors.

McNaughton (1977, 1985) also studied changes in precipitation as disturbances to the system. However, the differences in root depth between the different levels of diversity plots could have greatly affected the responses of the two community types to rainfall. The studies where McNaughton (1977, 1985) looked at the interaction of precipitation changes and grazing will be even more complicated because of the interactions of these two factors as described in the paragraph above. Silvertown et al. (1994) also looked at changes in precipitation but the regular fertilization additions and the community composition used in the study complicate the results.

The interplay of vegetation diversity, vegetation species composition, foraging behavior, and precipitation form a complex relationship in the Serengeti (McNaughton 1993). However, this complex relationship cannot simply be summarized as supporting the diversity-stability hypothesis. The coevolution among grasses species, large grazing ungulates, and precipitation, makes artificial and misleading the results that diversity begets great ecosystem stability and resilience (McNaughton et al. 1982). The upshot from further understanding of the ecosystem makes the use of the above-described disturbances inappropriate for testing the diversity-stability hypothesis in these communities and ecosystem. Only disturbances that are unbiased should be used.

Disturbances in future studies have to be unbiased to test the idea that diversity itself, and not some confounding factor, is related to ecosystem stability. In addition, the more natural the disturbance, the more likely that the response of the system will resemble a true response and not an artifact of the disturbance. In cases where this is difficult, long-term observations of natural systems could be used to study the relationship between diversity and stability. Thus, natural and unbiased disturbances, as opposed to artificial and selective disturbances, should be favored in future studies; they are less likely to complicate the results.

3.2.1.5.4 *Limitations of the study system*

One of the greatest limitations presented by the above studies is that they were all conducted in grassland ecosystems. Most likely, this system was used because it is relatively easy to manipulate, subject to disturbance, and determine species composition on short temporal and spatial scales. However, since every test of this hypothesis was conducted in the same ecosystem type, the general applicability of the results is questionable. This serious limitation of the findings is often not recognized (e.g., Kareiva 1994); however, it needs to be recognized, in light of the fact that grasslands are unique ecosystems (Johnson et al. 1996).

One unique characteristic of grasslands is their disturbance regime (Bazzaz and Parrish 1982). The main disturbances that affect the ecosystem are drought, fire, rainfall, and grazing (Laurenroth 1979, McNaughton et al. 1989). Grazing plays an important role; the grazing food web filters 60% of the energy and nutrients that flow through the system compared to approximately 1–5% in other terrestrial ecosystems (Golley 1971, Wiegart and Owen 1971, McNaughton et al. 1982). Coevolution between grasses and disturbances has created complex relationships among these variables and is an important component of how these systems function (McNaughton 1993).

Another characteristic that makes grasslands unique are the characteristically strong competitive interactions among species populations (Bazzaz and Parrish 1982). The species are more affected by competition than plant species with close evolutionary relationships and similar life history characteristics (Billing 1978, Bazzaz and Parrish 1982). The most commonly used niche separation mechanisms by perennial grasses is seasonality; the species exhibit phenological differences or differences in timing requirements for resources (Johnson et al. 1996).

Species coexistence occurs via niche separation on the beta-scale, alpha-scale, and through regeneration characteristics (Bazzaz and Parrish 1982). The one study that used annual as opposed to perennial grass species found no relationship between diversity and stability (McNaughton 1968). This might have resulted because these different types of grasses have different natural history characteristics. Although all the studies were in the same ecosystem type, this example illustrates how even within the grassland ecosystem, the species used in these studies can greatly affect the diversity-stability relationship.

Since ecosystems are holistic and integrative ecological constructs (Tansley 1935, Golley 1995), the interactions within ecosystems are system-dependent (Steele 1991, Johnson et al. 1996). Therefore, the choice of study ecosystem is likely to influence the diversity-stability relationship. Similarly, plants and plant communities differ in their life history characteristics, physical structure, and chemical contribution to the system. Therefore, the choice of plant study species will influence the diversity-stability relationship, as well (Frank and McNaughton 1991, Swift and Anderson 1993). Although the results described above may be in general agreement, they describe only the relationship between diversity and stability for one particular community and ecosystem type. The nature and magnitude of this relationship is likely to be different in different systems (Johnson et al. 1996). Both plant species composition and ecosystem type are important in defining the diversity-stability relationship and should be considered, as well as varied, in all future studies of this hypothesis.

Finally, all of the studies to investigate the relationship between diversity and stability were conducted using small spatial scales compared to the scale of the ecosystem function of interest and the size of the ecosystem and community. Clearly, ecosystem processes occur over various spatial and temporal scales and, therefore, measuring changes in ecosystem functioning using small spatial and temporal scales may bias the results compared to results that might have been taken for the whole ecosystem.

3.2.2   Human values integral to certification

3.2.2.1   *Social and natural science integration in natural resource management and assessment*

Joyce K. Berry, Kristiina A. Vogt

Human values are being integrated into natural resource management and evaluation and are embedded in the definitions of terms such as sustainability, sustainable development, and ecosystem management. All of these terms attempt to explicitly balance the sustainability of the social and natural systems. As part of the social sciences, human values become an important mechanism by which people and their activities are being linked to the natural resources (Vogt et al. 1999ab). Since sustainability is defined not only for the natural system, but also for the social system (see 3.1.2.2), certification has to deal with how to integrate both disciplines. Human values are a critical component of the social science side of frameworks attempting to integrate people as essential parts of influencing natural resource resilience and resistance.

Historically, forestry was often thought of as primarily a biological and economic enterprise and the study of people was peripheral to the study of trees and resources. Even though scarcity in the supply of forest products has been documented for hundreds of years (see Chapter 2), the importance of the social side of the equation, aside from economics, was not considered central to the management of forests until quite recently. When these past periods of shortages occurred, they were dealt with as a problem that required locating more wood sources (Vogt et al. 1997a). During that time, when supplies diminished in one area, the perception was that it was a matter of finding new sources, since unlimited timber supplies existed elsewhere (Vogt et al. 1997a). Shortages were not examined as an issue where consideration needed to be given as to how to manage forest resources in a sustainable manner. Once sustainability of a finite resource became the issue, it focused discussions on how to integrate the ecology or natural sciences of the resources, their management, and social issues revolving around people's utilization and values for these resources. Discussions along this line became more common during the last three decades, when it became apparent that ignoring the social side of the management equation resulted in unsustainable natural resources management (Decker et al. 1996).

Many changed perceptions towards natural resource management can be documented to the 1970s. At that time, the environmental movements began to have a major impact on natural resource management, especially on public lands. The public's changing mood was driven by several factors:

- The perception that the environment was degrading and that this would diminish their quality of life and directly affect their health
- The perception that human activities were changing the global environment in a historically unprecedented manner
- A decreased willingness of people to tolerate anthropogenic activities considered to be producing unacceptable environmental impacts

Because forestry was not sensitive to the changing mood of the public and what they would tolerate, it became caught in the political "crossfires" of the 1980s and 1990s (Salwasser 1990). Forest managers realized that publicly owned forests could not be managed without an understanding of what people valued and wanted from these forests. In the early 1990s, the connection between people and forests became a concept firmly established and formalized when ecosystem management became the accepted forestry and natural resources paradigm (Robertson 1992, unpublished manuscript; Christensen 1997, Vogt et al. 1997a, Larson et al. 1997, Berry et al. 1998). The changing philosophy on what was publicly acceptable forest management was reinforced by other international events (see 2.1, 2.2). These international events further focused attention on the principles of sustainable development and the sustainability of both natural and human communities. Two specific events were crucial in mobilizing public thinking on these issues: the publication of the *Brundlant Report* (World Commission on Environment and Development 1987) and the 1992 Earth Summit and the subsequent launching of Agenda 21 (Keating 1993).

By the late 1990s, the integration of social and natural systems has become a well-recognized foundation for ecosystem management and sustainable development, although the methodology or tools to implement these approaches have been difficult to articulate (Vogt et al. 1997a). There are four principles that define ecosystem management and are central to certification and its implementation:

- Ecosystem management focuses on the *sustainability of systems,* rather than efficient production of resource outputs, such as timber, animal units per month, or recreation days (Gordon 1994, Grumbine 1994, Knight and Bates 1995).

- The complex and dynamic nature of ecosystems requires a *holistic understanding* of all parts, linkages, and feedbacks within the system and between the system and the landscape within which it is embedded (Vogt et al. 1997a). This requires an understanding of the entire system, but at the same time, an ability to identify the minimum information needs that are critical in controlling the functioning of a system (Vogt et al. 1997a).

- *Longer-term temporal and spatial analyses* of systems are crucial to utilize when assessing system sustainability because of the legacies that control and constrain the current functioning of the system (Vogt et al. 1998).

- *People value and interact with ecosystems in multiple ways* and these interactions can promote, constrain, or reduce the sustainability of the system. This identifies the importance of being able to document and evaluate people's values, and to be able to determine when they become driving variables or constraints on how a forest ecosystem can be currently managed. These values can be economically, environmentally, aesthetically, socio-culturally, or politically driven (Kellert 1996, 1997, Decker et al. 1996).

In past attempts to understand the environment and human impacts on the environment, the social side of the equation was mainly perceived to be an economic analysis that drove the policy-formulating process (Decker et al. 1996). This type of approach failed to consider the socio-cultural components of the management environment (e.g., traditions, values, norms, religions, and public and resource manager philosophies) (Decker et al. 1996). The way in which a manager or the public interprets economic, ecological, and political information is based on the socio-cultural framework particular to their region (Decker et al. 1996). Some have concluded that it is "social scientists" who have been ineffective in conveying the social significance of this field outside their discipline. In addition, social scientists have exacerbated this problem by not learning the more technical language of forestry that would enable them to communicate better with this group (Luloff et al. 1996). However, the same statement can also be made for forest managers. Until quite recently, forest managers felt that all they needed was good silviculture to manage their forestlands and that the social sciences had nothing to contribute to the management of these resources.

In the last few years, many models have been developed for the involvement of people in natural resource management that include, or are predominantly comprised of, the social system. Most of these models are either conceptual or general in nature. They were not designed to be predictive or to identify the mechanistic links between the social and natural systems. Some of these models are presented below:

- The Human Ecosystem model developed by Machlis et al. (1997) is mainly driven by the human social system (e.g., social institutions, social cycles, social order). The human social system links to critical resources such as natural resources (e.g., energy, land, water, materials, nutrients, vegetation), socioeconomic resources (e.g., information, population, labor, capital) and cultural resources (e.g., organization, beliefs, myths).

- The protocol developed by Stanford and Poole (1996) integrates research and public opinion in adaptive management.

- An integrative methodology for sustainable management was developed by Armitage (1995) that includes the identification of structural and functional attributes of abiotic, biotic, and cultural systems.

- A "lacing model" for ecosystem management was developed by Bormann et al. (1993) that select goods, services, and states desired by society, and determines ecosystem patterns and processes thought to be needed for these goods and services.

- An integrative model that incorporated ecological, economic, and social principles was developed by Kaufmann (1994) and includes a decision model with an ecological filter and implementation, modeling, and feedback components.

- The analysis model developed by Driver et al. (1996) provides guidelines for integrating social and biophysical components in ecosystem management analysis and monitoring. Their model specifies biophysical objectives, defines key biophysical and social issues, makes biophysical and social assessments, defines the range of feasible biophysical and social conditions, completes and implements a plan, and then monitors.

- The Legacy model developed by Vogt et al. (1999ab) is based on the assumption that the natural and social systems are continuously connected by legacies that are either social- or natural-science driven. This model develops mechanistic links between the social and natural systems that allows weighting of information needs for any site. The analysis stage first identifies prior events or conditions (*legacies*) that constrain the current system and the driving variables (*spatial, temporal, disturbances*) that can be social- or natural-science based. This analysis identifies the desired endpoints (management, human values) distinct from the non-value-based endpoints that are then used to identify the indicators (values, in this case, are science-based and not advocacy-based) (Vogt et al. 1999ab). This is not a totally familiar approach for those coming from a management or planning background, where issue and endpoint identification are often the first step in the process. This approach saves researchers from data overload and the need to collect large amounts of uniform data for each site (see 5.1).

Recently, three different conceptual approaches have been proposed by scientists to link the natural and social systems (Figures 3.1a-c). One approach consists of two overlapping circles (Figure 3.1a). The area of overlap is the region where biological conditions and social needs integrate to create sustainable ecosystems (Bormann et al. 1993). As part of this conceptual framework, the circle on the left depicts what the current generation desires for themselves and for future generations, while the circle on the right implies what is biologically and physically possible in the long term. In the second conceptual approach (Figure 3.1b), economic and social needs are presented as separate circles and the area of overlap between all three circles is where ecosystem sustainability can occur (Kaufmann 1994). Kaufmann (1994) concluded that the economic and social spheres need to be shifted, because the physical and biological spheres were considered less flexible.

The third conceptual approach (Figure 3.1c) is represented by the "egg analogy" where there are no overlapping circles, since the social and natural system circles are embedded within one another (Vogt et al. 1997a). This last approach suggested that "no part of the natural system is isolated (or functions independently) from the social system" (Vogt et al. 1997a). This approach also reminds us that even if were are able to identify a "range of natural variation" in ecosystems, primitive or pure natural areas no longer existed in the

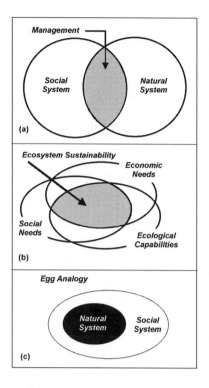

FIGURE 3.1

Three different views of social and natural system integration: (a) the ecosystem management model of Bormann et al. (1994), (b) the ecosystem sustainability model of Kaufmann et al. (1994) and (c) the egg analogy of Vogt et al. (1997).

environment. It also shows us that many human influences are indirect and appear at first glance to not be measurable (Vogt et al. 1997a). Which conceptual approach most appropriately represents the links between the social and natural systems becomes important, since each defines different assessment criteria and indicators to monitor system health and sustainability.

The problems with many natural resource or land management frameworks or models can be summarized as follows (Driver et al. 1996, Vogt et al. 1999ab):

1. Lack of incorporation of sensitivity analysis
2. Lack of identification of mechanistic links between social- and/or natural-science driving variables
3. Lack of identification of the relevant and necessary data needs for essential variables that result in data heavy needs and costly analyses
4. Failure to include social science variables as drivers or constraints on ecosystem functions
5. Conduct separate natural and social analyses
6. Lack of using a holistic approach

Machlis and McKendry (1996) concluded that interdisciplinary models have been slow to develop because of the lack of necessary data, the problem of data interpretation, and unspecified or ignored driving variables. These problems are the same as those facing certification, because of the need to conduct a holistic analysis of a site and the need to assess social- and natural-system sustainability.

A significant impediment to certification is the lack of several frameworks with different structures that could be used by a resource manager. Frameworks are important because they identify key components of a system and their relationships. Frameworks are needed which will allow the mechanistic feedback to be identified between the natural and social systems. For certification, a framework must allow an understanding of the system to be developed and the important points of linkage between the social and natural systems. This inability to determine the stages at which these links are relevant will be a problem inherent to any framework where mechanistic links have not been collaboratively developed for the social and natural system prior to the analysis process. Lack of mechanistic links will push the development of frameworks that link at every process stage the social and natural sciences. This linking of the natural and social systems at every stage suggests the need for the public to potentially be involved in, or to participate at, every stage of the analysis process. This requirement is a potentially cumbersome requirement for a forestland owner (see 7.3). This type of approach contrasted the approach introduced by Vogt et al. (1999ab) in their legacy model. In Vogt et al. (1999ab), the importance of the social or natural science legacies are identified early in the analysis process and the mechanistic links are identified at this stage. This means that public participation would occur early in the analysis process.

The lack of development of mechanistic links between management impacts (see Chapter 6) and their indicators also cause problems during certification, because it results in the need for large data sets as part of assessments (see 4.3.5, 5.1). Without this type of approach, the problems identified with the assessment of Toumey Forest by the Yale students will continue to occur in other assessments (Chapter 4). In this case study, there was an inability to document that good forest management was occurring on the site. This occurred because the approach was more structured to collect volumes of data without articulating an approach to weight the information needs for the site. Certification of a site was more dependent on being able to show that all that data were available and not whether the data had any relevance for the site. This type of approach does not ask the certifier to determine what data are pertinent from the large amount of required information that, in fact, reflects the sustainability of the human and ecological systems. This topic will be further discussed in Section 4.3.5.

3.2.2.2   Values of certifying organizations, society, landowners, and the desired forest condition

Kristiina A. Vogt, Joyce K. Berry, Andrew Hiegel

The types of forest ecosystems desired by the public external to the management unit, the people living within the management unit, the different certifying organizations, and owners of the forests frequently conflict with one another. When these values conflict, often the forestland owners are unable to harvest products from their land, as the values of society override the values held by the owners. In some cases, societal values for particular forest conditions cannot be achieved in today's landscape because of past human legacies or the inherent constraints of the natural system (see 5.2.1). It is not unusual to find that local values are overshadowed by international values generated in a country external to the management unit (i.e., the conflicts between developed and less developed countries) (see 2.1). However, forest management should not be approached or resolved by denying one set of values over another without understanding the implications of giving precedence to one or two values.

It is important to understand the role of values in controlling natural resource management because they determine the:

- Type of forest condition that is acceptable to the public
- Type of forest condition and forestlandscape pursued by management
- Types of forest management practices that will be allowed
- Types of regulations used to control natural resource uses
- Structure of current assessment approaches
- Definition of what is sustainable or good forest management

The primary goals of certifying organizations are not clear, because no single value appears to dominate the structure and implementation process. The following represent some of the different values that have been presented as goals for certification:

- Achieve good forest management
- Raise the bar for industrial forestry practices
- Function as economic incentives
- Inform the consumer
- Support ideals of social, economic, or environmental sustainability
- Create a desirable forest condition

Since the values of the certifying organizations (reflecting values of particular sections of society) influence the structure of the assessments, they end up defining acceptable forest management practices. Because of this, it is important to understand when values dictate protocol structures. These values are important because they directly or indirectly influence the selection of criteria and indicators that are used in the forest assessments (see 4.2.1). Since criteria and indicators are the core elements of certification, they will determine the failure or success of a company or individual to achieve certification. Because values drive an assessment process, value-derived information may be utilized in an assessment that may inadequately reflect site condition. This may happen even when considerable scientific information has been collected as part of the assessment.

Forestry and forest management has not been isolated from historical changes in societal values. Major changes in how forestry has been practiced in the past are a result of changing social values and interests (Vogt et al. 1997a). In forestry, most of the changes in public values have been expressed as conflicts over public land management. Currently, certification can be perceived as a mechanism to implement existing societal values that are held by certain sectors of society. Societal values are typically reflected in laws and regulations that are implemented at that time. The recent interest and growth in the use of certification has been generated by public values for how forest resources should be managed. Certification is an assessment tool that has evolved to respond to the public need to determine whether societal values are being realized in managed forests. Since the public often is not qualified to assess forest management, certification is a tool to give the public confidence (a label) that society values are being maintained. Certification is, therefore, an expression of social values that predominate at this time. Because public values, like natural ecosystems, are not static, certification has to be adaptable and structured holistically or it may not survive as a tool of the future. This suggests the need for certification approaches to understand not only that values are driving the process — but also whose values, the strength or intensity of those values and their representation in relation to broader society values. Approaches also need to be flexible in structure so they are capable of reflecting future public values.

Changing public values can be seen from an historical analysis of timber use around the world. When there was a need for wood to build houses and the railroad after World War II, major public impetus asked for increased timber harvest from public lands in the United States (Williams 1989, Vogt et al. 1997a). An enhanced production of a predictable supply of timber became a goal of policy when society viewed forests primarily as an economic resource to provide jobs and material goods. In the late 1940 and 1950s, the practices used to maintain this timber production (e.g., clearcutting, road building) were acceptable to society when the supply of material goods was the primary social value. Starting in the 1970s, the values of the public changed. This was reflected in the growth of the environmental movement. With these changes in values, clear-cutting and road building become unacceptable on public lands.

Forestry was slow to respond to these changing values and only changed their approach to management after considerable conflicts over land management had occurred. Eventually, the response of forestry groups to society's new values (e.g., saving species and habitats) was the implementation of New Forestry and adoption of ecosystem management as the guiding management paradigm in the early 1990s (Vogt et al. 1997a). This shifting view is expressed clearly by Franklin (1990) "as traditional forestry practices, such as clearcutting, shelterwood cutting, and selection cutting have focused on the regeneration of trees and not on the perpetuation of complex ecosystems" (Franklin 1990). These practices were thought to change many of the "natural linkages" that occurred in naturally regenerating forests and resulted in a public perception of degraded and less-healthy forests. "The New Forestry utilizes the concepts of ecosystem complexity, biological legacies, and viable landscapes to retain ecological values in managed forests" (Franklin 1990). What history documents so effectively are the changing values that society has for natural resources. It also shows the changing public perception of how land management should occur on public lands. Additionally, it shows that public values and the resulting regulations reflecting these values occurred much earlier than the managers' understanding of these changes (Vogt et al. 1997a).

Currently, the public (especially in developed countries) appears to value "naturalness," which drives what they define as acceptable aesthetics for a forest. This has strong implications for certification, since it defines the forest condition that the public would like forest management to imitate. The public frequently considers forests that present a visual range of color to the human eye to be quite aesthetic. In the United States, this can be seen in the recurring autumn scene, where busloads of people travel to view the changing foliage colors in the forests of New England or the aspen stands in the southwestern part of the country. Public attitude toward forests is also driven by their personal belief of how the ecology is affected by the silvicultural treatment and the motives that they attribute to the foresters and forest managers (Brunson 1996). In the latter case, when the public perceived that the timber industry was exploitative and non-responsive to the amenities that the public desired out of those lands, they opposed clearcutting (Brunson 1996).

The public also perceives that if their value for "naturalness" is followed in forest management, then their environmental values for saving species and certain habitats will also be satisfied. However, public values and the ecological basis for these values are not always compatible. Ecosystem degradation can occur when attempting to maintain public values in forests (Vogt et al. 1999ab).

Aesthetics may conflict with other amenities being desired from the same given piece of land. The public does not like the appearance of large, cleared-over areas where trees have been removed (Brunson 1996). These attitudes influence which forest management practices are considered to be acceptable by the public. For example, clearcutting generally is not acceptable to the public, even if it could be documented that this practice is needed to keep the forests healthy and free of insect pests or pathogens. In a comprehensive review of silvicultural practices and which types of practices society finds acceptable, Brunson (1996) stated that "high scenic beauty is associated with the presence of large, mature trees; moderate tree densities; grass/herb cover; color variation; and multiple species." Brunson and Shelby (1992) found that even-aged stands of trees were not considered acceptable by the people they surveyed. In an assessment of Indian Forests and Forest Management in the United States (IFMAT 1993), Native Americans told researchers that clearcuts looked like "death" — expressing their widespread opposition to the use of clearcuts. The general public does not find clearcutting an acceptable forestry practice even when it can be scientifically documented not to damage ecosystems in all locations (see Chapter 6).

Today, the public values of how they define healthy forests are integrally combined with their values for biodiversity and opposition to forest fragmentation. Interestingly, these values are also reflected in the federal laws (see 5.3.2, 5.3.3) that private landowners must satisfy — again showing that laws and regulations are a reflection of society values. In addition, these values are found as integral elements of certification protocols developed under the umbrella of the Forest Stewardship Council (see 4.1.3.2). An approach that focuses on species biodiversity and fragmentation reduction, however, may be inappropriate when it uses them as indicators of the system health and its sustainability. When reviewing the effectiveness of species diversity to reflect the stability and productivity of ecosystems, it becomes clear that *monitoring* species diversity to determine when forest ecosystems are degrading is not plausible (Johnson et al. 1996, see 3.2.1). This does not negate the fact that the public value for species is real and important to incorporate into management. However, public values should not be used as the primary indicators for assessing a system's condition.

The values held by the public can have a strong influence on which management practices they consider to be acceptable. However, even different values espoused by the public are frequently not compatible with one another. For example, Brunson (1996) stated that the values for biological diversity may conflict with the aesthetic and management values for that same forest. In this case, those forests with high species diversity are the same forests that have a high number of standing dead trees and logs decaying on the ground. However, the presence of large amounts of decaying wood is not considered aesthetic by most of the public. In another example, disease or insect outbreaks may be a natural part of the system, but people do not value the aesthetics of a forest where many trees have been killed by insects. Hollenhorst et al. (1991) reported that the negative aesthetics of a gypsy moth attack in North American forests could be mitigated by the significant increase in mountain laurel in the understory of a forest canopy. However, forests managed for timber selectively remove mountain laurel in the understory, resulting in possible conflict between aesthetic values and management practices. In an another example, also in the United States, the 1988 fire in the Greater Yellowstone ecosystem initially resulted in a forest condition not considered acceptable by the public (Knight 1991). It was not until later that the public realized that fire is a natural and important part of maintaining these ecosystems in healthy conditions into the future.

In addition to the societal values generated external to the management unit, the values of the landowner strongly influence their ecological, economic, and social objectives. Small nonindustrial forestland owners do not always have the same values as environment groups for the same piece of land. For example, in one study conducted by Rickenbach et al. (1998), landowners were questioned about their reasons for owning land in Massachusetts, U.S.A. In that study, only 12.8% of the respondents owned their land for mainly conservation or antidevelopment purposes, while only 9.4% were interested in owning land for wildlife, or wildlife habitat benefits or improvements. Environmental values are frequently lost as landowners attempt to achieve financial return from their land.

Society may expect private forestland owners to provide environmental services not desired by the landowner. In the United States, this expectation is growing because public forestlands are inadequate to maintain endangered species. Federal, state and local government forestlands comprise only 28% of the total forestland in the United States (Haynes 1990). This means that commercially owned land (15%) and nonindustrial private landowners (57%) (Haynes 1990) are needed for conservation efforts, and to maintain societal values from forests, especially as ecosystem management focuses on larger management units, such as landscapes or watersheds.

In the United States, other problems exist in achieving societal values from private forestlands since these lands do not easily lend themselves to maintaining forest conditions conducive to supporting endangered species because of their distribution and small size. Much of the existing forestland also occurs as a mix of private and public lands so forestlands are not contiguously owned by one individual or organization and are more difficult to manage as a unit. For example, 58% of the private forest owners had a total acreage of 1–9 acres (0.4 — 3.6 ha) and 27.9% owned land in the 10–49 acres (4–19.8 ha) range (NRC 1998).

Certifying-organization values are apparent when examining their goals for what they consider to be acceptable forest conditions and what indicators are used to assess the forest condition (see 4.2.1). The certification protocols under the FSC label tend to reflect environmental values by aiming for a forest condition that is minimally impacted by humans. For example, Drescher (1998) stated that we need a good reference point with which to compare the current system and that should be a condition that pre-dates human-induced disturbance. Drescher (1998) stated that we should be managing to duplicate nature since nature knows best how systems function. This has been defined as forest conditions that pre-date European colonization in the United States. Using this concept, management would be required to maintain species, genetic, and structural diversity at all levels within a forest similar to that of pre-European times (Drescher 1998). Drescher (1998) suggested that the most diagnostic of all of these indicators is forest structure (e.g., dead wood, canopy height, etc.) since it can be most easily measured.

Managing forests in conditions with minimal human influences raises many questions which certification assessments are unable to answer, and create subjectivity in the process. For example, what is the model for a certified forest? How far back in time should we go to identify the desired forest condition? Can reconstructed forests from the past exist under the current global climate-change scenarios or in the legacy landscapes produced by humans? How do we determine when our values identify a forest condition that does not represent past forest condition, but our current aesthetic values for forests?

In many parts of the world, it is difficult to identify indicators of forest conditions that go back more than one-hundred years. For example, what forest condition should be used as a model in the Maya region of Central America where forests grew back after being harvested more than a thousand years ago? The same can be asked about the tropical forests in the Amazon that have been managed for hundreds of years by the indigenous people living within these forests (Pinedo-Vasquez 1995). What is a "natural" forest in landscapes that have been managed by people for long periods?

In theory, it sounds like a good idea to use pre-European colonized forests in the United States to determine what forest condition to manage for, or to model. The main problem with using forest conditions that are more than one-hundred years old is the lack of data available to demonstrate the forest conditions that existed during that time. Forest conditions are also dynamic and move naturally through many different developmental stages (Bormann and Likens 1979, Oliver and Larson 1996).

It has been difficult to obtain consensus on the composition and structure of pre-European colonized forests in the United States. For example, arguments have occurred in the Pacific Northwest region of the United States on how much old-growth forest area should be maintained in face of past timber harvesting, and the current development and expansion of urban communities into these forests. These arguments are being played out in the policy arena where inadequate data of historical dominance of old-growth forests are being used to determine what should be the current desirable forest composition and structure. For example, no consensus has been reached on whether 20%, or up to 80% of the forested landscape was in old-growth forests prior to European colonization of the United States. However, these arguments are probably being driven more by public values for a forest condition and their aesthetics.

Silviculture will have to manage for "naturalness" if it is going to satisfy public values of forests (Brunson 1996). However, this raises an interesting question of what is naturalness. Frequently, the public perceives a natural forest to have no visible impact of humans and where the aesthetics of the forest match their values. Since it is extremely difficult to find a forest anywhere in the world that has not been managed by humans, the public is valuing an artifact of their own creation when they demand a forest to be "natural."

Values of the different certifying organizations can be gleaned from an examination of their required indicators and how good forest management is assessed. Few certifying organizations directly assess forest management. In some cases, there is no assessment of management in the field. For example, according to Mankin (1998), when criticizing the weakness of the ISO 14000 Series approach to forestry, it does not need to see the forest to be able to assess it (Mankin 1998). The focus of ISO is to develop environmental management systems. Therefore, their value is that the institutional structure and process can create a sustainable management system in the field — a definite social science approach. This contrasts with the focus of the FSC approach to certification that does not ask about forest management specifically, but assumes that management is acceptable if certain standards are met (Mankin 1998) (see 2.1, 2.4.1).

The FSC-type certification approach has strong environmental values and tries to satisfy current society values by emphasizing biodiversity and reserve establishments (see 4.2.1). Social sustainability is also a value expressed in the FSC-type certification. This can be seen in the need for information on land tenure, ability of communities to organize in unions, health of the workers, and safety of the workplace (see 1.2). However, the information needs in the FSC protocols are not site-based and they do not link the social and natural science information (see 4.3.5).

Any discussion of values must consider how compatible the values are of the certifying organizations and the people living in or around the management unit. This is a problem especially when people live within the forests (e.g., wet tropical forests, but also forest dwellers in boreal regions). The FSC approach to certification strongly considers social sustainability of indigenous communities even if they do not allow local values to dominate in assessments. The approach of most of the FSC certifiers is to incorporate into their protocols all the values which could possibly influence the management of an area. They attempt to keep all their criteria and indicators uniform so that local values are not specifically considered. Although FSC does not advocate a particular philosophy, many of the members of FSC have had, and continue to maintain, strong advocacy roles (WWF, Greenpeace, World Wildlife Fund for Nature, etc.) (Mankin 1998).

In general, most of the certification protocols do not specifically seek to determine local values and use them determine which indicators to use in an assessment. The exception is the Canadian Standard Association's Sustainable Forest Management system. This system explicitly uses local values to drive the process of identifying what the goals, indicators, and objectives of their management criteria should be (CSA 1996ab). As part of the process of converting a criterion to a performance objective, the public participates in the identification of local values that must be integrated into the forest management plan.

An example of how this works was given in CSA (1996ab) as follows:

> When conservation of biological diversity criterion is selected, the local value is identified to have a healthy population of woodpeckers. This produces a goal of increasing the necessary habitat (standing dead trees). The indicator is the actual increase in the relevant habitat (standing dead trees), and finally, the objective is to increase the standing dead trees in a specific watershed by a given amount by a specific date.

However, the Canadian Standard Association approach is quite specific in addressing one value at a time without considering how it exists within the context of all the other values that might exist on that site (see Vogt et al. 1999ab). In the Canadian system, the six different sustainable forest management criteria may conflict with one another. In addition, a mechanism for weighting the different criteria is not given which can cause problems during assessments.

4

Case Study and Evaluation of the Dominant Certification Protocols

Kristiina A. Vogt, Anna Fanzeres, Daniel J. Vogt, Bruce C. Larson, Jennifer L. O'Hara, Glenn Allen, Allyson Brownlee, Luisa Camara, Eva Cuadrado, J. Scott Estey, Alex Finkral, Brett Furnas, Jennifer Heintz, Andrew Hiegel, Heidi Kretser, Indah Kusuma, Jessica Lawrence, Marie-Claire Paiz, Peter A. Palmiotto, Brooke A. Parry, Christie Potts, Brian Rod, Manrique Rojas, Joe Taggart

In this chapter, eight protocols will be reviewed that have been developed by some of the most active certifiers. The choice of protocols for this exercise was made on the basis of availability and of their applicability to temperate forests. The information will be quite detailed in order to inform the reader about the protocols' basic structures and information needs (e.g., their criteria, indicators, and implementation approaches). This will be followed by a discussion of the goals and values of each certifying organization and, finally, the student evaluation of the protocols. The student evaluation will identify whether Yale Toumey Forest was certifiable based on their assessment; the reasons for the decision to certify or not to certify Toumey; what criteria were satisfied during the assessment; and finally, the strengths and weaknesses within each protocol that determined how each could be operationalized. This discussion will conclude with an analysis for each protocol of the amount and types of data needed (e.g., generality and uniformity of data, weighting of data); how values for a desired standard forest condition drive the choice of indicators; how social factors were integrated in the protocols; and how each was structured to deal with small vs. large landholding sizes.

4.1 Relevance of Toumey Forest to assess certification protocols

The purpose of this exercise was to begin to identify:

- The advantages and disadvantages of each protocol
- Their similarities or dissimilarities in how they approach the certification process
- What factors would create problems in implementing each protocol
- Recommendations for which elements work and do not work as part of any certification process

The certification protocols were not implemented following the exact procedures stated by the certification organizations. This was appropriate since the purpose of this exercise was not to specifically compare and contrast them to one another to determine if one protocol was better than another, but rather to understand them generally. The one- to two-week time period typically required for the actual site analyses by certifiers was not used, nor was a team formed consisting of an ecologist (e.g., wildlife, plant) or silviculturalist, an economist, and a social ecologist. The field assessment was conducted by a team of three-to-five graduate students (with a mix of social and natural science backgrounds). Each student group received two-to-three different certification protocols that they needed to use in the Toumey Forest assessment. All information that could have identified the protocol was removed so that the students were not biased by knowing the source of the protocol. The assessments were conducted during the first three weeks of class prior to any discussion of certification and before the guest speakers were brought in. Due to time limitations, the field assessments were conducted during a one-day period. However, many opportunities existed to go back to the field for ground-truthing information, and to interview adjacent landowners and the forest managers of Toumey Forest.

The purpose of this case study was to assess the difficulties that could arise from taking a non-preconditioned group of smart students who were merely reading guidelines to identify the advantages and constraints of implementing the different assessment protocols. The rationale for this was that many of the people hired to certify forestlands probably have abilities and scientific understanding similar to these graduate students. The purpose of this exercise was to determine the impediments to implementing each certification protocol and to begin to identify the relevance of their different information needs.

Another focus of this exercise was to determine what type of information (especially missing information) would prevent a forest from being certified. The purpose was to identify what information is relevant to determine social- and natural-system sustainability and to identify the minimum information needs for any assessment.

The comparison of the different protocols utilized information that was available in January 1998 (when the comparisons were made). Undoubtedly, these protocols have evolved since that time. However, since the purpose of this exercise was to determine which elements are needed to successfully determine forest sustainability and good forest management, the details inherent to any protocol are not as important as their general structure and data needs. The purpose of this exercise was not to critique specific protocols, nor was it to state whether one protocol was better than another.

In addition, there was interest in determining how the different certification protocols would assess a forestland that had similar characteristics and problems faced by small to mid-sized landholdings pursuing certification. In some cases, Toumey is an unusual forest because of its unique status as primarily a tool for research and teaching. Some of the requirements of the certification protocols are therefore not relevant for this forest. For example, issues related to satisfying "worker requirements" for certification are not relevant for this forest, since the workers are students employed by the university to manage the school's forestlands. Therefore, wage issues, insurance, adequate housing, good forest manager supervision, etc. are covered by Yale University and will not be part of any management plan developed by the school for the forest. Students are employed in order to obtain the training they need as part of their education requirements, rather than to earn a steady income. Labor unions are not a consideration for the employees of the school forests, nor is the requirement for annual continued educational opportunities for the employees. Student training is a continuous process in this forest.

Since the primary objective of school forests is education and research, it does not have a defined annual cut. Income generated from these lands is a luxury for the school, and is not needed to maintain the school's continued operations. The objectives at Yale Toumey are to maintain self-sufficient operations and to teach students how to manage tracts of land. Since Toumey is acknowledged to be well managed, it is appropriate to test these different approaches to certification there. Useful insights can be garnered from this type of study to determine how sensitively each certification protocol reflected the actual field conditions. By understanding what factors determine whether Toumey is, or is not, certifiable, a better understanding should occur of those elements important to be incorporated into any assessment.

4.2 Analysis of the elements comprising the dominant certification protocols as structured in January 1998

An analysis of the general attributes of the certification approaches examined as part of the Toumey case study suggested the following generalities (Table 4.1, Boxes 4.1–4.8):

- Most of the organizations certifying forestlands started work in this area or were founded in the early 1990s. Because of the relatively young nature of certification initiatives, changes are still expected in the language of criteria, indicators, and in the implementation of approaches.
- Several certification protocols (CSA, SFI, SILVA, Acadian Regional FSC Standards) have not been used to certify any forestlands as of early 1998, while a few have been quite active in certifying forestlands since 1991 (Smartwood, SCS, Soil Association) (see 2.1, 2.5).
- Most of the certification approaches initiated in response to strong environmental values are performance based, while the industry-developed initiatives are a mixture of performance-based and systems-based approaches.
- Most approaches do not define the minimum land size required for certification but assumed it to be the land area of the forest being assessed for certification.
- Most of the certification approaches initiated in response to strong environmental values required large land areas (> 1,000 ha), while initiatives developed by the forest industry assessed lands as small as 16 ha (see 2.5, 7.4).
- Most of the approaches required a few criteria (3–18) to be satisfied, but a high number of indicators (28–262) needed to be assessed.
- Initiatives developed by the forest industry required the lowest number of indicators, while non-forest industry approaches called for the highest number of indicators.

4.2.1 Criteria, indicators, and their implementation for protocols used in the case study

The presentation of the information on criteria and indicators follows the same format for each of the eight certification approaches. The principles are listed using roman numerals and are boldfaced. The criteria are listed using alpha letters and are written out using capital letters. The indicators are listed using arabic numbers below the criteria (sometimes given as the total number of indicators immediately following each criterion and not individually listed).

TABLE 4.1

Certifiers, their general characteristics and approaches examined for the Toumey case study based on their structure in January 1998.

Organizations*	Year Formed	Types **	Approach	Minimum Land Size Needed	Number of Criteria and Indicators
CIFOR Project on "Testing Criteria and Indicators for the Sustainable Management of Forests" — Austrian set	1994	Forestlands	Performance based	Forest Management Unit (FMU)	4 Principle Objectives; 10 Critical Elements; 50 Indicators
SFI (AF&PA)	1996	Forestlands	Systems based	FMU	4 Principles; 12 Principle Objectives (Criteria); 8 Sub-Objectives; 28 Performance Measures (Indicators)
Smartwood (Rainforest Alliance) FSC accredited in 1996	1989	Forestlands; Resource Managers; Chain of custody; Certifiers; Plantation evaluation; Companies process, manufacture or sell products made from certified wood**	Performance based; Needs regional standards; Uses universally consistent certification standards (minimizes liability)	FMU	9 Criteria; 92 Indicators
SCS (Forest Conservation Program — FCP) FSC accredited in 1996	SCS founded 1984 FCP founded 1991	Forestlands; Environmental claims; Compliance evaluation program; Life-Cycle assessment and Environmental performance evaluation; Chain of custody; Plantation evaluation	Performance based	FMU	3 Principle Objectives; 17 Sub-Objectives (Criteria); 108 Indicators

SILVA (Silva Forest Foundation) Applying for FSC accreditation	1992	Forestlands: Stand level, Whole forest level***	Performance based	FMU and landscape	18 Principle Criteria; 149 Indicators
Woodmark (Soil Association) — FSC accredited in 1996	1992	Forestlands; Resource managers, producers, manufacturer; Processing facilities (mills); Importer or distributor	Performance based	FMU	6 Principles; 14 Criteria; 188 Indicators
Acadian/FSC Regional Standards	1997	Forestlands	Performance based	FMU	10 Principle Objectives; 10 Criteria; ~242 Indicators
CSA	1993	Forestlands — none certified using CSA, as of April 21, 1998 (LaPointe 1998)	Systems and Performance based	FMU	6 Criteria; 21 Critical Elements; 83 Indicators

* CIFOR = Center for International Forestry Research; SFI = Sustainable Forestry Initiative (AF&PA — American Forests and Paper Association); SCS = Scientific Certification Systems; CSA = Canadian Standards Association

** Types of certification defined:
- Resource managers (e.g., foresters, loggers, landowners, anyone who manages land) (e.g., group certifications of dispersed properties)
- Chain of custody (e.g., mills producing certified wood products)
- Certifiers — implement certification protocols

***Stand Level Certification: verification that timber is grown and managed according to the principles of ecologically responsible forest use. Applies where the land steward or forest manager has decision-making control over only a forest stand or patch.

Whole Forest Certification: verifies timber is grown and managed according to the principles of ecologically responsible forest use. Applies where the land steward or forest manager has decision-making control over the landscape or part of the landscape and has decision-making control over the forest stands that make up that landscape.

4.2.1.1   Center for International Forestry Research (CIFOR) (Prabhu et al. 1996)

Many organizations are involved with developing criteria and indicators that will assess whether sustainable forestry is occurring. Although, CIFOR is not a certifying organization, it is included in these discussions because it is developing and testing criteria and indicators. CIFOR has tested hundreds of existing criteria and indicators and, from that amalgamation, has selected criteria and indicators that have appeared to successfully assess forest sustainability (Prabhu et al. 1996). As part of their study, CIFOR used criteria and indicators from Smartwood (Rainforest Alliance, U.S.), Initiative Tropenwald (Germany), Woodmark (Responsible Forestry Standards, Soil Association, UK), Lembaga Eko-label Indonesia (Indonesia), and the Deskundigenwrkgroep Duurzaam Bosbeheer (the Netherlands). CIFOR's testing of criteria and indicators was conducted in five locations around the world. The effectiveness of the different criteria and indicators, and eventually the identification of those attributes that were considered relevant, were used to develop the CIFOR approach (Prabhu et al. 1996). The information needs of this approach are summarized in Box 4.1.

Box 4.1
Principles, criteria, and indicators for CIFOR (The Austrian set)

I.   General and Organizational Requirements

    A. General Information
        1) Identification of the areas subject to management; 2) Management; 3) Product identification; 4) Compliance with legal standard within the framework of forest management and fulfillment of regional standards

    B. Basic Data on Production and Framework Conditions
        1) Ecological and biodiversity data; 2) Protection areas and areas designated for specific forms of use; 3) Necessary economic and production data; 4) Product diversity and forest areas' basic information

    C. Management Concept
        1) Management objectives; 2) Ecological aspects; 3) Economic aspects; 4) Social measures; 5) Documentation and monitoring

II.   Ecological Aspects

    A. Quantity and Quality of Ecosystem Components
        1) Soil; 2) Forest area, forest fertilization and soil tillage; 3) Growing stock size of cutting, logging and log transport; 4) Water; 5) Drainage facilities, shoreline designs; 6) Logging and log transport; 7) Road construction; 8) Forest protection and protection areas; 9) Biodiversity, structural and age-class diversity, genetic diversity

    B. Vitality, Health, and Productivity
        1) Forest protection and preventive measures; 2) Stress factors and risk assessment

III.   Economic Aspects

    A. Forest Product and Forest Functions
        1) Spatial distribution and extent

    B. Profitability
        1) Financial productivity; 2) Investment of capital and personnel; 3) Utilization/management contract; 4) Silviculture and forest protection

C. System
1) Regeneration and tending measures; 2) Seed trees; 3) Forest protection and forest hygiene; 4) Harvesting and skidding; 5) Damages from harvesting; 6) Product wastage; 7) Transport: road and skidding track network; 8) Timber production; 9) Stand regeneration; 10) Rotation period; 11) Species diversity; 12) Planning and control mechanism; 13) Non-timber forest function/services

IV. Socio-Economic Aspects

A. Design of External Relations
1) Right and participation of the local/traditional/indigenous population; 2) Cultural heritage; 3) Job and integration of the local/traditional/indigenous population; 4) Consultation of the local/traditional/indigenous population; 5) Effects of local timber processing on the regional economy
B. Internal Aspects
1) Working conditions, safety provisions, and health insurance; 2) Education and further training; 3) Freedom of organization of the employees

---

4.2.1.2 Sustainable Forestry Initiative (American Forests & Paper Association) (AF&PA 1995)

This protocol has four principles followed by 12 principle objectives (comparable to criteria). Each of the principle objectives is accompanied by a set of specific "performance measures" (comparable to indicators). The principles and criteria needed for this protocol are outlined in Box 4.2.

Box 4.2
Principles and criteria for the Sustainable Forestry Initiative

I. Practice of Sustainable Forestry on AF&PA Members Forests

A. Broaden sustainable forestry practices using an array of scientific, environmental, and economically sound practices for forest growth, harvest, and use (*2 indicators*)
B. Promptly reforest harvested areas to conserve forest resources and maintain their long-term productivity (*1 indicator*)
C. Protect water quality using riparian protection measures and use EPA-approved best management practices for all forest-management operations (*3 indicators*)
D. Enhance wildlife habitat — promote habitat diversity and plant and animal conservation in forests (*2 indicators*)
E. Minimize visual impact of harvesting (*3 indicators*)
F. Special consideration of unique sites (*1 indicator*)
G. Promote biodiversity by enhancing landscape diversity and providing a variety of habitats (*2 indicators*)
H. Improve forest utilization efficiency (*1 indicator*)
I. Make prudent use of chemicals to protect forests while protecting employees, neighbors, the public, and sensitive areas (*1 indicator*)

II. **Involve Loggers and Other Landowners Procuring Wood and Fiber in Implementing Sustainable Forestry**

    A. Involve non-industrial landowners, loggers, consulting foresters, and company employees in the practice of sustainable forestry (*6 indicators*)

III. **Public Reporting and Involvement in the Practice of Sustainable Forestry for AF&PA Member Companies**

    A. Publicly report AF&PA progress in fulfilling its commitment to sustainable forestry (*3 indicators*)

    B. Provide opportunities for the public and the forestry community to participate with AF&PA members in implementing sustainable forestry (*3 indicators*)

IV. **Promote AF&PA public policy goals for sustainable forestry on all private and public lands in the United States (8 criteria)**

4.2.1.3   Smartwood (Rainforest Alliance) (Smartwood 1998)

The Smartwood Program (Rainforest Alliance) was initiated in 1989 and was the first forest certification program in existence (Smartwood 1998). The information needs for the Smartwood certification approach are given in Box 4.3.

Box 4.3
Smartwood criteria and indicators

    A. General Information (*21 indicators*)

    B. Forest Security
        1) Secure land tenure for defined land area and legal timber harvesting; 2) Landowner dedicated to long-term forest management; 3) Clearly defined pool of land established; 4) Maximum 10% of land logged annually, records kept justifying reasons; 5) No "single harvest only" individual forest ownership

    C. Management Planning
        1) Sufficient information for a forest management plan; 2) Long-term monitoring for adaptive management; 3) Forest management objectives and prescription documented; silvicultural prescriptions current; 4) Existence of forest management plan; 5) Forest management plan field-implemented; 6) Existence of maps; 7) Non-timber forest products considered and managed

    D. Sustaining Forest Production and Resource Quality
        1) Allowable cut with non-declining sustained yield of forest products; 2) Prevent disturbance to regeneration; 3) Management strategies prevent over-harvesting; 4) Management strategies emphasize improving long-term stand quality; 5) Management addresses restoration of degraded or low-quality stands; 6) Planting incorporates ecology and restores high-graded forests to natural mix of species or cover; 7) Planting uses native stock, not exotic species

    E. Forest Operations
        1) General (*4 indicators*); 2) Pre-harvest activities (*4 indicators*); 3) Tree felling (*2 indicators*); 4) Skidding, yarding, and hauling (*3 indicators*); 5) Post-harvest activities (*2 indicators*)

F. Environmental Impacts
    1) General (*1 indicator*); 2) Measures to conserve and enhance biodiversity (*8 indicators*); 3) Reserves and special management zones (*3 indicators*); 4) Water and soil resource protection (*2 indicators*); 5) Control of chemicals (*3 indicators*); 6) Product processing (*1 indicator*)
G. Social
    1) Community relations (*6 indicators*); 2) Employee relations (*2 indicators*)
H. Economic Viability
    1) Long term investment in forest (*4 indicators*); 2) Enhancing resource potential (a) In the forest (*3 indicators*) and (b) In production and sales (*2 indicators*)
I. Tracing and tracking (*2 indicators*)

Implementation Approach Guidelines:

Six steps must be followed for certification with the Smartwood approach (Heaton 1994, Donovan 1998). The first step consists of submitting an application and required documentation. This is followed by a discussion between the applicant and Smartwood. The third step is a field assessment by an interdisciplinary team (e.g., forester, ecologist, and social scientist) that can last from one to three weeks and includes the write-up of a report and its follow-up analyses. The fourth step is a peer review of the report to be conducted by three of the 13 members of the Smartwood advisory board. The fifth step is approval or rejection of certification; this decision is made by the director of the Smartwood program based on the written report and feedback from the peer-review process. The sixth step consists of annual and random audits of the applicant's operations to ensure adherence to the principles of Smartwood.

4.2.1.4   Scientific Certification Systems (SCS) (SCS 1995, 1998)

Founded in 1984, Scientific Certification Systems is a multi-disciplinary scientific organization based in Oakland, California. Its Forest Conservation Program was established in 1991. The types of information used in the evaluation criteria for the SCS protocol are given in Box 4.4.

Box 4.4
Principles, criteria, and indicators for the SCS protocol

I.   Timber Resource Sustainability

A. Harvest Regulation
    1) Actual yields per acre compared to predicted yields; 2) Current and projected merchantable inventory volumes per acre for stands to be harvested within the next thirty years; 3) Target age of crops and trees under selection management; 4) Rotation lengths relative to stands' ages approaching maximum mean annual increment; 5) Extent to which current harvest levels are justified by allowable cut effects; 6) Annual harvest levels compared to planned levels; 7) Species composition by volume of annual harvests, compared to planned levels; 8) Annual softwood harvest volume as percent of total annual harvest, compared to softwood inventory volume as percent of total inventory volume; 9) Size class distribution of stands, stratified by species classes; 10) Historical rates of stand-type conversion by valued types; 11) Average annual harvest levels compared to growth levels

B. Stocking and Growth Control
   1) Design and execution of stand treatments and consistency with projected yields; 2) Harvesting priorities at stand and individual-tree level; 3) Stocking levels and species composition of young stands; 4) Extent original diversity of natural forests maintained with silvicultural prescriptions; 5) Extent to which field foresters possess and apply current silvicultural knowledge; 6) Extent to which prescriptions are tailored to individual stand conditions and markets; 7) Extent to which expedient prescriptions are routinely applied; 8) Extent and effectiveness of pre-commercial and commercial stand release treatments; 9) Damage to residual stand during partial harvest entries; 10) Adequacy of residual stocking after partial harvests

C. Pest and Pathogen Management Strategy
   1) Pre- and post-harvest softwood species's composition, particularly the extent to which high-risk species and stand conditions are systematically reduced, using harvesting priorities and plantation composition; 2) Explicit efforts to manage for natural pest predators (e.g., bird species) and modification of prescriptions to increase structural diversity to provide favorable habitat for natural predators; 3) Tendency of foresters to rely on future insecticide spraying as principal strategy for surviving the next epidemic

D. Harvest Efficiency/Product Utilization
   1) Condition of landings and log decks; 2) Incidence of sound logs not being trucked out of woods; 3) Frequency of excessive falling damage to harvested trees; 4) Extent of trees damaged during harvesting operations; 5) Appropriateness of end uses of harvested logs; 6) Harvesting decisions driven by short-term, low-value product realization at expense of long-term productivity

E. Management Plan and Information Base
   1) Breadth, depth, and currency of forest plan; 2) Written guidelines for avoiding or minimizing environmental impacts; 3) Extent to which plan is used by field foresters; 4) Extent to which aggregate harvesting activities are reconciled to forest plan; 5) Extent and accuracy of field data; 6) Monitoring procedures for acquiring information on plan attainment and resource conditions; 7) Adequacy of planning response to natural catastrophes; 8) Adequacy of log-marking system for tracking all logs coming from fee and non-fee lands

F. Forest Access
   1) Average miles of haul roads per acres; 2) Average area accessed per mile of new spur road; 3) Number of times lack of access limited desired management prescriptions; 4) Road right-of-way widths; 5) Condition of culverts, water bars, and roadway surfaces; 6) Road-bank vegetative management; 7) Runoff drainage patterns during storms

II. Forest Ecosystem

A. Forest Community Structure and Composition
   1) Seral stage distribution across watershed; 2) Age, size, and species distribution of trees within a stand; 3) Presence/absence and diversity of indigenous herbs, shrubs, and non-commercial trees; 4) Degree of "green retention" after harvesting; 5) Vertical diversity of stand; 6) Use of exotic species and genetically engineered organisms

B. Long-Term Ecological Productivity and Health
   1) Length of managed rotations to ecological rotations; 2) Management efforts designed to maintain nutrient capitals; 3) Extent of soil damage during harvesting; 4) Extent and appropriateness of whole tree logging; 5) Excessive exposure of soils to harsh micro-climatic stress

C. Watercourse Management Policies and Programs
   1) Effectiveness of design and execution of watercourse buffer policies; 2) Extent and effectiveness of stream-restoration projects; 3) Frequency and nature of Land Use Regulation Commission violations; 4) Effectiveness of design and maintenance of stream crossings; 5) Frequency of stream crossings within harvest areas; 6) Location and layout of roadways near watercourses; 7) Road-bank vegetative management near watercourses; 8) Extent of observable roadway rainfall-runoff into watercourses

D. Ecosystem Reserves Policies
   1) Extent of ecologically significant areas protected; 2) Permanence of set-aside areas; 3) Observations of ecologically important areas being substantially altered through timber harvesting

E. Wildlife Management Actions, Strategies, and Programs
   1) Regular involvement of wildlife biology experts; 2) Extent of data on wildlife populations, habitat conditions, species requirements; 3) Degree of integration of wildlife concerns into management prescriptions; 4) Degree of retention of desirable habitat features; 5) Status of working relationships with state wildlife officials; 6) Extent and condition of wildlife-oriented special management areas

F. Pesticide Use, Practices, and Policies
   1) Frequency of pesticide use and reasons for use; 2) Extent to which silvicultural methods minimize need for pesticides; 3) Effectiveness of use; 4) Use of targeted versus broadcast aerial insecticide spraying; 5) Policies and procedures for proper use and disposal of hazardous materials

III. *Financial and Socio-Economic Consideration Criteria*

A. Financial Stability
   1) Ownership structure and vertical integration where log requirements of a mill owned by a company might dictate land-management decisions; 2) Stability of ownership structure; 3) Cash flow demands of the company; 4) Accounts payable performance or other financial-performance data; 5) Review of company's annual financial statements; 6) Evidence that financial considerations drive land-management decisions; 7) Management philosophy of corporate officers (interviews, written statements)

B. Community/Public Involvement
   1) Company policies to encourage employee participation in community programs; 2) Corporate contributions to charitable causes; 3) Employee participation in local, state, and regional professional and natural-resource organizations; 4) Employee participation in ad hoc and standing public/private committees dealing with forestry and management issues; 5) Efforts to hire from within the local and regional workforce; 6) Effort to support local business when making decisions about the sale of wood products, or in purchasing decisions; 7) Procedures for identifying and protecting areas of special cultural, economic, or religious significance

C. Public Use Management
   1) Policies guiding the extent to which the general public has access to company lands for recreation; 2) Barriers and inducements to public recreational use; 3) Management of public use to control resource damage; 4) Efforts to minimize avoidable opportunity costs, in terms of foregone timber-production capability, because of recreation or other uses

D. Capital and Personnel Investment
   1) Average annual expenditures on pre-commercial silvicultural prescriptions; 2) Expenditures on, or commitment to, ongoing employee training and education; 3) Financial support or investment in improved harvesting machinery

E. Employee/Contractor Relations

1) Employee wages and benefits compared to industry norms in region; 2) Average tenure of workforce; 3) Employee work attitudes and general morale; 4) Opportunities for employee participation in management decisions, policy formulation; 5) Contract harvest/hauling rates compared to regional norms; 6) Average daily compensation of woods crews relative to regional industry norms, and to pre-service contract era; 7) Safety records of employees and contract woods crews; 8) Contractor attitudes about the company; 9) Stability of relationships with wood contractors

Implementation Approach Guidelines:

The operational approach utilized by the Scientific Certification Systems is to:

- Determine the scope of the potential project.
- Conduct a preliminary evaluation.
- Sign a contract between SCS and the forest products company.
- Assemble the evaluation team.
- Determine the scope of the evaluation.
- Collect and analyze data.
- Consult with regional stakeholders.
- Weight the criteria and modify scoring guide as necessary.
- Assign a numerical performance score.
- Specify conditions for certification, if necessary.
- Solicit and respond to the client's review comments.
- Solicit and respond to peer-review comments.

The evaluation team is comprised of SCS staff and contract/consultant field-level personnel who have expertise in the relevant disciplines needed for the evaluation (e.g., forestry, biology-ecology, economics, and wildlife). The field teams are selected based on having a balance in the relevant disciplines, field experience, and regional experience, as well as regional credibility.

4.2.1.5 Silva Forest Foundation (SILVA) (SFF 1997)

The Silva Forest Foundation began work on forest certification in 1992, and in 1994 published their standards for ecologically responsible forest use. SILVA certification protocols require information in five categories followed by what are described as standards (a combination of criteria and indicators) within each category. This organization incorporates the concept of landscape level in their analysis. The principles, criteria, and indicators for this protocol are summarized in Box 4.5.

Box 4.5

Principles, criteria, and indicators for the SILVA protocol

I. ***Landscape-Level Standards***

A. Planning — Landscape-Level Plan

1) Description and condition of applicants land (*4 indicators*); 2) Mission and objectives ecosystem based; 3) Timeframe of plan is at least 150 years; 4) Identified ecological limits are accurate and ecosystem-based; 5) Description and maps adequate for on-ground activities; 6) Social and economic needs of local communities are analyzed; 7) Minimum landscape-level requirements for stand-level applications met (*2 indicators*); 8) Monitoring impacts of management activities established

B. Planning — Protected Landscape Network

1) Protected riparian ecosystems (*3 indicators*); 2) Ecologically sensitive areas protected (*2 indicators*); 3) Old-growth nodes and old-growth recruitment needed (*6 indicators*); 4) Cross-valley corridors needed (*3 indicators*); 5) Protect entire watersheds when over 20,000 ha in size (*5 indicators*)

C. Planning — Ecologically Responsible Forest Use Zoning

1) Forest allocation is ecologically responsible and balanced (*3 indicators*); 2) Roads and landings are designed to maintain full forest functioning (*2 indicators*); 3) Large protected reserves designated when management unit over 100,000 ha in size (*6 indicators*)

II. ***Stand-Level Standards***

A. Planning — Stand-Level Plan

1) Description and condition adequate to guide management (*7 indicators*); 2) Timeframe of plan is at least 150 years; 3) Mission and objectives clear and ecosystem-based; 4) Prescriptions achieve objectives; 5) Description and maps adequate to guide on-ground activities; 6) Descriptions of protective measures for soil, water, and permanently reserved composition and structure are clear; estimation of impacts due to timber management accurate (*6 indicators*); 7) Social and economic needs of local communities analyzed and provisions for them described; 8) Plans for monitoring management clear

B. Planning — Protected Network of Ecosystems

1) Protect riparian ecosystems (*3 indicators*); 2) Protect ecologically sensitive areas (*2 indicators*); 3) Protect rare, threatened, and endangered species and ecosystems from logging

C. Planning — Permanently Reserved Composition and Structure (*7 indicators*)

D. Operations — Protection of Soil and Water

1) Permitted logging in riparian ecosystems designed to maintain full functioning of the system (*7 indicators*); 2) Pesticides, synthetic fertilizers, and other chemicals not used, except in special cases (*2 indicators*); 3) Roads and landings designed to maintain full forest functioning (*12 indicators*); 4) Skidding or yarding systems designed to maintain soil and water structure, and function (*6 indicators*); 5) Wastes properly handled and disposed of (*3 indicators*)

E. Operations — Cutting Patterns and Cutting Levels

1) Logging designed to mimic/maintain natural disturbance regimes (*5 indicators*); 2) Cut over not to remove more than 80% of growth for the same period; 3) Selection of trees to be cut maintains natural biodiversity, improves tree quality, and develops viable economic resource (*3 indicators*)

    F. Operations — Tree Regeneration

        1) Natural regeneration preferred (*2 indicators*); 2) Tree planting designed to maintain natural biodiversity (*3 indicators*); 3) Brush control, site preparations maintain full forest functioning (*4 indicators*)

    G. Operations — Salvage Logging (*4 indicators*)

III. Social and Economic Standards

    A. Indigenous Peoples (*3 indicators*)

    B. Local Communities (*4 indicators*)

    C. Workers' Rights (*4 indicators*)

    D. Financial Means and Investment (*3 indicators*)

    E. Resource Utilization (*2 indicators*)

IV. Restoration Standards

    A. Reestablishment of Natural Composition, Structures, and Functioning (*2 indicators*)

    B. Ecological Impacts of Restoration Activities (*2 indicators*)

V. Monitoring

    A. Monitoring (*3 indicators*)

Implementation Approach Guidelines:

The Silva Forest Foundation protocol guidelines have a series of additional documents which accompany their approach: Practical Methodology for Landscape Analysis and Zoning (35 pages) and Ecologically Responsible Timber Management (13 pages). These attachments cover the following topics: how to evaluate the criteria presented as part of the protocol; which tools to use to obtain that data; and the purpose of the data being requested. The Ecologically Responsible Timber Management section compares current timber management approaches to ecologically responsible timber management as defined by the organization.

4.2.1.6   Soil Association (Woodmark 1994)

The Soil Association, founded in 1946, began by working on organic food certification in the U.K. Their work on forest certification (Woodmark program) started in 1992. The principles and criteria integral to Woodmark are given in Box 4.6.

Box 4.6
Principles, criteria, and critical elements of the Woodmark program

I. Environmental Impact: forest management minimizes negative impacts on biodiversity, soils, water, and the landscape of the forest and adjacent areas

    A. Planning for Conservation of Biodiversity (*7 indicators*)

       • This critical element explicitly outlines 3 recommendations and 1 prohibited activity for forestland management and timber production.

    B. Roads and Firebreaks (*11 indicators*)

       • This critical element explicitly outlines 7 recommendations, 1 restriction, and 3 prohibited activities for forestland management and timber production.

C. Harvesting (*8 indicators*)
  - This critical element explicitly states 2 prohibited activities for forestland management and timber production.
    1) Felling (*2 indicators*)
      - This criteria explicitly states 3 recommendations, 2 restrictions, and 1 prohibited activity for forestland management and timber production.
    2) Extraction and collection of logs emphasizing soil management (*7 indicators*)
      - This critical element explicitly states 3 recommendations, 2 permissible activities, and 2 prohibited activities for soil management with timber production.
D. Site Selection and Preparation for Plantations (*9 indicators*)
  - This critical element also explicitly states 1 recommendation, 1 permitted activity, and 1 prohibition for forestland management and timber production.
E. Planting (*9 indicators*)
  - This critical element explicitly states 2 recommended activities, 1 permissible activity, 2 restricted activities, and 1 prohibited activity for soil management with timber production.
F. Pollution Control (*6 indicators*)
  - This critical element explicitly states 3 recommended activities, 8 permissible activities, 2 restricted activities, and 7 prohibited activities for soil management with timber production.

II. *Sustained Yield: yield of forest products and services are sustainable in the long-term*
  - This principle has 8 indicators
  - This principle has 2 recommended activities

III. *Land Rights: legal land rights of indigenous and traditional peoples are established and enforced. Customary use rights to the forest are maintained.*
  - This principle has 5 indicators
  - This principle has 1 prohibited activity

IV. *Local Control, Consent and Benefit: indigenous and traditional communities control forestry activities on their lands. Forestry operations must receive full and informed consent of local communities, enhance their long-term social and economic well-being, and not reduce in any way their ability to use the forest*
A. Control and Consent (*7 indicators*)
  - This critical element explicitly states 4 recommendations and 3 prohibited activities for forestland management and timber production.
B. Cultural Heritage (*1 indicator*)
C. Labor, Health, and Safety (*7 indicators*)
  - This critical element explicitly states 3 recommendations for forestland management and timber production.
D. Economic Gains (*2 indicators*)

V. *Economic Potential: forest management encourages an optimal and efficient use of all forest products and services in order to ensure a wide range of environmental, social, and economic benefits*
  - This principle has 4 indicators.
  - This principle has 2 recommended activities for forest management and timber production.

VI. Management and Monitoring

    A. Management Plan (*4 indicators*)

        1) Biological/ecological context (*4 indicators*)

        2) Socio/cultural context (*7 indicators*)

- This criteria has 1 permitted activity for forest management and timber production

    B. Monitoring (*5 indicators*)

- This criteria has 2 recommended activities for forest management and timber production

---

Implementation Approach Guidelines:

This protocol has very explicit operational requirements and a detailed outline for implementing this protocol. The types of records required include: purchase, processing, stock, sales, and chemical uses. The application procedure requires:

- Upfront documentation on the description of the forest.
- Details on the forest-management activities.
- A description of restocking activities and preparation for plantations.
- Details on how forest operations (in management plan) minimize impacts on the forest.
- A description of the pollution-control methods being used.
- A statement of proposed yield of forests products.
- A copy of the management plan.
- Description on monitoring procedures.
- Specifics on the individuals and organizations consulted.

4.2.1.7 Acadian/FSC Regional Standards (unpublished draft)

The Acadian FSC Regional Standards set of principles, criteria, and indicators are given in Box 4.7. This document was kindly given to us for use by Francis Raymond despite still being in a draft form.

Box 4.7
Principles and criteria from the Acadian/FSC Regional Standards

I.   Compliance with Laws

    A. Forest management respects all applicable laws of the country, and international treaties and agreements to which the country is a signatory (*10 indicators*)

II. Tenure, Use Rights, and Responsibilities

    A. Long-term tenure, use rights to the land, and forest resources are clearly defined, documented, and legally established (*5 indicators*)

III. Indigenous Peoples' Rights

A. Legal and customary rights of indigenous peoples to own, use, and manage their lands and resources, will be recognized and respected (*10 indicators*)

IV. Community Relations and Workers' Rights

A. Forest-management operations maintain or enhance long-term social and economic well-being of forest workers and local communities (*34 indicators*)

V. Benefits from the Forest

A. Forest management operations should encourage the efficient use of the forest's multiple products and services to ensure economic viability, and a range of environmental and social benefits (*26 indicators*)

VI. Environmental Impact

A. Forest management should conserve biological diversity and associated values; water resources; soils; and unique and fragile ecosystems and landscapes; and thereby maintain the integrity of the forest (*69 indicators*)

VII. Management Plan

A. Management plan and supporting documents should provide management objectives; description of forest resources to be managed; description of silvicultural and other management systems; rationale for rate of annual harvest and species selection; provisions for monitoring forest growth and dynamics; plans for identification and protection of rare, threatened and endangered species; maps describing forest resource base; description and justification of harvesting technqiues and equipment used (*38 indicators*)

VIII. Monitoring and Assessment

A. Monitoring which is appropriate to the scale and intensity of forest management to assess the condition of the forest; yields of forest products; chain of custody; and management activities and their social and environmental impacts (*12 indicators*)

IX. Maintenance, Conservation and Restoration of Primary, Natural and Semi-natural Forests

A. Primary, natural and semi-natural forests and sites of major environmental, social or cultural significance to be maintained, conserved, or restored. Management should not modify the character and function of the forest in any major way, and should not reduce the complexity of the forest ecosystem (*14 indicators*)

X. Plantations

A. Plantations must be managed following all principles and criteria given. Plantations should only complement the management of, reduce the pressures on, or promote the restoration and conservation of natural forests (*24 indicators*)

Implementation Approach Guidelines

Guidelines are not given with these regional standards, but are similar to other FSC-accredited guidelines.

4.2.1.8   *Canadian Standards Association (CSA 1996ab)*

The Canadian Standards Association (CSA) program consists of 23 coalition members from industry and environmental groups (CSA 1996ab) and requires public participation so that local needs are satisfied (LaPointe 1998). The Canadian Standards Association used the International Organization of Standardization (ISO) to help develop the standards for CSA, which were accepted in 1996 (CSA 1996ab). The CSA protocol was not structured on the ISO format, but was used to develop CSA standards. For example, ISO 14001 has no public participation as part of its protocol, no mechanism for continued improvement of the protocol, no third-party audit on-ground performances, nor any forecasting (CSA 1996ab).

The Canadian Standards Association approach identifies six criteria that must be addressed and managed for within a sustainable forest management (SFM) system (CSA 1996ab), and they present four methods for addressing these criteria. Each of the critical elements listed below is required to have at least one indicator that will be used to assess that critical element. In 1996, 83 indicators were approved by the Canadian Council of Forest Ministers. The principles and criteria for the Canadian Standards Association are summarized in Box 4.8.

Box 4.8
Principles and criteria for the Canadian Standards Association

I.   *Conservation of Biological Diversity*
  A. Ecosystem diversity (*4 indicators*)
  B. Species diversity (*3 indicators*)
  C. Genetic diversity (*1 indicator*)

II.   *Maintenance and Enhancement of Forest Ecosystem Condition and Productivity*
  A. Incidence of disturbance and stress (abiotic and biotic) (*8 indicators*)
  B. Ecosystem resilience (*2 indicators*)
  C. Extent biomass (biota) (*2 indicators*)

III.   *Conservation of Soil and Water Resources*
  A. Physical environmental factors (*5 indicators*)
  B. Policy and protection forest factors (*3 indicators*)

IV.   *Forest Ecosystem Contributions to Global Ecological Cycles*
  A. Contributions to global carbon budget (*9 indicators*)
  B. Forest land conversion (*2 indicators*)
  C. Forest sector CO_2 conservation (*3 indicators*)
  D. Forest sector policy factors (*5 indicators*)
  E. Contributions to hydrologic cycles (*1 indicator*)

V.   *Multiple Benefits to Society*
  A. Productive capacity (*5 indicators*)
  B. Competiveness of resource industries (timber/non-timber related) (*3 indicators*)
  C. Contribution to the national economy (timber/non-timber sectors) (*4 indicators*)
  D. Non-timber values (including option values) (*4 indicators*)

VI. *Accepting Society's Responsibility for Sustainable Development*

A. Aboriginal and treaty rights (*1 indicator*)
B. Participation by Aboriginal communities in sustainable forest management (*5 indicators*)
C. Sustainability of forest communities (*4 indicators*)
D. Fair and effective decision making (*3 indicators*)
E. Informed decision making (*6 indicators*)

Implementation Approach Guidelines:

There are four steps that must be adhered to for a forest to be successfully certified:

1. Commitment: commit to establishing an SFM system and work to apply the system over a defined forest area. As part of this, the applicant must:

   - Develop a corporate SFM policy that is communicated internally and externally to the company
   - Meet or exceed all regulations and laws at the federal, provincial, and local levels
   - Commit to engaging the public in its planning process.

2. Public Participation: establish a public consultation process for local people directly affected by the management of the defined forest area. The public consultation process must have the following elements:

   - Local sustainable forest-management values that are used to identify goals, indicators, and to set performance objectives.
   - A representative group of stakeholders who the forest operator consults throughout the process.
   - Incorporation of aboriginal peoples and their traditional knowledge when identifying sustainable forest management values and eventually when identifying indicators and performance objectives.
   - A clear set of rules and procedures for public participation in the certification process (these must be operational, developed from public participation and with a communication system developed for keeping everyone informed of what is happening at all times).

3. Implement an ISO-based management system: establish an SFM system that includes plans to achieve specified local objectives related to the critical elements and other identified values using the public participation process.

   - Earlier performance criteria are integrated with the management system elements.

4. Implement a strategy for continual improvement of the sustainable forest-management system.

   - This part compares the actual results to the forecast results implemented as part of the ISO-based management system.

4.2.2 Grading and ranking data within certification protocols

No consistent ranking system was used by the eight certification approaches examined in this case study. Some protocols did not stipulate how to rank the information required as part of the assessment, and for others, at the time of the field test, we did not have access to an existing system. For each protocol, the following information is presented:

- The ranking system
- What standard forest condition was identified as part of the assessment and used when grading criteria

Center for International Forestry Research (CIFOR)
Grading System:

1. Not relevant (not a certifying organization, but testing and developing criteria and indicators).

Sustainable Forestry Initiative of the AF&PA (SFI)
Grading System:

1. No grading criteria
2. Method of grading: scale of 1 to 5, pass > 3

Grading Criteria:
  Forests are graded on their ability to be managed in a sustainable manner which follows the principles of good stewardship ethics.

Rainforest Alliance (Smartwood Program)
Grading System:

1. No grading criteria
2. Method of grading: scale of 0 to 5, pass > 3 (0 = not applicable; 1 = strongly unfavorable; 2 = more unfavorable than favorable; 3 = neutral; 4 = more favorable than unfavorable; 5 = strongly favorable)

Grading Criteria:
  Smartwood sources are certified according to how closely they adhere to Smartwood principles and guidelines (Smartwood 1998). The following classifications are used in Smartwood:

- Sources operating with strict adherence to these principles, and having long-term data to support this, are classified as **sustainable**.
- Sources that can demonstrate a strong operational commitment to the principles and guidelines are classified as **well managed**. Forests certified as well managed lack long-term data documenting that they are being managed in a sustainable manner, and improvements are needed for the forest to be classified as sustainable (Heaton 1994).
- Smartwood also has a third category called **pre-certified**, where sources are not ready to be certified, but want to learn about the requirements for certification. "Once the applicant has satisfied a mutually agreed timetable of improvements, Smartwood will certify" (Heaton 1994). This certification category can be used to support forestry fundraising efforts but cannot be advertised (Heaton 1994).

- Smartwood has two types of certification for companies: **exclusive,** where only certified wood products are sold, and **non-exclusive,** where non-certified stock is available (Heaton 1994).

Scientific Certification Systems (SCS)
Grading System:

1. Examples are provided with grading criteria to show what would fail the test of sustainability.
2. Evaluation team is required to use collective judgment to evaluate.
3. Method of grading: scale from 0 to 100, pass > 80%.

Grading Criteria:
   As part of its forest conservation program, forest operations which score well in their indexing system are designated as **well managed** (Scientific Certification Systems pamphlet, undated).

SILVA Forest Foundation (SFF)
Grading System:

1. No grading criteria
2. Method of grading: scale of 3 to –3, need to exceed a 0 score to be certified. A positive score sum needed for the entire checklist.

Grading Criteria:
   The purpose of the SILVA forest certification system is **Ecologically Responsible Timber Management** (SFF 1997).

Soil Association (Woodmark Program)
Grading System:

   Not available

Grading Criteria:
   The purpose of the evaluation process for certification is to determine whether **good forest management** is occurring, whether **forest management is sustainable**, and to label wood from **well-managed forests**.

Acadian/FSC Regional Standards
Grading System:

1. No grading criteria.
2. Method of grading: scale of 1 to 5, pass > 3.

Grading Criteria:
   Supports the FSC philosophy of **Forest Stewardship** which "recognizes that forests provide more than just timber… Forest stewardship practices recognize these multiple values by maintaining the forest's ecological integrity, minimizing negative impacts to biological diversity, respecting the rights of forest-dependent communities, and conserving the forest's economic values." (FSC 1999a).

Canadian Standards Association (CSA)

Grading System:

1. No grading system

Grading Criteria:
The four required elements are:

1. A **sustainable forest management system** over a defined area that meets local objectives and values
2. An ISO-based management system
3. A **public consultation process** which promotes direct involvement by local people in management
4. A **strategy for continued improvement** towards a sustainable management system

4.2.3 Social integration in certification protocols

4.2.3.1   *Public participation in protocols*

Most of the FSC approaches to certification do not explicitly incorporate public participation as part of certification. The FSC approach is in contrast to that used by the CSA standards, where there is a strong emphasis on public participation throughout the entire certification process. This means that the FSC approach more strongly reflected the values and goals of the organizations conducting the certification than those of the communities living adjacent to forests being certified. The CSA approach explicitly required local values to be incorporated into the decision-making process for the certification assessment.

Of the eight protocols examined in the case study, the CSA approach had the strongest requirements — and the greatest number of procedures articulated — for achieving effective public participation that satisfied local needs. To meet CSA standards, a forestland owner must to demonstrate that there has been "a determined effort to solicit participation from local stakeholders and there is a clearly defined process to ensure their input" (CSA 1996ab). According to the CSA standards, the purpose of public participation is to identify the values, goals, objectives — and also the indicators — of sustainable management for each defined forest area being examined for certification (CSA 1996ab). According to the CSA approach, the explicit purpose of participation is to determine the public values and goals that are associated with a given forestland. As part of their approach, the public must be kept continuously informed about their participation in the evaluation process, and have external audit reports available to them at all times (CSA 1996ab). The goals and values of the forestland owner are not considered as part of this process, nor is how to resolve the potentially conflicting values between the public and private forestland owners. CSA assumes that these conflicts will not occur since the goal of certification is to achieve sustainable forest management — this higher goal would incorporate all relevant values.

The Sustainable Forestry Initiative of the American Forest & Paper Association is also structured to include public participation. However, SFI focuses on those individuals and organizations that are involved in the forestry industry. For example, SFI encouraged state groups to "sponsor training and education programs for loggers, employees involved in procurement and landowner assistance, contractors, and suppliers" (AF&PA 1995). The other mechanism used to promote interaction with the public is by requiring the AF&PA

to report annually to the public on how well its members are complying with the performance measures required by their protocol. The AF&PA also wants to be a vehicle to
facilitate interaction between the public and forestry communities to develop the principles and performance measures of sustainability and good forest management for private
and public lands. It is unclear how this goal will be achieved, but it entails AF&PA working
with Congress and other public agencies to implement ecosystem management on these
lands. Priority will be placed on helping to establish sustainable forestry initiatives on
public lands that have health problems (e.g., insect or pest outbreaks).

Smartwood considers public participation to be:

- Educational opportunities to inform the public concerning forestry practices and
  their sustainability.
- Paying local taxes on time.
- Employing local communities and businesses for employment.
- Including special or unique sites in protected zones.
- Considering public goals for the natural resources
- Following zoning laws.

The approach presented by the Smartwood certification does not directly involve the
public in the decision-making process for developing management plans, or in identifying
the values to be maintained in the forests. The approaches for obtaining public participation in forestry are not given — no framework or process is articulated for how this should
occur.

The SILVA protocol's approach to public participation in certification is also indirect
and similar to Smartwood's. The public is not incorporated into the decision-making
process, although the rights of the indigenous people and the protection of traditional and
customary rights are emphasized. In this protocol, the emphasis is placed on ensuring
that the rights of groups (e.g., indigenous people and workers) are sustained with the
decisions left to the organizations being certified.

4.2.3.2   Goals and values of certification organizations

The goals of the certifiers clarify which human values are integral to each protocol. The
values embedded within each certifying organization will be used by a forestland owner
to select which certifying organization they will contract with to assess their lands. Enough
certifying organizations exist to satisfy the different values that a forestland owner wants
to be obtained from their land.

An examination of the different groups that support, or are linked to, a certifying
organization helps explain the approach and the goals of that organization. Scrutiny of
one of the first forest certification organizations demonstrates this well. For example,
Smartwood is comprised of an international network of local, non-profit conservation
organizations which have been involved in implementing certification of forestlands (private, state, county, and city) (Donovan 1998). Smartwood became "involved in certification
because, at least in part, we also saw it as a practical field-oriented tool that might help
resolve conflicts between communities and forests; conflicts that were destroying both
forests and communities." The Smartwood program — part of the Rainforest Alliance —
stated that they provide independent, objective evaluation of forest management practices
that do not destroy forest ecosystems (Smartwood 1998). According to this program, the
role of certification is to provide environmentally responsible landowners and wood-
product businesses (mills, lumber distributors, furniture makers, etc.) with an opportunity

to market their products to consumers with a "green" label. This label indicates that the product has been sustainably harvested. The members of Smartwood feel that certification has been a valuable tool to improve or verify the quality of forest management using environmental, silvicultural, and community-stability criteria (Donovan 1998).

The following section summarizes the goals and values of the protocols examined during the case study:

CIFOR

Goals or objectives (Prabhu et al. 1996):

- To develop a methodology that evaluates and generates criteria and indicators.
- To generate a minimum number of cost-effective and reliable criteria and indicators based on iterative and comparative field evaluations.
- To evaluate the sustainability of forest management based on the recommended criteria and indicators.

SFI

Certifier goals (AF&PA 1995):

- To practice sustainable forestry that allows the needs of the present to be satisfied without compromising the future ability to meet these needs by implementing a land stewardship ethic.
- To use and promote practices that are environmentally and economically responsible.
- To maintain long-term forest health and productivity by protecting forests from wildlife, pests, diseases, and other damaging agents.
- To protect special sites (e.g., those which have biological, geological, or historical significance).
- To continuously improve how forestry is practiced and to monitor, measure, and report how satisfactorily members are achieving the goals mentioned above.

Smartwood

Certifier goals (Smartwood 1998, Donovan 1998):

- To "provide independent, objective evaluation of forest management practices, forest products, timber sources, and companies, enabling the public to identify products and practices that do not destroy forests" (Smartwood 1998).
- To change how forestry is practiced using certification as a practical, field-oriented tool to resolve conflicts between communities and forests.
- To promote credible natural resource managers in the eyes of the general public.
- To use certification as an educational tool in order to improve the relationship between people and forests.
- To create an atmosphere that rewards certified operations with commercial incentives.

- The values which the forest is expected to provide now and in the future
- The time horizons of the values
- The spatial aspects of forest values
- The management strategies needed to sustain values
- The relationships between the desired and actual outcomes based on forest values

4.3 Student evaluation of protocols at Yale's Toumey Forest

There is tremendous value in comparing the different approaches used to assess forest ecosystems for their social and natural system sustainability or for good forest management. Recent efforts by CIFOR have focused on comparing the criteria and indicators of several different protocols (Hahn-Schilling et al. 1994, Prabhu et al. 1996) and are important contributions to our continuing improvement of evaluation systems. The purpose of the Toumey Forest study was to expand the analyses beyond the criteria and indicators level; to scientifically analyze many of the assumptions behind the certification approaches; and to examine the different frameworks that certifiers have developed.

4.3.1 History of Toumey Forest

David M. Smith

During the 1730s and 1740s, European immigrants settled in the area known as the Toumey Forest. Many of the sandy glacial outwash areas probably had natural, pure stands of white pine that had been induced by fire, although most of the area had hardwood-hemlock-pine forests. The settlers converted almost all of the land to agricultural production. As was the case elsewhere in New England, most agricultural use ceased during the last half of the 19th century. The old grassy fields revegetated naturally with pure stands of fast-growing white pine.

The region became the center of a pine-box industry that thrived until the 1920s. Many old-field pine stands were clearcut because the industry used trees as small as five inches in diameter. In 1913, Professor James W. Toumey of the Yale Forestry School began the gradual acquisition of cutover land for the Yale Research and Demonstration Forest. One of his primary purposes was to show that timber could be grown on a sound financial basis. Its location was chosen because it was on a main highway that connected urban areas to the White Mountains. At the time, there was virtually no intentional silvicultural practice underway anywhere in America.

The objective of management was to grow white pine and other conifers for sawtimber. Until the 1950s, there was no market for hardwoods and most of the soils of the forest were too dry to grow good hardwoods. Most of the land that was acquired was either in recent cutovers or young old-field stands. Many of the white pine stands that had been clearcut had been so young that little advanced regeneration of shade-tolerant hardwoods or hemlock had started beneath them so they reforested naturally with pioneer gray birch. White pines often seeded in beneath the gray birches. Red or white pines were planted beneath gray birch stands which did not have enough pine seedlings, and on remaining open fields.

During the 1920s and 1930s, many pines were released by selling gray birch and other hardwoods for fuelwood to people who came from Keene to cut it. The quick-burning gray birch was used for summer fuel to avoid overheating kitchens. Not many sawlog timber sales were made since most of the stands were too young, and it was rarely possible to get loggers to do anything but clearcut. One area was deliberately clearcut and planted.

Until the 1930s, the main objective was the fast growth of "box-board pine" whose logs did not need to be straight or have much clear lumber. It became apparent that the wooden-box industry was fading, so the emphasis changed to growing pines of higher quality. A tree-pruning program was instituted and there was a small amount of improvement cutting. In 1938, a hurricane blew down many of the stands of old-field white pine; coincidentally, there was an abundant cone crop which restocked the blowdown areas with pine seedlings. However, the stands that had been blown down were old enough to have had abundant red maple and oak advanced regeneration which grew faster than the white pines and resprouted after it was cut. For the next two decades, attention turned to releasing white pines on the blowdown areas by hand cuttings which had to be repeated two or three times. Herbicides had not been adequately developed until late in this program when hardwood saplings and surplus pines began to be killed with injected herbicides. The only use of foliar herbicides was to clear brush from some of the woods' roads.

After World War II, it was possible to get local loggers to engage in and become educated about partial cutting. For the next 30 years, nearly all of the cuttings were improvement thinnings in the immature, white-pine stands that made up most of the forest. The primary objective of these thinnings was to eliminate crooked white pine dominants that had been deformed by the white pine weevil, an insect that feeds on and kills only the tender terminal shoots that have quickly grown under direct sunlight. The object was to release straight, unweeviled, co-dominant trees of the kind that had been artificially pruned. The harvested trees supplied the dying pine-bucket industry which did not require straight logs since the bucket staves were cut out of the clear stem internodes. These cuttings proceeded only as fast as the last remaining bucket mill needed the wood. There was no other outlet for small or poor pines, so the process of getting rid of malformed dominants took too long and the stands were not thinned heavily enough to get good sawlog growth. Funds were not available to supplement the harvests with pre-commercial thinning. The last bucket factory closed in 1976, but by then it had become possible to continue the improvement thinnings by harvesting conventional sawlogs. Now that old enough stands exist, shelterwood regeneration cuttings are starting.

People in the area have continually utilized the forest for non-timber products: a spring beside Route 10 supplies much non-chlorinated drinking water; hunters regularly use the forest, and many people pick berries. The part of the forest adjacent to the Ashuelot River protects major groundwater resources. This large island of pine forest has most of the biota peculiar to such habitats. It includes some of the last samples of naturally pure white pine forests that once flourished on glacial outwash sands in New England.

In the early 1900s, the forest was set up by Yale as a demonstration forest to show that trees could be harvested profitably and as a research/teaching laboratory. Today, most of Toumey forest is comprised of eighty-to-ninety-year-old stands and of stands which formed after the 1938 hurricane. This has produced a forest that today is comprised of many different age classes of stands. The emphasis in the forest has been on timber management and the development of tools for good management of timber using an ecological approach. The primary silvicultural method used in the forest is shelterwood cutting.

4.3.2   Information and site description for Toumey Forest

The Yale-Toumey Forest is located between the towns of Keene and West Swanzey in Cheshire County, New Hampshire. The forest is currently approximately 583 ha (1,440 ac) in size. The forest is bordered on the south by the Ashuelot River and stretches north across the Keene town line. Route 10, one of the area's main north-south arteries, runs through the forest along with several other smaller public roads.

The annual average temperature in Keene is 46.8°F (8.2°C), with a wide daily range in temperature during the growing season. The annual average daily minimum temperature is 34.8°F (1.6°C) and the average daily maximum temperature is 58.8°F (14.9°C). Although the first heavy frosts in the autumn generally occur around the middle of September, they may occur in any month of the year. The growing season for most crops falls between April and September. Precipitation is evenly distributed throughout the year with snow occurring mainly between December and April.

Cheshire County is over 80% forested and although suburban development has spread throughout the area, the Toumey Forest is surrounded almost entirely by forest. The majority of the Toumey Forest's stands are single-aged and dominated by eastern white pine (*Pinus strobus*), which exist in pure or mixed stands on sandy soils. Stands of red pine (*Pinus resinosa*) are also common on glacial outwash soils. Hardwood stands consisting of several birch species, sugar maple (*Acer saccharum*), red maple (*Acer rubrum*), northern red oak (*Quercus rubra*), and white oak (*Quercus alba*) generally occur on rocky elevations and upland glacial till soils. Eastern hemlock (*Tsuga canadensis*) is scattered throughout most stands.

The Yale-Toumey Forest is used by faculty and students of the Yale School of Forestry and Environmental Studies as a laboratory for teaching and research. A member of the faculty serves as director and all management is carried out by students working as interns or managers. The forest is maintained as a *working forest*, which includes selling timber and other products and services. Students working on the forest not only gain experience as land managers, but also learn the social aspects of management, such as relationships with neighbors and compliance with myriad local and state regulations.

Most research on the forest involves experimentation on the effects of management, including timber harvesting. Past and current research projects include wildlife ecology, hydrology, silviculture, and social studies. The forest is also used for field trips by many courses, workshops, and demonstrations to show techniques in wetland management, protection of rare plants, and forest-stand development pathways.

As part of the student assessment of Toumey Forest, the information given below was made available to each student evaluation team. Reviews of Yale-Toumey were conducted through an analysis of existing forest records, discussions with persons involved with or responsible for past and/or existing management practices, and a one-day field trip to the site.

The following information was used in the analyses:

- State Best Management Practices for New Hampshire. *Best Management Practices for Erosion Control on Timber Harvesting Operations in New Hampshire*, April 1996

- History of the Yale-Toumey Forest

- Yale Bulletin 1916

- Toumey, J.W. (1932) *The Yale Demonstration and Research Forest near Keene, New Hampshire.* Yale University School of Forestry, Bulletin No. 33

- Letters from David M. Smith, director of school forests from 1981–1991 and reports from Toumey Forest
- Forest stand maps
- Inventory data for average acreage (1957, 1967, 1976, 1985)
- CIF inventory data on commercial species
- Generic agreements used for timber sales
- Contract between Yale and the loggers
- Management guidelines for the Yale-Toumey Forest
- Toumey Forest, timber sales annual summary
- Timber sales data of the school forests, annually from 1981–1991
- Timber harvest reports
- Recent sales maps (Junkyard [Compartment 20], Lost White Pine [Compartments 12 & 13], Red Pine Sale [Compartments 14 & 15], and the 81 South Sale from 1983 [Compartment 5])
- Aerial photographs, maps of Toumey Forest
- Soil survey book
- Other miscellaneous documents in the Toumey records
- Interviews with the forest manager (Alex Finkral) and director of the school forests (Bruce Larson)
- Site visit on January 31, 1998

The overall approach of each protocol's assessment was to first determine if the main criteria were being recognized and considered at Yale-Toumey. Second, to assess whether or not sustainable forest management (SFM) was occurring at Toumey Forest according to each protocol. The purpose of this exercise at Toumey was not to exactly duplicate the process followed by each certification organization, but to identify the elements of each protocol that were positive and those that were restrictive to their implementation. For example, there was only a one-day site visit to the field area during this exercise although most assessments can take 3–10 days in the field (Richard Donovan, Smartwood, seminar at Yale, February 26, 1998). Similar to most of the certification procedures, the students had access to all the information available for Toumey prior to the field assessment. Certifiers do not measure or conduct inventory analyses during their typical field assessments but use this period to cross-check available documented data. Similarly, the Yale graduate students did not collect primary field data during their field excursion but used this period to verify existing data for Toumey.

4.3.3 Class evaluation of Toumey Forest using the different protocols

A synthesis of the graduate student evaluation of the eight approaches is presented below. Each protocol evaluation is based on the results of three individual teams that did not consult with one another during the assessment. The following information will be summarized for each of the eight protocols:

- State whether Toumey Forest was certifiable, using each certification approach
- Summarize the reasons for scoring Toumey Forest as to its certifiability ranking
- Enumerate the criteria that were satisfied during each assessment

- Enumerate the existing strengths of each certification approach which make it operational
- Give the existing weaknesses identified for each certification approach that make it difficult to operationalize
- Give any comments particular to a certification approach

Following the summarization of the student evaluations of each protocol, several aspects of each protocol will be discussed in more detail in Sections 4.3.5 and 4.3.6.

4.3.3.1. CIFOR

CIFOR SCORE: Not certifiable (conclusion of three independent teams)

CIFOR: Reasons for denial of certification

General Site/Ecology Information

- Missing information on a formal, written management plan identifying all areas of the forest, vegetation, inventory surveys, and future operations.
- A set of management guidelines exists, but is too general to fit the criteria called for in the assessment guidelines.
- Protected areas and areas designated for special use are marked on a map, but clear boundaries are not demarcated in the field. Only includes wetland and protected areas/old-growth stands, but not water catchment areas.
- Documentation inadequate to assess many of the indicators (e.g., pest control, protected areas, chemical control, silvicultural statistical data, etc.).
- Quantitative data lacking to support all silvicultural, economic, ecological, and social practices.
- Needed information on the legal size of area and mechanisms to protect fragile ecosystems, high diversity, water catchment areas, erosion abatement, wetlands.

Forest Management Information

- No spatial planning for logging; no special wood storage areas prepared.
- Basic information on product diversity and forest areas was incomplete. For example, timber harvesting occurs every year, but without volume projections.
- Management objectives are narrow with emphasis on timber production. Science research in forestry and biology/hydrology has low level of implementation of results.
- Lack of current information, quantitative information on forest management practices and land use, and nebulous management guidelines.
- Needed current inventories, predicted rotation intervals, maps, and harvesting practices and silvicultural strategies.
- Missing data on usage, storage, and treatment of fertilizers.
- Needed data on the targeted growing stock, rotation lengths, and volume predictions.
- Needed documentation of current silvicultural practices.

SCS

Certifier goals (SCS 1995, SCS 1998):

- To "spur the private and public sectors toward more environmentally sustainable policy, planning, product design, management systems, and production operations."

SFF

Certifier goals (SFF 1997, SFF 1998):

- To promote ecologically responsible timber management which "recognizes that the current social, economic, legal, and political contexts...do not allow total compliance, regardless of a steward's sincere desire to achieve this."

Woodmark

Certifier goals (Woodmark 1994):

- To promote good forest management worldwide by supporting those already managing their forests responsibly and to encourage others to improve their practices so that they can become certified.
- To provide independent assurance to consumers and timber traders that products sold under the Woodmark label are harvested from forests that satisfy particular environmental, economical and social needs. In addition, to assure that the processing chain, from the forest to the sale of wood products, has been audited to verify the source of the timber products.

Acadian/FSC initiated regional standards

Certifier goals:

- No guidelines given, but should be similar to other FSC-accredited organizations

Canadian Standards Association

Certifier goals (CSA 1996ab):

- A voluntary tool to assist responsible forest organizations to move towards sustainable forest management.
- To "maintain and enhance the long-term health of forest ecosystems, while providing ecological, economic, social and cultural opportunities for the benefit of both present and future generations."
- "Canadians and customers at home and abroad want assurances that the Canadian industry can continue to provide a stable supply of wood and forest products while safeguarding the social and ecological values of forests."
- A commitment to maintaining and conserving biodiversity.

As part of the CSA, forest values are explicitly incorporated into certification by requiring the forest manager or owner to determine:

Biodiversity/Wildlife Information

- Missing information on biodiversity and wildlife status/surveys.
- Managing ecological aspects to maintain age classes of stands but not for biodiversity.
- Harvesting priorities (preferred species being pine) do not contribute to enhancing biodiversity.

Ecosystem Health Information

- No documented evidence of forest vitality, health, and productivity activities (based on managing for natural processes to maintain these).
- Lacked documentation that satisfied regulations involved with the management and application of pesticides.
- Applied herbicides to suppress competitors of desired trees.
- Lacked data documenting annual tree damage/mortality due to insect infestation or natural disasters.

Social/Cultural/Economic Information

- Large emphasis on cultural heritage and concerns of indigenous peoples that were not satisfied.
- Economic aspects of management focus on managing growing stock, logging operations and transport, and road construction, but little effort for afforestation and forest-protection management.
- With respect to economic aspects and profitability, the financial productivity was low and the investment in capital and personnel was low.
- No evidence of consultation with or participation of the local/traditional/indigenous populations.
- No evidence of integrating the local communities by hiring them at Toumey, or of local timber processing, since so little wood is sold from the forest.
- Needed documentation on compliance with legal standards (ILO, ITTA, CITES, Convention of Biodiversity, Forestry Law, land-use planning, property rights, 'other' ecological rules, 'other' economic regulations).
- Quantified data on secondary forest uses, phyto-pharmaceuticals, and non-timber revenues.
- Needed financial statement specific for this property.

CIFOR: Criteria satisfied during the assessment

- Many indicators for assessing the site are available in the two Yale Bulletins from 1916 and 1932 (however, this information is outdated).
- General information and identification of property are satisfactory.
- General ecological data exists.
- General management plan exists.

- Clear information exists on the geographic location, boundaries and borderline, the size, and the area's land-use functions.
- Evidence that the forest is managed based on Best Management Practices, and follows state, regional and international regulations.
- Soil data, stands structure data, inventory of the species, climate, and wetland information well documented. Growing stock data documented (but not continuously). Good documentation and monitoring going back to early 1900.
- Size of area cut during logging and log transport is very small and the gap size satisfies environmental standards.
- Uses Best Management Practices (regional standards) and laws to guide field activities for drainage facilities, wetland protection, and logging and log transport.
- Protection of old growth and archeological site.
- No logging damage or product wastage observed.
- Stand regeneration is well managed and natural.
- Non-timber forest functions and services are provided free to adjacent communities. Public uses are mainly recreational: berry collecting by local communities, and Christmas trees and boughs.
- Controls on the working conditions, safety provisions, health insurance, and the freedom of employees to organize are all covered as rules established by Yale University.

CIFOR: Strengths for operationalizing the CIFOR approach

- Requires multiple uses on lands to be shown in maps, and the presence of quantified data of the relevant information. Multiple uses include timber production, recreation, water, hunting, fishing, pasture, and extraction of non-timber forest products.
- Emphasizes the diversity of tree species and vast range of products to be managed on the forest for multiple uses, specifically the non-timber forest products.
- Emphasizes good communication with neighboring landowners and other interested parties in order to maintain a good dynamic within the community.
- Highlights cultural heritage and concerns of indigenous peoples.
- Very specifically looks at different types of silvicultural systems, identifying issues of concern, and management requirements for different types of harvesting.
- Completely addresses most of the pertinent issues in forest management.

CIFOR: Weaknesses for operationalizing the CIFOR approach

- Too much emphasis on cultural heritage and concerns of indigenous people, which is not applicable for all operations.
- Tended to be rather patronizing in its tone toward both the certifier and the landowner pursuing the certification. Assumed that landowners were not good

business people and reiterated many ideas that are common business practices. Being law-abiding was also strongly stressed, and specific laws were mentioned, which noncompliance with would involve jail time or large fines.

- Required some unusual specifications for issues that may or may not be relevant to a particular forest. Wording was a bit unclear and rationale for requirements was not supported. For example, it required that landowners set aside 5% of their land as a protected area if the land area was larger than 200 ha. This requirement existed regardless of whether or not there were any plant or animal species in need of protection. This set of criteria also specifically stated that no further drainage facilities were to be added to any existing road system. However, this does not state whether this limits the improvement of the existing road infrastructure or even the construction of new roads. This could create a problem for the landowner should they wish to improve their land.

- For issues of herbivore damage to regeneration, this set of criteria required that herbivore-effect exclusion areas be set aside for future study of the large-hooved herbivore impact. It seems as though a wildlife management study is required in order to comply with these certification procedures.

- Some criteria and indicators were too rigid, without much flexibility, while other criteria were too general making them difficult to objectively evaluate.

- Did not stipulate how a certifier could use the guidelines, indicators, or criteria to quantitatively evaluate a forest for sustainable management practices. Though guidelines, criteria, and indicators were developed to assist the certifier, no definitive scoring system was developed and incorporated into the rating system. Without the inclusion of a scoring system, it is difficult to make decisions on criteria.

- Assessment of forest was completely subjective and involved a yes/no rating to evaluate the compliance of Toumey with the designated indicators.

- Not clear if criteria required decisions to be made on an area-assessment basis only of Toumey forest, or through a comparison with other areas.

- Insufficient information provided about who the authority is for certifying these criteria, and how the certification should be applied to other regions.

CIFOR: Additional comments

- The guidance given by this protocol is too general to be effectively used — it needs to be made more specific. If the guidelines are not made more specific, it will be too subjective to be ascertained objectively in practice.

- In order to make it useful, this set needs to make suggestions on what should be achieved with the assessment. It needs to include guidance on how to quantify, measure, or find representative examples.

- There is a need to know who will be in charge of the certification, or who will approve whether it can be accepted nationally, regionally, or internationally.

- Certification procedure is directed at large forest operations with multi-purpose usage and is less conducive to smaller-scale operations.

4.3.3.2   SFI

SFI SCORE: Not certifiable (conclusion of three independent teams).

SFI: Reasons for denial of certification

General Site/Ecology Information

- No management plan.
- General lack of written information regarding activities occurring at Toumey Forest.
- Toumey Forest does not have management guidelines specifying its own policies, programs, and plans to implement and achieve the Sustainable Forestry Principles and Guidelines (Criterion 1a).

Forest Management Information

- No plan for regeneration after a harvest by planting or direct seeding within two years, or by planned regeneration within five years (Criterion 2a).

Biodiversity/Wildlife Information

- No research on wildlife habitat and habitat areas.
- No attempt to create diverse habitats for biodiversity of plants or animals.
- No analysis of biodiversity.
- Not specifically applying knowledge gained through research, science, and technology for conserving biological diversity (Criterion 7b).

Ecosystem Health Information

- No research on water quality.
- No documentation of herbicide use.
- Requirements of landowner to support research ranging from wildlife to ecosystem and landscape levels were not satisfied during the assessment. Requires funding for water-quality research (Criterion 3c); for improving the health, productivity, and management of forests (Criterion 1b); for wildlife (Criterion 4b); improve the science and understanding of landscape management, ecosystem function, and conservation of biological diversity (Criterion 7b).

Social/Cultural/Economic Information

- No tracing the "chain of custody," or following the forest products to the market.
- No information regarding to which mills the forest products go.
- No documentation of public meetings and forest demonstrations.
- No involvement of a state-level "panel" of experts to establish and help identify goals for implementing riparian protection measures for all perennial streams and lakes (Criterion 3b).
- Yale is not involved with an established state group (comprised of logging and state forestry associations and other agencies in the forestry community) to improve the professionalism of logging and to push for sustainable forestry (Criterion 10b).

SFI: Criteria satisfied during the assessment

- The Best Management Practices (BMP) for New Hampshire cover the criteria for careful timber removal near waterways and wetlands and Yale Toumey follows the BMP.
- The school provides some funding and facilitates other fund-raising activities for forest research on wildlife; supports understanding of landscape management, ecosystem function and conservation of biological diversity; and sees to the improvement of the health, productivity, and management of all forests (Criteria 1b, 4b, 7a).
- Meets or exceeds all established BMPs approved by EPA, all applicable state water quality laws and regulations, and the requirements of the Clean Water Act (Criterion 3a).
- Fulfills — but some verification is required — its policies, programs, and plans to promote habitat diversity (Criterion 4a).
- Silvicultural practices do not leave a negative visual impact. Toumey has its own policies, programs, and plans to minimize the impact of harvesting on visual quality (Criterion 5a), with timber-sales contracts specifying clean-up conditions for landings, informal policy to set back landing from public roads, and no large cuts.
- Requires that no harvesting cuts exceed over 120 acres (Criterion 5b). At Toumey Forest, clearcutting is not part of management operations and no cuts are over 120 acres because of logistic complexity. Small crop cuts are used to promote multi-cohort advance regeneration at final harvest.
- Criterion 5c required the adoption of a "green up" policy or more comprehensive method for providing age, habitat, and aesthetic diversity. Even though there is no policy in writing, silvicultural policy (both in management guidelines and in practice) promotes uneven aged stand with structural diversity.
- Yale Toumey manages its forest to encourage reforestation using Best Management Practices and a major objective for the forest (articulated in the 1932 Toumey report) was to demonstrate sustained silvicultural practices to private forest owners (through forums, an internet page, other outreach activities) (Criterion 10a).

SFI: Strengths for operationalizing the SFI approach

- Includes "performance measures" that pertain to specific actions that could be obtained if a defined management plan exists.
- Requires specific silvicultural practices.

SFI: Weaknesses for operationalizing the SFI approach

- Too general and brief. Often vague terminology, very subjective, few real guidelines. The entire guideline lacks clarification.
- No grading system.
- Too much research required. It would take years to compile all the information required.

- Most criteria do not have quantifiable terms. The criteria are written in such a way that it sounds as if they are the criteria for the certifier, and not the criteria for the certified forest.
- Does not specify flexibility for specific forests.
- Requires planning documentation, but provides no guidance on quality.
- Some criteria remain vaguely defined and subjective (e.g., Criteria 1a, 7b, 8a). For example, (1) Criterion 1a requires defining policies, programs, and plans to implement and achieve Sustainable Forestry Principles and Guidelines. (2) Criterion 7b requires applying knowledge gained through science, research, and technology, and field experiences for conserving biological diversity. This criterion is very broad and subjective. (3) Criterion 8a, employing appropriate technology, processes, and practices to minimize waste and ensure efficient utilization of trees harvested. This criterion needs greater specification of what is "appropriate technology" and "efficient utilization."
- Parts of a criterion can be satisfied, while other parts result in problems in weighting the information. For example, Criterion 6a requires the identification of special sites, managing them as appropriate to their unique features, and involving organizations with expertise in protecting special places to suggest how these lands should be managed. Toumey satisfies part of this criterion by cooperating with outside researchers interested in studying a significant anthropological site at Toumey, but does not participate in a state historical registration program because Yale University found it to be too much of a nuisance.
- As far as reaching an overall evaluation decision is concerned, the relationship between "objectives" and specific "performance measures" is not obvious. (e.g., Objective 9. Objective is prudent use of chemicals for certain purposes and the performance measure is mere legal compliance).
- Reference to regulatory compliance in some of the performance measures may be difficult to verify.
- Several requirements are given to establish, interface, report, or perform extension functions to other groups which are quite cumbersome and would require a considerable time investment by the landowner to satisfy them (Criteria 10b, 10e, 11b, 11c, 12a, 12b, 12c). For example:
  1. Criterion 10a requires landowners to work closely with logging and state forestry associations, appropriate agencies, and others in the forestry community to further improve the professionalism of logging by establishing state groups (where none exist) and by cooperating with existing state groups to promote the training and education of loggers. State groups will be encouraged to sponsor training and education programs for loggers, employees involved in procurement and landowner assistance, contractors, and suppliers.
  2. Criterion 10e requires supporting and promoting efforts by consulting foresters, state and federal agencies, state groups, and programs like the American Tree Farm System to educate and assist non-industrial landowners and to encourage them to apply principles of sustainable forest management on their lands.
  3. Criterion 11b requires the issuing of an annual report to the public on its membership performance regarding compliance with, and progress on, sustainable

forestry, including a listing of companies complying with the Sustainable Forestry Principles and Implementation Guidelines.

4. Criterion 11c requires the establishment of an advisory group of independent experts to assist in the preparation of the annual report, including validation of conclusions and the assessment of reported progress.

5. Criterion 12a requires supporting and promoting appropriate mechanisms for public outreach, education and involvement related to forest management (e.g., toll-free numbers, environmental education, and/or private- and public-sector technical assistance programs).

6. Criterion 12b requires the establishment of appropriate procedures at the state level to address concerns raised by loggers, consulting foresters, and employees regarding practices that appear to be inconsistent with the Sustainable Forestry Principles and Implementation Guidelines.

7. Criterion 12c required the establishment of a national forum of loggers, landowners, and senior industry representatives, including CEO representation, that will meet at least twice annually to review progress towards the Forestry Principles and Implementation Guidelines.

SFI: Additional comments

- The protocol needs to develop an explicit grading system that is not subjective.
- Protocol needs to develop region-specific evaluation criteria.
- This approach is designed for use by large organizations or large landowners but not small landowners. For example, Criterion 10d requires commitment to the Sustainable Forestry Principles to be communicated throughout all levels of a company — particularly to mill and woodland managers, wood-procurement operations, and field foresters. This criterion is not relevant for Yale since organizational levels do not exist and would not be appropriate for small landowners.
- This protocol has too many criteria that need to be less vaguely defined (e.g., "appropriate technology" for waste minimization).
- For this protocol, the relationship between general "objectives" and specific "performance measures" needs to be made more explicit (i.e., chemical use objective presents guidance on prudent use for specific purposes, but the performance measure only covers legal compliance).
- Reference to regulatory compliance in some of the performance measures may be difficult to verify in practice, so it would be better to specify the lack of any enforcement actions.

4.3.3.3. Smartwood

SMARTWOOD SCORE: Not certifiable (conclusion of two independent teams); Not certifiable but close to being certified (conclusion of one independent team) — not as "sustainable," but as a "well managed" forest. This protocol states "Sources that demonstrate a strong operational commitment to these principles, but are lacking long-term supporting data, will be classified as "well managed" (Smartwood 1996).

Smartwood: Reasons for denial of certification

General Site/Ecology Information

- No management plan. Missing information on a formal, written management plan identifying all areas of the forest, vegetation, inventory surveys, and future operations.
- A set of management guidelines exists, but is too general to fit the criteria called for in the assessment guidelines.
- Supporting documentation was sparse. Several bits of required information were available from interviews, but were not documented (e.g., "Merchantable log loss and waste in the forest is minimized").
- Some criteria of approach were not fulfilled.

Forest Management Information

- Required specifics on harvesting and production information which was not all readily available.
- Required operating and harvesting plans and information on pre-harvest silvicultural conditions which was not available.
- "Sustained yield" not defined or specified in guidelines.
- No documentation of acreage in plantations and regenerated forest stands.
- No separate plans for plantation forests.
- Cutting to eliminate red pine (i.e., to create a monoculture of white pine) might be considered "unsustainable."

Biodiversity/Wildlife Information

- Missing information on biodiversity and wildlife status/surveys.
- No research on wildlife, wildlife habitat, or habitat areas.
- No analysis of biodiversity.
- Exotic species not documented.
- No attempt to create a diverse habitat for biodiversity of plants or animals.

Ecosystem Health Information

- Undocumented use of herbicides.
- No GIS maps or comprehensive hydrology information.

Social/Cultural/Economic Information

- Large emphasis on cultural heritage and concerns of indigenous peoples.
- No tracing of forest products to mill and beyond.
- No use of, nor documentation of, non-timber forest products or recreation uses.
- No documentation of social aspects, such as town meetings or community interactions.

Smartwood: Criteria satisfied during the assessment

- Silvicultural practices do not leave a negative visual impact.

Smartwood: Strengths for operationalizing the Smartwood approach

- Can serve as a general guide for managers.
- Tries to account for local variability.
- May provide incentive for managers to keep better records and stay abreast of current research and recommendations.
- Easily understood set of criteria and tone assumed that the certification procedure involved knowledgeable landowners sincerely looking to improve their land.
- Required both the management plan and inventory surveys to be examined together to cater to the needs of wildlife and special areas to be maintained.
- Seemed more realistic in combining the goals of sustainable forestry and maintaining ecological diversity. Acknowledged that each operation is going to be different and gave the certifiers flexibility in deciding whether an operation complies with certification guidelines. This set did not have the same depth in some areas as the other protocols, but allowed the certifier to consider situations that might not be normal. A knowledgeable certifier would be required to determine the degree of sustainability.
- Mentioned the use of a certified forest as a training/educational tool.
- Criteria are clearer and less subjective.
- Includes documentation on non-timber forest products.
- Gave more specific guidelines for conserving and enhancing biodiversity, i.e., gaps (in the plantations protocol).

Smartwood: Weaknesses for operationalizing the Smartwood approach

- No explicit criteria weighting procedures (assumed equal importance and all had to be met).
- Most criteria do not have quantifiable terms.
- Questions are vague and open to interpretation. Many two-part questions are difficult to answer definitively. Often vague terminology, very subjective, few real guidelines.
- Data analysis approach suggests a "fill in the blank" approach, supplemented by a prose analysis of findings for each major category. Ranking from 0 to 5 is used, but the ability to determine what ranking should be recorded is not supported, making evaluation ranking subjective. No way to determine relative importance of criteria.
- Unclear instructions in the Forest Plantation Section (Part II).
- No real designation of areas of special concern within the procedures.
- Best Management Practices only mentioned in passing and not integrated into the protocol.
- Too much research required, it would take years to compile all of the information required as part of the protocol.
- Does not specify flexibility for specific forests.

4.3.3.4   SCS

SCS SCORE: Not certifiable (conclusion of two independent teams); Not certifiable, but close to being certified (conclusion of one independent team).

SCS: Reasons for denial of certification

General Site/Ecology Information

- Some activities to maintain habitat and ecological landscape but also indications of developing a homogenous species site (e.g., white pine).
- Watercourse management policies and programs only geared to wetlands.
- No management plan.
- Management plan is vague and too broad to assess against certification standard. Not documented that all management actions are based on statistically accurate information.
- No documentation of landscape-level planning initiatives.
- No ecological reserve system.
- No identification of ecologically significant areas that are non-managed or protected.

Forest Management Information

- For timber-resource sustainability criteria, no specific regulation strategy, no rotation and target volume, and no long-term strategic plan for harvesting.
- No written yield target for timber.
- Lack of historical data and research findings that support target rotations and predicted yields of volumes and values.
- No data regarding the full stocking of each species.
- Most silviculture and harvesting data are out of date (1916–1932).
- Most information related to rotation lengths is outdated, and current research not being conducted to determine target rotations that will meet MAI standards. Lacking maximum mean annual increment values for all tree species.

Biodiversity/Wildlife Information

- No evidence of wildlife management actions, strategies, or program.
- Full access to data concerning wildlife requirements for food, shelter, water, and space is missing for this forest.
- No data on wildlife population inventories, habitat conditions, or dynamics.
- Diversity and structure not being maintained since suppressing hardwoods because of selection for white/red pine.

Ecosystem Health Information

- No evidence of pest and pathogen management strategy (i.e., to salvage firewood when disease identified).
- No records kept on long-term ecological productivity and health.

- Priorities not well documented regarding the use of biological controls vs. pesticides.
- Water-protection guidelines not available.
- No philosophy on pesticide use (i.e., when applied, how applied, etc.).

Social/Cultural/Economic Information

- No evidence that management is maintaining financial stability because the production scale is small compared to the land size.
- No public involvement in the financial and socio-economic considerations.
- No evidence that management invests a rational capital and personnel investment.
- No records kept on harvest efficiency/product utilization.
- No documentation of worker/contractor relations or long-term commitment to stability.
- No clear documentation of company policies and long-term strategies throughout all levels of organization.
- No budget analysis available of Yale-Toumey income and expenditures.
- No list of items purchased for improved silvicultural practices.

SCS: Criteria satisfied during the assessment

- Good implementation of ecosystem reserves (maintaining old growth and wetlands). Ecologically significant areas identified (e.g., swamps), protected and non-managed.
- Good implementation of public-use management since area is open for ecotourism or recreational use and the production activities maintain the aesthetic externality.
- Best Management Plan developed, although the information within is weak.
- Pesticide use, practices, and policies follow the Best Management Practices for New Hampshire.
- Compliance with state water-quality-protection regulations (however, water quality not checked regularly) and maintenance of optimum water quality and riparian habitat conditions based on BMPs for New Hampshire.
- Forest being managed at an ecological rotation relevant for maintaining ecological productivity and health, and not based on economic rotation. Ecological considerations are valued over market demands for product.
- Forest community structure and composition not affected by management; management is at a landscape level and full range of seral stages of stand development exist and is part of the research focus on this land.
- Continuous forest inventory conducted every 10 to 12 years.
- A long-term forest regulation strategy exists with respect to species and harvest.
- Evidence that harvest regulation strategy can absorb unforeseen events (e.g., pests).
- Actual harvest levels are sustainable and in compliance with management plan.

- Current merchantable growing stock vigorous.
- Have an assessment every 5 to 10 years of all products and potential products with monetary value.
- Management decisions based on empirical data and experience (but not well documented).
- Minimal loss of forestland to road system and rights of way.
- Minimum ecological impact (e.g., erosion, degradation) due to road system.
- Long-term goal of company is clear in management plan.
- Communication with (certified) wildlife official is maintained and recommendations are incorporated into management. Provisions are in place to maintain wildlife habitat structures in forest (e.g., unfragmented area, cavity trees, mast).
- Financial stability is strong and cutting occurs on as-needed basis.
- The company is concerned with the effect of forest management in the surrounding community and employs local loggers. The company is also involved in trading lands locally to achieve their environmental obligations and to maintain community interactions.
- With respect to public-use management, there are regulations controlling trespassing and conflict, clear statement of open operations to public, and written annual report.
- Contractors harvesting timber must have liability insurance; Yale-Toumey meets the state regulations on safety issues, and guarantees appropriate stipend and housing support for student internships.

SCS: Strengths for operationalizing the SCS approach

- Each criterion lays out clear examples of failure.
- The contents of this procedure are lucid and well organized.
- The separation of criteria and examples of non-sustainable practices helps to clarify ambiguous statements which helps to ensure consistent interpretation by land managers and certifiers.
- Does not tie the hands of forest managers with quantitative restrictions and allows for enough subjective interpretations for a site-specific assessment.
- Each criterion is practical and reasonable for even small landowners to implement.

SCS: Weaknesses for operationalizing the SCS approach

- The weighting of the criteria ends up treating all criteria as being equivalent and the types and number of types of information listed below each criterion will determine the final ranking. This appears to be good, except there is no way of knowing how to score each criterion.
- The use of a mathematical scoring system where criteria are weighted differently and each scored on a 100-point scale with 80 points as a failure threshold appears to be quantitatively rigorous. However, this system is problematic because no instructions are given on how to score each criterion, combine its components, or judge the final, weighted average score. Grading system is quite subjective and inconsistent even though the procedure outlines grading thresholds and optimum management strategies.

- The numeric weighting system scoring has to be assigned subjectively by the certification crew which will result in large variability between different evaluation teams.
- Many of the criteria are vague, highly subjective, perhaps not scientifically valid, and difficult to ascertain in practice (e.g., "all stand treatments are designed and carried out so that future yield targets are met"; "Species and habitats maintained in a manner such that ecosystem components are similar to those predating European settlement…species composition…is unaltered or restored…"; "Company takes a *progressive* position with respect to balancing landowner rights with public trust obligations").
- Questions are vague and open to extensive interpretation.
- Terms are not defined and left open to operator interpretation.
- Some esoteric ideals are included that may not be practical for implementation.
- Criteria focus on "original diversity of natural forests" and "maintaining habitats components similar to those predating European settlements" which are not conducive to plantation management or research forest practices. The emphasis here is on preservation rather than on active management for commercial timber. The values of the certifying organization are strongly expressed by this requirement.
- Assumes unlimited resources and time.

SCS: Additional comments

- This protocol has to prorate each criterion before ranking and scoring the results.
- This certification procedure promotes a holistic view of sustainable management in forest ecosystems.
- This certification protocol is applicable to both commercial and non-commercial forestlands with minor modifications.

4.3.3.5. SFF

SFF SCORE: Not certifiable (conclusion of two independent teams); Certifiable (conclusion of one independent team). Flexibility existed to examine other evidence of sustainability (i.e., the implied management regime being implemented on the ground). Therefore, even though no management plan existed, certifiers could substitute other information to satisfy this requirement.

SFF: Reasons for denial of certification
General Site/Ecology Information

- Lack of a formal management plan and all it entails (clear objectives and methodologies).
- Absence of information on natural disturbances.
- Lack of information on ecosystem types.
- No demarcation of ecologically sensitive/protected areas (e.g., riparian zones, wetlands).
- No road design (to maintain full forest functioning).

Forest Management Information

- Timeframe of management (mostly long-term).
- Plans to mimic natural disturbances in harvesting operations.

Biodiversity/Wildlife Information

- Lack of information on wildlife distribution and abundance.
- Lack of information on non-woody plants.
- No list of endangered or protected species.

Ecosystem Health Information

- No system to monitor the impacts of management.
- No explicit protective measures for soil and water conservation.

Social/Cultural/Economic Information

- Social criteria could not be addressed according to the requirements of the protocol. For example, many of the variables were not relevant to Toumey Forest since the forest uses student trainees (i.e., summer crew) instead of locally hired, permanent employees. Students do not fit into the same category or meet standards associated with housing, unions, etc.
- No community involvement in the management of the forest (no "town meetings," or reporting to locals), although locals are permitted to use the land in a variety of ways.
- Social and economic needs of local communities are not considered.

SFF: Criteria satisfied during the assessment

- A robust and long-term strategy for managing exists, so the implied management structure could substitute for a written management plan.

SFF: Strengths for operationalizing the SFF approach

- The "fatal flaw" sections include some guidelines that cannot be overlooked during an assessment. These sections are particularly useful.
- It is short, easy to understand, and well organized. It is clear what is needed for certification assessment.
- Straightforward methodology.
- Includes an explicit, tangible set of requirements.
- Concise ranking system to weight information collected that provides a grading system requiring all sections to have a positive value across all criteria. The use of a point system gives flexibility.
- Versatile because it emphasizes that the evaluation teams have the authority to modify standards to suit specific characteristics of the sites.

SFF: Weaknesses for operationalizing the SFF approach

- Has vague indicators (Section 8.2 "Cutting over any period removes no more than 80% of the growth for the same period" — what period is meant?); statements (Section 14.2 "Financial decisions do not dictate management decisions" makes it seem as if a certifiable forest cannot be run for profit), or financially unrealistic expectations (Section 18.1 "Management reviews ecological, social, and economic impacts of activities immediately after each entry and thereafter on at least an annual basis" is unrealistic, and attempts in this direction will be extremely expensive).
- Some of the listed "fatal flaws" chosen should not necessarily disqualify a forest from certification.
- A 150-year management plan is not realistic (in the southern United States that would be three rotations and realistic economic or even ecological forecasts are difficult to project that far in advance).
- Corridor requirements are arbitrary and can have potentially negative impacts on the forest (edge effects, disease risks not considered).
- Requiring permanently reserved structure at each felling does not take into account economic, silvicultural, or disturbance (e.g., hurricanes) information of unmanaged systems.
- Requires too much information.
- The economic and social sections are not very clear, and several sections overlapped in the information needs.
- Requires the present condition to be compared with the natural range of variability, but places no time frame on the natural range of variability.
- Requires the landowner to inventory rare, threatened, and endangered species.
- Ambiguity in protocols requiring the certification team to interpret the evaluation process.
- Many criteria appear to conflict with one another, making it impossible to satisfy these criteria at the same time.
- Many criteria are not relevant for the site (very specific to an area). For example, old-growth nodes and recruitment will provide habitat, but this does not exist in New England.

SFF: Additional comments

- Significant attention given to the public interest on how it feels the forest should be managed and not on how it is being managed. Public participation should not be expected throughout the entire process of developing a management plan. For example, a certification procedure can include checks on biodiversity, but does not need to hold public meetings to confirm that this is part of the forest management objectives or for it to be managed sustainably. Public participation is costly — both in terms of time and money.
- Requested information needs to be made less vague and unrealistic, and fewer criteria should be required.
- Boundaries and edges of the management unit should be considered (external influences on the forest under management).

- Focus of certification procedures is on the management plan; however, it should be on the organization of the forest management, or on how the operation is running on the ground.
- Values of the certifiers should not be imposed on the management of the forest if it is to be certified.
- In a developing country, it would be difficult to complete all of the ecosystem-related information because of the expense and time required to collect all this information, especially in areas where little prior research exists.
- Needs an explicit grading system that leaves little room for subjectivity.
- This very regionally constructed evaluation protocol will not readily transfer to other locations.

4.3.3.6   Woodmark

WOODMARK SCORE: Not certifiable (conclusion of three independent teams).

Woodmark: Reasons for denial of certification
General Site/Ecology Information

- Missing information on a formal, written management plan identifying all areas of the forest, vegetation, inventory surveys, and future operations (clear objectives, methodologies, annual allowable cuts and methods of calculations, rotation length, and method of calculation). A set of management guidelines exists but these are too general to fit the criteria called for in the assessment guidelines.
- No designation of conservation areas.
- No description of biological or ecological features.
- No description or identification of water resources.
- No designation of biological corridors.

Forest Management Information

- Source of red pine planting stock not identified.
- The purposes for establishing red pine plantations were not given, so it was difficult to determine whether they were socially and ecologically appropriate.

Biodiversity/Wildlife Information

- Missing information on wildlife status/surveys (distribution and abundance).
- No biodiversity inventory exists.

Ecosystem Health Information

- No information on the quantity and total area to which pesticides have been applied.
- No monitoring of the environmental impact of forest management (e.g., soil compaction, quality of water resources, damage to residual stand, etc.)

- No explicit policies that promote watershed protection.
- No assessment of whether culverts obstruct the migration of fish in the stream.
- No explicit statement of fire and pest control.

Social/Cultural/Economic Information

- Considerable emphasis is placed on local communities and land rights — which are not very applicable to New Hampshire. For example, uses of student trainees (e.g., summer crew) instead of locally hired, permanent employees (students do not meet standards associated with housing, unions, etc.) make much of this section irrelevant for Yale-Toumey.
- No community involvement in the management of the forest (no "town meetings," or reporting to locals), although locals are permitted to use the land in a variety of ways.
- Large emphasis on cultural heritage and concerns of indigenous peoples which is not considered during forest management.
- No information on local people's involvement in management decisions.

Woodmark: Criteria satisfied during the assessment

- None given during the student evaluation process.

Woodmark: Strengths for operationalizing the Woodmark approach

- Explicitly states the requirements for certification.
- Is a holistic approach. Economic, cultural and social issues are clearly presented (much more than in other protocols). Integrates the multiple disciplines of sustainability (e.g., social, economic, biological, etc.).
- The set has a glossary.
- The principles are well thought out.
- Based on the information required and the detail to which it is taken, this certification procedure requires the landowners to take a much more active role in the forestry operations taking place on their land.
- Requires continual monitoring of the forest and periodic updates of the management plan to reflect new conditions and growth on the forest, again forcing a more active role by the landowner.
- Regarding the issues of a continuous supply of raw material, this set requires an annual cut designated in a management plan.
- Requires areas of potential concern to be managed as special units and set aside. This process occurs at the discretion of the certifier and practicing forester, which may increase the flexibility for following the certification procedure.
- Looks for evidence of economic return from non-timber products. This requirement will help to ensure the sustainability of these alternative products.
- Has good guidelines for handling relations between landowners and traditional users.

Woodmark: Weaknesses for operationalizing the Woodmark approach

- The set is hard to follow — very dense and complicated.
- Many statements come in the form of "must...if appropriate," which is very vague.
- Protocol is too rigid and does not have much flexibility.
- Does not give an explicit ranking system for each point. Rankings can be subjective. There is no point system or systematic way to decide whether the principle is being upheld.
- Structured to be specific to management locations with emphasis on local communities and land rights as being being important values driving the certification process.
- Combines landowners, processing facilities, and contractors in the certification scheme when they might not be relevant for many operations. For example, Yale bids out timber sales so the processing facilities are not linked to them but becomes relevant for the contractor awarded the bid.
- A lot of attention paid to details when looking at record keeping for a certified unit. It lists the forms that should be used to maintain accurate records and labeling requirements. These requirements have the potential to create a headache of paper trails that a landowner may or may not decide are worth the benefits of being certified.
- Tends to be patronizing in its tone to the landowner. One should be able to assume that a landowner will be looking for the most cost-effective means of sustaining an income off their land.
- Emphasis on protection of biodiversity is a subjective process and no clear procedure exists to determine what level of biodiversity is appropriate for a given site.
- Terminology and some of its implications are unclear. For example, what do the standards mean when they refer to local opposition, local rights, and culturally appropriate practices?

Woodmark: Additional comments

- Public participation is not needed on all aspects of achieving management objectives. A certification procedure can include checks on biodiversity, but does not need to hold public meetings to confirm that this is part of the forest management objectives or for it to be managed sustainably.
- Requested information needs to be made less vague and unrealistic.
- Boundaries and edges of the management unit should be considered (external influences on the forest under management).
- Focus of certification procedures is on the management plan, but should be focused on the organization of the forest management system or how the operation is running on the ground.
- Significant attention given to the public interest on how it feels the forest should be managed, not on how the forest is being managed. Public participation should not be expected at each point in the management development process.

Obtaining public participation at all these stages is costly (e.g., time and money) to implement.

- Values of the certifiers should not be imposed on the management of the forest if it is to be certified.
- Decrease the amount of paperwork required for certification.
- Decrease the amount of time spent on issues that are merely good business practices and assume that the landowner has the ability to make decisions regarding the most cost-effective manner of managing their land.
- This approach is mostly relevant for large landowners and too many constraints are imposed on smaller landowners.
- Should be applicable to tropical areas because of its structure. This plan is probably more geared to those locations where indigenous people require more consideration. This makes this protocol more site specific.

4.3.3.7 Acadian/FSC Regional Standards

ACADIAN/FSC SCORE: Not certifiable (conclusion of three independent teams).

Acadian/FSC: Reasons for denial of certification

General Site/Ecology Information

- Unable to evaluate. Too much information required that was unavailable.
- Lack of a formal management plan.
- No documented information on buffer zones. Buffer zones not implemented for wildlife and recreational uses.
- No riparian buffer zones established for different slopes.
- No plans detailing watersheds, fisheries, and ecological studies.
- Protected area network not implemented since Toumey is not large enough. Did not satisfy the need to have 5% of representative ecoregion land contributed to network of protected areas if the ecoregion is underrepresented.
- No monitoring or assessment documentation.

Forest Management Information

- No documentation of silvicultural goals, including species to be cut and desired regeneration.
- Forest regeneration plans not explicit.
- No salvage plans.
- Management plans do not differ for primary, natural, and semi-natural forests.
- No separate plans for plantation areas.

Biodiversity/Wildlife Information

- No wildlife surveys. Lack of information on wildlife distribution and abundance.
- No inventories of all plants and animals.
- No documented information on endangered species.

- Cutting to eliminate red pine (i.e., create a monoculture of white pine) might be considered "unsustainable."
- No attempt to create a diverse habitat for biodiversity of plants or animals.

Ecosystem Health Information

- No environmental impact assessment.
- No documentation of waste material disposal and clean-up.
- No specific information for soils and soil conservation.
- No documented information on chemical use.

Social/Cultural/Economic Information

- Could not satisfy the social requirements since they are not relevant at Toumey. For example, use of student trainees (e.g., summer crew) instead of locally hired, permanent employees means that most standards associated with housing, unions, etc. cannot be addressed.
- No community involvement in the management of the forest (no "town meetings," or reporting to locals), although locals are permitted to use the land in a variety of ways.
- No clear evidence for handling interactions with local communities, especially in case of a conflict.
- Few job opportunities provided to communities neighboring Toumey Forest.
- No documentation on recreational uses of land or non-timber uses by communities adjacent to Toumey Forest.
- Lack of financial records.
- Lack of documentation on compliance with laws.
- No tracing "chain of custody."

Acadian/FSC: Criteria satisfied during the assessment

- Silvicultural practices do not appear to leave a negative visual impact.

Acadian/FSC: Strengths for operationalizing the Acadian/FSC approach

- Detail is one of its strong points (although its length makes it unwieldy). Explicitly sets forth particular criteria by which to evaluate each principle along with an underlying philosophy.
- The set is very thorough.
- Considers differences between regions (Principle 5) related to landscape, stand, and market conditions.
- Dedicated set of criteria (Principle 10) for plantations and recognizes that it is a significant part of timber management (even though it discourages their use).
- Section 7.1.f.6 states that "If the management unit is larger than 500 hectares, a landscape level plan is required in addition to a stand level plan." This is good because it pushes managers to think on a landscape scale, not just about their management unit.

- Considers past land uses.
- Calls for long-term monitoring and assessment.

Acadian/FSC: Weaknesses for operationalizing the Acadian/FSC approach

- Establishes "pre-colonization forest conditions" as goal of forest condition.
- Principles and criteria are often stated as philosophy or unreasonable requests (i.e., pre-colonization forest condition).
- Mixes subjective and objective criteria (i.e., prescribes a "cookbook" 50 ft. riparian buffer requirement; states that preference must be given to locals when conducting business).
- Length makes it unwieldy. Too detailed, and the details are not quantifiable.
- Information on wildlife needed, and environmental impact assessment and/or inventory of plants and animals. This is prohibitively expensive and time-consuming for each landowner to carry out. Expertise may not exist in system.
- Although the set is very thorough, many of the subsets to the principles are repeated in different areas. This creates problems in how to weight the information needs, so that some types of data are not given more weight because of being repeated throughout the set.
- Many parts of the set are criteria for the buyers, which is not relevant to certifying forestlands.
- Impossible to enforce most of the criteria after certification.
- Uses words without commonly accepted definitions, and uses many subjective terms.
- Restrictions are not appropriate and do not make sense for some criteria.
- Many criteria are vague, but demanding, in their information needs.
- The values of the certifiers are imposed on the managers (i.e., 4.2a: "Management policies and practices should confirm a commitment to regard investment in human employment as preferable to investment in equipment." Also see 5.1, 6.4, 9.5).
- Some stipulations ignore the ecology or silviculture particular to a region. For example, Section 6.3.a states "management, at all levels, should favor the growth of natural seedlings of underrepresented species." At Toumey, this would require more planting of hemlock which acidifies the site and is being attacked by an introduced insect pest (e.g., the wooly adelgid).
- This set of criteria requires a gradual return to "natural diversity" (Section 6.3.b.6), but is not clear about what natural diversity is and how it can be achieved on a managed forest.
- This set requires the protection of representative samples of existing ecosystems (Section 6.4). This is a considerable burden on the landowner. It also does not consider the landscape-level problems of maintaining smaller matrix land units and whether they function in a similar way as when they were present as part of a larger landscape.
- The requirements for the management plan reiterate the first half of the criteria and are too repetitive.
- Criteria are more appropriate for larger operations.
- Does not allow for flexibility for specific forests.

Acadian/FSC: Additional comments

- Two levels of criteria are needed to satisfy local and regional differences that could be difficult for an evaluator to integrate into one analysis.
- Two-pronged approach to evaluation process: set of comprehensive (regional) criteria defined in quantitative terms (e.g., 50-ft. buffer zones) and a subjective evaluation of those same criteria by the evaluator. The evaluator should have authority to override a failure to comply with a quantitative criterion, but this can result in different evaluation results depending on who the evaluator is.
- Public participation is not needed on all aspects of achieving management objectives. A certification procedure can include checks on biodiversity, but does not need to hold public meetings to confirm that this is part of the forest management objectives or for it to be managed sustainably.
- Requested information needs to be made less vague and unrealistic.
- Boundaries and edges of the management unit should be considered (external influences on the forest under management) instead of focusing on the management unit by itself.
- Focus of certification procedures is on the management plan. This focus does not address whether the management organization is sustainable or whether its operations on the ground reflect good forest management.
- Significant attention given to the public interest on how they feel the forest should be managed and not on how the forest is being managed. Public participation should not be expected during all phases of developing the management plan. Public participation is costly to obtain both time wise and money wise.
- More appropriate for international forest and large tracts of land.
- Requests need documentation of past and present chemical uses.
- Recommends a post-harvest assessment of snags, gaps, and woody debris.

4.3.3.8   CSA

CSA SCORE: Not certifiable (conclusion of three independent teams).

CSA: Reasons for denial of certification
General Site/Ecology Information

- Main reason for not certifying Toumey was a lack of information and public participation in the development and implementation of all stages of a management plan.
- The management plan does not specifically address how each objective will be achieved.
- The requirement for revising the management plan every ten years has not been satisfied.
- Lack of formal process to assess whether continued improvement is occurring on the management of forestlands and whether there is a mechanism to detect changes.
- No information on the impact of management on global ecosystem functions (i.e., contribution to global carbon budget), or formal procedures for its assessment.

- No inventory available on the status of selected indicators.
- No information on ecosystem, species, and genetic diversity.
- No documentation of the incidence of disturbances and stresses on the site.
- No documentation of ecosystem resilience.
- No documentation of existing biomass.
- No documentation of water quality, trends in stream-flow events, distribution, and abundance of aquatic fauna.
- No documentation of forest areas protected.

Forest Management Information

- None resulting from the assessment.

Biodiversity/Wildlife Information

- No formal plan to assess either existing forest biodiversity or to monitor its change.
- No information on non-woody plant species and animal populations.

Ecosystem Health Information

- Criteria requiring conservation of soil and water resources cannot be assessed. Best Management Practices are adhered to but there are no means of assessing existing conditions or for monitoring whether the conditions are changing.
- No future monitoring of changing forest conditions.
- No existing assessment protocols for evaluating the condition or monitoring changes in soil and water resources.
- No environmental-risk assessment conducted on these forestlands.

Social/Cultural/Economic Information

- No direct public input into the management system and its implementation. No defining of the management system by incorporating values, goals, indicators, and objectives through the public-participation process.
- No consideration of the trends in global market share.
- No consideration of the forestland contributions to GDP.
- No data on the economic value of non-market goods and services.
- No information on the sustainability of forest communities.
- No information on whether informed, fair, and effective decision making is occurring.

CSA: Criteria satisfied during the assessment

- The need for multiple benefits to society are satisfied due to the sustained level of timber harvest, accessibility of the forest to the public for recreational activities, as well as for non-Yale research.

- Forest managers generally recognize that there is a range of social values associated with the forest. Management activities are carried out in such a way as to minimize potential conflicts with those values (although no formal public input is solicited).

- The management plan entitled "Management Guidelines for the Yale-Toumey Forest" established a defined forest area and a policy statement that defined the vision, mission, guiding principles, and the codes of management practices.

- Information and map is available on forest types and age classes at Toumey Forest.

- Information exists on forest inventories.

- Management guidelines exist and the New Hampshire BMPs are followed.

- Yale has a commitment to disseminate information to the public.

- Documentation exists on an aboriginal archeological site that is protected.

CSA: Strengths for operationalizing the CSA approach

- Because of its strong focus on public participation during all steps of the certification process, this protocol is designed for regions/countries with significant public ownership of forestland, and where industrial operators are provided access to the resource through a cutting-rights or tenure-based system.

- This sustainable forest management system (SFM) explicitly details required procedures, documentation, criteria, and indicators that are necessary for certification by this method.

- Goals and intentions of this sustainable forest management system are honorable.

CSA: Weaknesses for operationalizing the CSA approach

- Existing management strategies and/or practices are not in conflict with the underlying principles of this certification approach. However, could not assess them because of the requirement for a certifiable process that would require the creation of values, goals, indicators, and objectives as they related to the six main criteria.

- Emphasis on the required level of public participation made this protocol extremely cumbersome. This requirement will create a significant roadblock for private landowners (especially small landowners) to manage conflicting objectives and absorb additional costs associated with administering the public-participation process.

- Requires a large amount of information that is expensive to measure.

- No specific, simple, quick, and relatively cheap methodology is given to measure all the variables.

- The protocol is very abstract and therefore appears to have little applicability.

- Insufficient attention is paid to social issues, except for public participation.

- This sustainable forest management system (SFM) is quite vague regarding evaluation or analysis of all the variables needed for certification.

- Criteria include an exhaustive list covering ecosystem, trade, policy, public involvement, education, and timber aspects that will be extremely difficult for a landowner to obtain. Has unrealistic expectations on the types and amount of data needed.

- This approach would require the expertise of almost the entire faculty at the School of Forestry and Environmental Studies, supported by student crews and several full-time managers, to address and fulfill each of the required criteria.

- Even when information was available, the system did not provide a gradient to judge success or failure. Often the audit asks for "evidence" of something, but almost every criteria or indicator requires a subjective judgment to be made.

CSA: Additional comments

- This protocol is only appropriate on publicly owned lands.
- It is impractical because of its high data needs and requirement for public participation at all stages of assessment.
- Serious concerns exist regarding the feasibility and practicality of applying this to certify any forest.
- This protocol targets lands held by governments due to need for information on national and international economies, and "the commitment of governments to sustaining global ecological cycles" by evaluating forest-sector policies.
- This system was designed for large corporate landowners or governmental agencies that have staffs of researchers to support their projects.

4.3.4 General summary of the assessment of Yale's Toumey Forest

The use of the eight protocols did not produce agreement on whether Toumey Forest should be certified. In general, the students concluded that this forest did not satisfy the requirements for certification when using some of the protocols (e.g., CIFOR, SFI, Woodmark, Acadian/FSC Regional Standards, CSA) (see 4.3.4). However, three exceptions to this conclusion resulted and are summarized below:

1. Smartwood protocol: two teams deemed that the forest could not to be certified but one team found it close to be certified as a well-managed forest.
2. SCS protocol: two teams found the forest to be not certifiable, but one team concluded that Toumey Forest was close to being certified.
3. SFF protocol: two teams found the forest to be not certifiable, while one team concluded that it was certifiable.

Each protocol (except for the CSA in the Forest Management Information category) required information in five areas. This information was either unavailable for Toumey Forest or the students did not know to ask for other information that supported or verified that Toumey did satisfy a particular requirement. The categories of information that the student teams identified as missing were surprisingly similar for the different protocols. In each protocol, information was identified as missing in the following categories: general site/ecology; forest management; biodiversity/wildlife; ecosystem health; and social/cultural/economic. All student teams concluded that the following information was unavailable at Toumey Forest and, therefore, it could not be certified:

- Lack of a formal written management plan and all that it entails (clear objectives, methodologies).
- Lack of documentation of long-term timetable of management and identification of all areas in the forest and how they will be managed as part of the future operations of the forest systems.
- Lack of specific maps including legal documentation, identification of corner markings, boundaries, etc. as well as vegetation inventories for each stand, basal areas, ages, etc.
- Lack of wildlife surveys.
- Lack of information on biodiversity; wildlife status and surveys; non-woody plant species; and list of rare, endangered, and protected species in the forest biodiversity surveys.
- Lack of demarcation of ecologically sensitive/protected areas (i.e., riparian zones, wetlands, etc.) and habitat for biodiversity.
- Lack of systematic monitoring plan for the forest (much of the monitoring is currently informal), and for monitoring impacts of management.
- Lack of systematic description of pre- and post-harvest silvicultural treatments required for the future sustainability of the forest stands.
- Lack of description of, and issues related to, non-timber forest products to be found in the forest (presently, firewood cuts and blueberry patches are informally noted by the managers, but are not recorded in permanent documentation).
- Lack of evidence of explicit protective measures for soil and water conservation.
- Lack of documented information on whether pesticides and herbicides are being applied to the management unit (records of the type and quantity of chemicals used).
- Lack of evidence of social and economic needs of local communities (i.e., local community participation as employees, or in identifying values or concerns of indigenous communities in management).

The fact that so many similar variables were found to be missing by student teams is unexpected considering the different philosophies underlying each certification approach and their fundamentally different structures (see 2.4). The similarity of which data were identified as missing is also noteworthy because the protocols did not produce the same results regarding the certifiability of Toumey Forest (see 4.3.3). It is disconcerting that the different protocols did not produce similar results. Although they are not structured in the same manner, they should produce similar outcomes. If they do not, there is a need to ask why, and to determine which aspects of each protocol were responsible for the different results. Although the general groupings of information required by the eight protocols did not vary much (see 4.2.1), each protocol did emphasize different amounts of data collection within each category and the weighting or non-weighting of that information varied dramatically. Both these factors help to explain the different outcomes that were produced with each assessment.

In general, Toumey Forest was not certifiable because there was a marked absence of much of the written information that most protocols required. Even though Toumey Forest has a history of being well managed and has been one of the premier historical training grounds for forest managers around the world, there was no way to prove that by using the existing protocols. Toumey Forest is classified as being well managed

TABLE 4.2

The indicators and results of good forest management at Yale-Toumey Forest.

- More inventory information than ever before
- No areas of significant soil erosion or compaction
- No loss of wetlands
- All wetlands managed separately, including buffers
- More species diversity and diversity of stand age classes than ever
- No significant insect or disease problems
- Active recreational use by neighbors for biking, hiking, and mushroom collecting
- No citations or expressed concerns from local conservation commissions
- Consistent production of high-quality sawtimber

because of the indicators shown in Table 4.2. Toumey Forest uses an ecological/ecosystem perspective in timber production and economic factors do not drive management. However, because many of the options for management are in the heads of the forest managers rather than in written documentation, the management approach could not be verified. Even though this forest has been — and is being — well managed, the required indicators were unable to verify the ecological approach to management for most of the certification approaches.

Interestingly, three individual teams did determine that Toumey Forest was or was close to being certified (see 4.3.3). This ability to classify Toumey as certifiable was based mainly on the ability to assess that this forest followed the principles of good forest management. This conclusion was reached by some protocols despite the lack of the other documented data listed above for this forest. What this result shows is that most of the data listed as missing do not assess whether good forest management is being practiced. This also suggests that if good forest management were the goal of certification, less information would be required as part of the assessment. Lower data requirements would reflect the decreased priority given to achieving all the values from a forestland.

The criteria for evaluating a sustainable management plan for a forest similar to Toumey was challenging. The criteria required explicit documentation regarding harvesting practice, forest/wildlife health, and finances. The management plan for Toumey Forest was quite general and lacked statistical documentation and the specific guidelines which were required by the assessment tool. Due to a dearth of quantitative information, Toumey could not be certified. The managers of the forest contend that the plan is designed to be somewhat indefinite to allow it flexibility as a research forest. The Toumey Forest is unique in that its primary role is to provide an area for research and study which allows experimentation that may include unorthodox methods of silvicultural practices and management strategies. The forest is managed under the general philosophy of sustainable management. However, this forest will not meet certification criteria until its managers provide detailed information and guidelines regarding harvest regulation; pest and pathogen management; statistical data availability; wildlife inventories; and pesticide usage.

This case study assessment of Toumey Forest by the students has produced several questions that need to be examined in order to explain the mixed results produced. This is not simply a case of identifying those protocols which satisfy one's values for a forest. It is important to determine how the structure of each protocol is unique. Should a generic protocol or consistent process be implemented across all types of ownership and different goals for that land ownership? Another relevant question is: what should determine the structure of any certification protocol and how much should our values determine the type of approach used in a certification protocol? It is questionable whether all our values should be satisfied as part of a certification (see 3.2.2, 7.2, and 7.3). The other question is:

how do we know what data should be required as part of assessment, and is there a road map that allows one to filter through all that information? How does one identify the indicators that will effectively assess social and natural science integration and the political variables affecting certification? How are the social and natural sciences linked in these protocols and do they need to be linked?

4.3.5   Amount and types of data needed for certification

4.3.5.1   *Generality and uniformity of data*

Smartwood stated that there is a need to have universally consistent standards in order for certification to be successful and as a means of minimizing liability (Dovonan 1998). Uniformity in the type of data to be collected sounds logical on initial analysis. However, the high data requirements of most certification protocols are a result of this need to collect the same data for each site. Instead of identifying the most important information for each site, the development of a generic set of data means that more information needs to be collected to ensure that all the potential variables are included. This need for so much data creates problems in determining how to weight the importance of the information collected during the certification process. It also masks the "real" driving variables that control the sustainability of a site in response to different management activities. The driving variables are obscured because they are embedded in a matrix of data that is not as critical in predicting how the site will respond. This may produce a situation where the appropriate indicators are not used to assess the site. Therefore, the certification process will be unable to really state what the impacts of disturbances are on that site. The real driving variables end up being a minor part of the analysis and do not drive the decision-making process. Ultimately, the results of any analysis of a management unit should be suspect. The opposite result may have occurred if the real driving variables were used to assess the site. This is where a framework similar to the Legacy Framework approach would be extremely useful as part of certification (Vogt et al. 1999). Such a framework would help to identify relevant indicators from the myriad indicators that could exist for a system.

Although the different protocols appeared to be relatively similar to one another, differences did exist. For example, the Smartwood protocol was the only one that spoke of certification as an educational tool. The CSA appeared to be the most cumbersome because of the amount of information required and the number of steps involved. Of all the protocols, the CSA approach would create the greatest burden for any landowner attempting to satisfy the requirements of certification. Success at passing the CSA certification was not dependent on how well-managed a forest was but on how well-documented the owner was. Much of the information requested by CSA would not be made available by a big timber company since they would not want this information to be shared with their competitors. At the same time, the CSA was the best protocol for minimizing the value of the certifier, but the amount of public involvement required was too much.

The amount of information required by some of the protocols (e.g., CSA) is probably not available for any managed forest. Even if forest management plans were available, they would not have all the data needs identified with this certification protocol. The required data is more likely to be found in a biological research station than in a working forest. The biological requirements are so strict that it will be extremely difficult to ever certify a working forest. Some of the data requirements for certification are so high that acquiring sufficient information may necessitate long-term ecological monitoring and storage of data in a regional data bank. This raises the issue of the type of information that is needed and when too much information is being required.

Even though it is perceived that credibility is achieved by designing assessment systems with uniform data, the system of indicators needs to be more flexible to deal with the unique characteristics of each site being certified. A common set of indicators will not be able to detect the changes across all ecosystems. The FSC-accredited certifiers have attempted to deal with this issue by developing both a regional and a local set of standards. The idea of having two sets of standards to adhere to is good, and will help to alleviate the problems that regional standards produce. However, although local conditions are taken into consideration using this dual approach, the choices of indicators are similar and are not weighted to select those which best reflect the site. When no weighting of indicators occurs, the values of the organizers (see 3.2.2.2) will probably dictate indicator selection, so that an unbiased analysis of site sustainability will not be reached.

Part of the inability to certify Toumey Forest using several of the protocols was due to generality of the criteria. Many of the protocols had criteria that were too general to be objectively evaluated in practice. The criteria sets could be vastly improved by the incorporation of more specific "legal" language. In many cases, no guidance was provided on how to conduct measurements, quantify values, find representative samples, or make concrete evaluations based on general criteria.

> Because of their generality, implementation of any of the sets would lead to subjective judgments resulting in the undesirable phenomenon in which different investigators reach different conclusions using the same data. This is exactly what happened with the Toumey Forest case study.

In order for these certification sets to be viable, they should be made much more specific, incorporating quantitative thresholds where feasible. There should be more attempts made to incorporate aspects of certification programs developed for industry, which include requirements to document the reasons why certain indicators were chosen and the thresholds that were used (see 2.4.2). In the protocols, even the references to regulatory compliance were difficult to verify in practice. It may be better to specify the lack of any enforcement actions (any actions brought against the company) because that will tell if the individual or company is complying with the regulations.

For several of the protocols (e.g., SFI, SFF, and Acadian/FSC), many of the criteria were not explicitly documented.

> This lack of documentation can result in what might be called a "Technical Failure" to pass the stated criteria, although the criteria may be well implemented on the ground.

For example, the Acadian/FSC Principle 7 required several aspects to be included in a management plan. Since Toumey has no formal management plan, this section received a failed score during the analysis. This failure occurred despite the fact that Toumey Forest is well managed and follows many of the criteria stated in Principle 7. In another example, the SFF protocol calls for a 150-year management plan under several criteria sections. Although Toumey does not possess a long-term, written management plan, the implied management regime implemented on the ground at Toumey Forest is robust and long-term in nature. In fact, the usefulness of using a 150-year management plan is questionable. Even if a 150-year management plan is developed by landowners, the time frame forecasts so far into the future that it does not mean much on the ground. Any forecast that reaches beyond a one-rotation cycle is impossible to assess because of our poor ability to predict at these time scales.

4.3.5.2 *Weighting of data*

The inability to weight the importance of the different data needs (e.g., treating most data on an equivalent basis) and the inability to exclude data not relevant for a site or region produces problems for a landowner pursuing certification. Because of the requirement for a large amount of equivalently weighted data, there is a strong likelihood that a forestland owner cannot satisfactorily meet all the data requirements, and therefore will not be certified. In general, the guidelines encourage subjectivity because the criteria, as detailed as it might be, does not give a specific system for assigning grades that uses objective measures. Indeed, some protocols (e.g., SFI, Acadian/FSC, etc.) had no grading system at all. The different protocols had varying approaches for ranking the information collected (see 4.2.1).

When scoring within the SFI, no weighting system was given. The ranking system of SFI protocol had a scaling of points that varied from 1 to 5, with a *pass* being greater than 3. During one team's analysis of Toumey Forest using the SFI protocol, 4 out of the 12 principle objectives (3, 5, 6, 9) were ranked as *passed*, 3 out of 12 were ranked as being *neutral* (4, 7, 8), and 5 out of 12 as *failed* (1, 2, 10, 11, 12). For SFI, the overall ranking was *failed*, since only 73 out of 135 points were achieved. This overall ranking resulted in an average grade of 2.7 which is below the acceptable ranking of higher than 3. As part of another team's assessment of Toumey Forest using SFI, 7 criteria were ranked clearly as *pass*; 6 as probably fulfilling the criteria, but better verification was desirable; 6 had insufficient information, or unclear criteria prevented evaluation; and 8 instances where Toumey violated the criteria. If one summed up the potential ranking from this last analysis, 13 criteria were satisfied and 14 criteria could not be satisfied for whatever reason. The outcome of this was that 48% of the criteria could not be satisfied so, therefore, Toumey Forest was not certifiable. In this example, 12 out of the total 27 criteria were ranked by the student evaluators as unclear or requiring greater verification.

The Acadian/FSC protocol also had a ranking system from 1 to 5, with a *pass* ranking required to be greater than 3. In one team's analysis of Toumey using the Acadian/FSC, 5 out of the 10 principle objectives were ranked as *passed* (1, 2, 5, 6, 9), 2 out of the 10 were ranked as *neutral* (3, 10), and 2 out of the 10 *failed* during the grading (4, 7). The summation of this analysis resulted in an overall *failed* ranking, since 688 out of a possible 1185 points produced an average ranking of 2.9 (again below the >3 required for a *pass*).

The SFF protocol seemed to be the most versatile because it provided a grading system (3 to –3) that required all sections to have a positive value across all criteria (see 4.2.2). It also emphasized the fact that the evaluation teams had the authority to modify standards to suit specific characteristics of the sites. With the SFF protocol, the grading criteria ranked from 3 to –3. In one team's analysis of Toumey, 11 out of 16 principle objectives were *passed*, 1 out of 16 was *neutral*, and 3 out of 16 were considered *non-applicable*. The result of this analysis was an overall *pass*. This protocol allowed the right to accept the implied management structure at Toumey and Yale in lieu of a written management plan. This flexibility is why Toumey was deemed certifiable using this protocol.

The SCS protocol developed a procedure where each category was worth 100 points, and these categories were weighted to calculate a total score for each section (see 4.2.2). It was assumed that the evaluation team determined the normalized weight for each category. The 2 to 5 criteria under each category were also given equal weight (based on 100 points), depending on the number of criteria. (This is not really weighting the quality of the different information being requested as part of this protocol). Full credit is given when a criterion is explicitly addressed in the management plan and is implemented.

Points are subtracted from the score when ambiguity is observed in the management plan, when there is an absence of statistical data when required, and there is complete absence of information on a set of criteria. Although the procedure outlined grading thresholds and optimum management strategies, the grading system was felt to be subjective and inconsistent.

In the SCS protocol, the following information resulted in Toumey Forest not being certified by one of the student teams. In the Timber Resource Sustainability section, harvest regulation (there was no specific regulation strategy, no rotation and target volumes existed and no long-term strategic planning for harvesting) failed to be satisfied by the weighting system designed for this protocol. In the Forest Ecosystem criteria, using the same protocol, Toumey failed by not having wildlife management actions, strategies, or programs. In the Financial and Socio-economic Consideration criteria, Toumey failed 3 out of the 5 types of criteria. It failed because it was unable to show financial stability, community/public involvement as part of its activities, nor enough evidence that management was investing in capital and personnel.

At the outset, SCS appeared robust because of the incorporation of a mathematical scoring system whereby criteria are weighted differently. However, despite the appearance of quantitative rigor, the scoring system failed because there are no instructions on how to score each criterion (and sub-criterion), or to judge the final, weighted average score. The types and quantities of information required to satisfy the criteria also determine how successful a certification will be. For example, in the SCS protocol, more information was requested in the Timber Resource Sustainabilty and Forest Ecosystem criteria (the sections that Toumey Forest was less able to satisfy) than in the Financial and Socio-economic sections.

In addition to understanding which data caused Toumey not to be certified, it is worthwhile to know whether some data should have been excluded from the analysis and whether other data could have satisfied the data requirements during certification. For Toumey Forest, some data requirements needed to be excluded from the assessments because of their irrelevance. Some of the irrelevant data included issues with indigenous peoples, employee relations, processing guidelines, and many of the plantation issues — especially those related to the introduction of non-native species. However, since they were part of the evaluation and scoring system, it was not an option to eliminate these information needs.

Toumey did have certain information that was quite useful for the certification process, although it did not satisfy the data needs of most of the protocols. For example, good records existed for all of the timber sales that have taken place on the Toumey Forest since the early 1900s. Because of its role as a teaching and research forest, Toumey had a good compilation of every management issue related to all of Yale's forests. In addition, past timber-sales contracts were a good source of written evidence demonstrating that care was taken to maintain the diversity and structure of the forest during harvesting procedures (i.e., skid trails were laid out, dates were set so harvesting only took place in the winter, etc.).

Bruce Larson, the director of Yale School Forests, stated that the wrong type of documentation was requested for the certification of Toumey (Larson, personal communication). For example, there is no study of organizational behavior in the certification protocols. Most protocols are incapable of dealing with the fact that the school is buying badly managed land to upgrade this land. Many questions were raised related to public participation in management at Toumey Forest; the public has been actively involved in management issues at Toumey Forest (i.e., issues related to deer hunting), but these are not documented. Because the protocols explicitly define the types of data to be collected, the student teams did not pursue other information which could have verified some of the information needs.

The types of data needed to assess a forest will also change depending upon the goal of certification. Depending on whether the goal is sustainable systems or good forest management, the data requirements need to be adjusted to achieve each one. Certification must consider the data demands for each protocol and whether they are realistic or unrealistic in terms of the landowner. It is easier to demand that landowners satisfy the following criteria: conserve biological diversity; maintain and enhance forest ecosystem condition and productivity; maintain a landscape that is resistant to pests and pathogens; and conserve soil and water resources. However, it is not realistic to ask landowners to contribute to ameliorating degradation of the global ecological cycles; to provide multiple social benefits; to alleviate poverty; and to satisfy society's responsibilities for sustainable development. This raises the question of who should be responsible for — or pay for — additional benefits that forestland owners are being asked to provide.

In several places throughout the protocol, the high data needs also resulted in inconsistencies and reiteration of some indicators. Many criteria appear to conflict with one another and cannot be satisfied at the same time. The following examples are from the Acadian/FSC-initiated regional standards. For example, Principle 6.3 requires at least 25% of the pre-logging number of snags to be retained, but then states that at least 50% of large-diameter snags are to be retained. Principle 8.3 of the same protocol requires the selection of cut trees to maintain the natural biodiversity of the stand. However, at the same time, it requires improvements in tree quality that would require planting selected tree species, since the natural biodiversity may not have these qualities. However, natural regeneration is preferred over planting (9.1), and no exotic or genetically engineered species stock are to be planted (9.23). Therefore, neither requirement can be satisfied. In another example, Principle 10.4 states that salvage logging should remove less than 50% of the fallen trees and less than 50% of the snags remaining after the disturbance; however, Principle 15.1 states that product wastage is to be minimized.

4.3.5.3 *Value-driven indicators and desired standard forest condition*

Some of the protocols strongly express the values of the certifying organizations and are integrated into the criteria/indicators (see 3.2.2.2). Attempting to satisfy these values requires the collection of so much data that the protocols were judged by the student evaluators to be cumbersome. Interestingly, no supporting evidence justifying the information requirements has been articulated by most approaches (the exception being SFF). Some of the information needs appear to be based on collecting information on all components of an ecosystem, as if Toumey Forest were a research site with significant financial resources. For example, in the CSA protocol, information has to be collected on water quality, trends in stream-flow events, distribution, and abundance of aquatic fauna. Collection of this information is a typical graduate-student project that can take two or more years of research to accomplish. This information need requires a landowner to hire someone with the technical skills to conduct these types of measurements — and to pay for them.

It is questionable whether private forestland owners should be expected to acquire all the information currently articulated in the protocols. For example, CSA requires a landowner to collect information on the impact of management on the global carbon cycle. This requirement does not appear to be appropriate for assessing whether a forest is well managed. It would be nice to have that information, but it is unrealistic to ask a landowner to provide this type of information. The landowner having to consider how their management unit contributes to GDP and to trends in the global market share (e.g., CSA), says nothing about how well managed or sustainable a particular forest is. In many cases, information on forest communities and their sustainability (e.g., CSA) is an issue that needs to be addressed for public lands, but may not be relevant for small or large private landowners.

CSA also requires that operators document information on the economic value of non-market goods and services collected from a particular forest. This practice is questionable if a private landowner does not harvest non-timber products. However, if the landowners harvest non-timber products from their lands, then they need to address how their management activities affect forest sustainability. If the public collects non-timber forest products (as occurs at Toumey Forest) from private lands, then private landowners should be given financial assistance in obtaining information related to the sustainability of non-timber forest products.

Most protocols emphasize the protection of biodiversity. However, managing for biodiversity can be a subjective process, since no clear procedure exists to determine what level of biodiversity is appropriate for a given site (see 4.3). If a forest is being managed for a few commercial species that require suppression of non-commercial species (e.g., hardwoods at Yale Toumey), should certification protocols override the management considerations? These requirements raise the question of who should decide how many species are enough.

The expectations for certified landowners are not always clear, and sometimes go against the rights of private landowners and question the role of the public when they have no tenure rights. For example, who should decide what is culturally appropriate? What if local opposition is unjustified, but their values have to be incorporated into the protocol (e.g., Woodmark)? Who will determine when one set of values should override other values? When the lands are in private hands (as in the case of small, non-industrial landowners), do local people have a say on management of that land, even when they have no tenure rights? Some certification protocols appear to impose economic constraints on the landowner (e.g., Woodmark) which can be a valid approach, but will depend on the site, the landowner, and the size of the property.

Some of the criteria are inappropriate, costly, or cannot be achieved. For example, the Acadian/FSC (Principle 6.3.c.4) asks operators to manage for a "pre-colonization forest condition." This is a value defining what type of forest condition is desirable for a managed forest. Implementing that value, however, is not easy, since we really do not know what a pre-colonial forest condition is. This is also an inappropriate criterion, as it does not take into consideration the changing mosaic of the region over time. Furthermore, it would be prohibitively expensive to determine. The SFF protocol requires that the "present condition is compared with natural ranges of variability." This places no time frame on the natural range of variability. One could argue that a natural range of variability includes temporary glaciating of a site every few thousand years. Other examples of this include the many criteria which ask for inventories of wildlife and endangered species. The SFF protocol requires that "rare, threatened, and endangered species are inventoried." This is not a landowner's responsibility. The cost of acquiring this data, especially on small holdings, would be prohibitive to certification.

Since most of the protocols are continuously evolving, the forest industry has expressed concern that even if their operations were certified, conditions will continue to be imposed on them by the certifiers. This situation is plausible, since a typical contract between the certifier and the landowner is for a fixed period, but annual audits are required. For example, Smartwood has a five-year contract with the certifying organization, but requires annual audits during this period. There is the potential for more data or for the satisfaction of new requirements which were not imposed during the first audit. In fact, Smartwood has stated that they do not guarantee an assessment (Donovan 1998) because other information may become known that may negate the results of their earlier analysis. Since some companies have had their certification withdrawn (see 2.5), this concern has some merit for individuals being certified.

For most landowners, it will be unclear which values are being introduced into the certification process by the certifying organizations themselves. In some cases, the vagueness of criteria and the poorly defined message that certification gives may result in public confusion about certification. The certifier may also be placed at risk of litigation if there appears to be no objective reason why one forest has been certified and another has not. Considering the globalization of business today, the lack of information about the certifying body is also a problem. It is not apparent how the criteria that Toumey needed to satisfy would be used to evaluate forests in other regions (on a local, national, or international scale) or how applicable the certification of a local Connecticut forest would be internationally.

It is obvious that certification can be equated with ecosystem management. The large volume of data needed for any certification suggests that certifiers are attempting to structure their protocols based on an ecosystem perspective. All the variables are included that can be identified as critical to understanding an ecosystem — it is a good list of variables. This is laudable and shows that certifying organizations are pushing a holistic approach. Several questions are then raised: is this the way to force landowners to utilize the ecosystem approach in the management of their land base? Are there other ways to assess the whole health of an ecosystem? What are the most appropriate tools for landowners to use? Does this mean that certification is relevant for public lands only because of the requirement to satisfy all the other values that society has imposed on landowners?

Are there other types of certification approaches that need to become part of our suite of tools for assessing landowner systems? A small landowner will not be able to satisfy the societal value for supporting all the goods and services that can come off the land or are particular to the land. This suggests that several different types of certification protocols are necessary in order to satisfy the different values that may exist on any landscape. Many of our environmental problems have led managers to recognize the value of a holistic approach to natural-resource management. A holistic approach does force individuals to consider the hidden factors that may affect the sustainability of their land — which a focus on simply the resource would not have. However, there is a need to be able to select which parameters are relevant to any site instead of requiring a large amount of uniform data that includes all possible variables for a generic ecosystem.

4.3.5.4   *Social integration in protocols*

Social information is required in all of the eight protocols examined for the case study. However, none of the protocols link social and natural science information (see 3.2.2, 4.2.3, 7.2) so a change in one would result in a change in how the other would respond (Boxes 4.1–4.8). Much of the social science information requested by the protocols consists of public participation in the process, using the public to identify local values, the goals of the certifying organizations and the desired forest condition.

The information required in the social sciences was typically a small part of each protocol's data needs (see Boxes 4.1–4.8). A summary of what proportion of the indicators required social science information (e.g., social and economic) follows:

- SFI: 14 out of 28 indicators were social science-based (50% of the total)
- Smartwood: 17 out of 92 indicators were social science-based (18.5% of total)
- SCS: 30 out of 108 indicators were social science-based (27.8% of total)
- SFF: 14 out of 149 indicators were social science-based (9% of total)

- Woodmark: 27 out of 107 indicators were social science-based (25% of total)
- Acadian/FSC: 85 out of 242 indicators were social science-based (35% of total)
- CSA: 35 out of 83 indicators were social science-based (42% of total)

Of course, no assessment should be approached with the idea of having an equivalent amount of information requested in the social and natural sciences. This type of approach ignores the fact that the relative importance of each will vary based on the ecosystem being assessed, and that there should be a functional rationale for the required data. The choice of whether social or natural science indicators are used in an assessment must be determined for each specific site. They should be selected based on which factors are driving the functioning of that system and what types of natural and social science-based legacies exist on the site (see 5.2).

> The important take-home message from this discussion is that social-science data needs are not integrated with the natural sciences, and much of the social-science information is treated as "human values."

Most of the protocols ask for social and natural science data as separate pieces of information. Many of these social variables reflect the fact that these protocols were initially proposed as a means to improve the livelihood and economic potential of indigenous people living in close association with the forest resources in the tropics (see 2.1). Certification was, therefore, a useful mechanism for society to stop unacceptable forest-harvesting operations and to improve conditions for people dependent on these resources for their livelihood. Therefore, the social and natural science variables are only linked when information is requested on both topics within one protocol. The different types of data do not drive one another or feedback to affect how they function within the ecosystem. For example, social factors are not measured as to how they modify the expression of the biological system. This in itself highlights how much of the social information is added — not as an attribute of the system — but actually to express other values that are considered desirable. If social factors are not linked to affecting the sustainability of the natural system, then including this type of information could be considered unnecessary for assessing whether sustainable management of the forests is occurring. If this is an end product of the certification protocol, then it may only be appropriate under certain circumstances (i.e., when people are intimately living with their resources or on publicly owned land).

Currently, most of the social and natural science links occur at levels of economics where information about the resources feeds into an economic model. A focus on economics still does not deal with the fact that other aspects of the social sciences can play a very important role in controlling the sustainability of natural resources. Public participation and values have strong controls on how natural resources are managed, as can be seen by the conflicts that occurred in the Pacific Northwest's old-growth forests (see 3.2.2.1).

4.3.6   Certification of large vs. small landowners

The certifying organizations, in most cases, realize that their protocols will not be applicable to small landholdings. For example, Smartwood states that a typical area that they certify ranges from 5,000 to 15,000 acres (Donovan 1998), which eliminates a sizable portion of the small landowners who own most of the forestlands in the United States (see 2.5). Certification should not systematically ignore a significant proportion of the forestland owners who own working forests. Part of the rationale for not certifying smaller landholdings is that the "chain of custody" cannot be tracked for small landholdings (see 7.7).

Even Toumey Forest approaches the size that the FSC-type of certification does not consider when certifying "working forests." Smaller tracts of land have been certified by the FSC-type certification (see 2.5); however, these were not "working forests" but privately owned reserves or parks.

This poses the question of whether Toumey should be used as a model for designing a protocol that will work for smaller landowners. Toumey Forest has characteristics that make it more similar to small forest landowners than to the larger landowners. Since much of the forestland is owned by small landowners in many parts of the world (see 7.4), protocols need to be adapted to allow small landowners practicing good forest management to pursue certification. In designing such a protocol, it would be interesting to tease apart the elements of the protocols that certified — or almost certified — Toumey Forest.

In comparing certification protocols which appear to be useful for small land holdings, it is apparent that some are impractical for use by small landowners. For example, SFI Criterion 1b had the following requirement, "Provide funding for forest research to improve the health, productivity, and management of all forests." It is unrealistic for a small landowner to support research related to the health of their land holdings. This only pertains to large industrial and non-industrial owners who have the capacity and have supported research on their land holdings for a long time. Also, this criterion does not allow a landowner to satisfy this requirement by utilizing research conducted within the region. If the size of the land is too small, the ability to conduct research on it is severely limited. It is unclear whether Yale-supported research conducted at other sites would satisfy this criterion and the other "research" criteria for SFI. In addition, Criterion 2a of SFI requires reforestation plans to occur within five years after the final harvest which is difficult for small landowners to accomplish. At Yale Toumey, even though there was no way to verify that regeneration occurred within five years, good natural regeneration was observed on all cut-over sites visited and there was no history of poor regeneration. In this case, a plan is required to satisfy this requirement of certification.

Although Toumey did not have a management plan, many small landowners meet or exceed the BMPs and are in compliance with all applicable water-quality laws and regulations and the requirements of the Clean Water Act. Toumey Forest exceeded the requirements of BMPs and complied with all the laws. This verification was based on verbal statements from the managers of this forest, and was partially verified by examining the conditions listed in timber sale contracts and observations made by the evaluators of harvested sites in the field. Other means of verifying that good forest management is being practiced do not exist and probably will not be something that small landowners will be able to produce. If the BMPs and laws are being complied with, should this be sufficient for certification of small landowners?

Another requirement that is extremely difficult for most small landowners to fulfill is the need to be involved in disseminating information and using panels of experts. For example, the requirement for using a panel of experts at the state level to help identify the goals for riparian protection of all perennial streams and lakes is a difficult criterion to fulfill (see Criterion 3b of SFI). Although riparian protection is a high priority on Yale lands, by not having a panel of state-level experts means that Yale has failed to satisfy this requirement. This conclusion occurred although other indicators of riparian-zone protection existed at Toumey Forest. For example, Yale has guidelines on riparian-oriented BMPs and the director of school forests directs supervision of skid-trail layout when trees are to be harvested. Other criteria (3c and 4b of SFI) are also unrealistic. Criterion 3c requires the landowner to provide funding for water-quality research, while Criterion 4b requires the provision of funding for wildlife research. Another criterion (4a of SFI) defines the need for landowners to define their policy, program, and plans to promote habitat diversity. This criterion was satisfied at Toumey in several ways and indicated that wildlife issues were factored significantly into management decisions. For example, the management plan lists the need for different habitats for wildlife studies as a reason for managing the forest for a variety of stand conditions; informal policies to leave snags after cuts; and the decision to preserve, for the time being, a white pine stand as a wildlife habitat. The ability to satisfy some of the suggested support of research on the landowner's property is satisfied at Toumey because it is a research forest. However, these criteria are more difficult to fulfill by small landowners.

Some of the protocols require landowner involvement in forming or interfacing with different organizations involved in information transfers. These criteria are not relevant to assessing whether sustainable logging or sustainable forestry is occurring on a given piece of land, but are striving to change how the entire profession practices forestry. Whether these criteria should be embedded in a framework evaluating a given management unit is questionable. Should certification strive to change the whole profession of forestry and are these values appropriate? In most cases, there is no supporting documentation given to substantiate the values which are embedded in the protocols by certifying organizations.

For example, SFI has at least seven criteria (Criteria 10b, 10e, 11b, 11c, 12a, 12b, 12c) that are involved with the requirements to establish, interface, report, or perform extension functions for other groups. These criteria are used as part of the evaluation process to determine whether a landowner can satisfy the certification requirements. These requirements comprise 7 out of the 27 criteria (26% of the criteria) for which information must be included as part of the evaluation. One criterion requires landowners to work with or establish (when nonexistent) state groups comprised of logging and state forestry associations and other agencies in the forestry community to increase the professionalism of logging by focusing on the education and training of loggers. Another criterion calls for supporting and promoting efforts by consulting foresters, state and federal agencies, state groups, and programs like the American Tree Farm System to educate and assist non-industrial landowners in applying principles of sustainable forest management on their lands. Another criterion requires supporting and promoting appropriate mechanisms for public outreach, education, and involvement related to forest management. Another of these criteria requires the establishment of a national forum of loggers, landowners, and senior industry representatives to meet at least twice annually to review progress towards the Forestry Principles and Implementation Guidelines.

The SFI would have to be modified so that criteria are not weighted and used to determine if a landowner is certifiable by relying on information particular to large, complex industrial organizations managing forestlands. For example, Criterion 10d calls for Sustainable Forestry Principles to be communicated — at all levels of a company — to mill and woodland managers. This is not relevant for a small landowner. Criterion 10f, which calls for ensuring that a mill's inventories and procurement practices do not compromise its adherence to the Principles of Sustainable Forestry, is not relevant for small landowners.

What type of system would be an effective assessment protocol? Criteria must be chosen so that they are relatively few, and are accompanied by an explicit grading system that leaves little room for subjectivity. Furthermore, the evaluation criteria must be region-specific. In some protocols (e.g., SFI, SFF, and Acadian/FSC) there existed multiple criteria which had little or no relevance to New England forests. It is important to delete the ambiguities which, barring interpretation on the part of the certification team, are crippling to the evaluation process. Many times, the wording of a specific criterion yielded a negative score for the landowner despite the fact that they might have satisfied the requirement. Many of the criteria seemed to be conflicting or counter-intuitive. Some criteria required that the forest be encouraged to maintain full functioning, or productivity, which would seemingly require planting or a change in species composition. Another criterion called for maximum utilization of salvaged materials, while another required that 50% of salvageable materials be left on site.

Protocols must be structured for the assessment goals instead of using one protocol for all systems. In some cases, the protocol should be used only when certifying government-held lands. For example, the CSA protocol requires that prior to certification the "registration applicant has all components of an SFM (Sustainable Forest Management) system in place and function." Thus, a landowner must commit many hours to gathering scientific and historical data, and prepare a comprehensive report demonstrating that appropriate components are "in place and functional" prior to being audited. One glance at the criteria and indicators requirements is enough to dissuade many landowners from even trying to certify their land using this method. In addition, the values incorporated into the CSA protocol address various spatial and temporal scales, from local community to global issues. They also address a broad range of scientific and historical criteria (including ecosystem integrity and health; social impacts and context; bio-diversity issues; long-term management strategies; conservation of soils and hydrologic cycles; and public participation). This system does appear to be targeting lands held by the government because of its reference to national and international economies, and its commitment to sustaining global ecological cycles by evaluating forest-sector policies.

It is important to discuss whether different types of certification protocols should exist. The demand for commitment to global ecological cycles that exists in the CSA protocol is unrealistic to require from either a large or small landowner. However, it is not unrealistic to require this commitment on federally owned land where the lands are held in trust for the public. In this case, the audience is probably quite clear. The CSA system appears to be overbearing because of the many pages devoted to the "process" and "documentation" aspects of successful certification. The detailed and rigorous procedural reporting requirements create hurdles that an individual landowner cannot likely overcome. The sustainable management system is cumbersome. Not only should questions be asked concerning who the audience is for certification, but also how the certification protocols are driven by different values. These values determine which protocols should be used for different situations. The values of the certifying agency should be clear.

5

Indicators Relevant for Inclusion in Assessments: Types, Minimum Number, and Those Derived from Non-Human Values

From an analysis of how successfully the certification protocols were able to assess Toumey Forest, the factor that appears to contribute the most to unsuccessful certification is the inability to satisfy many of the indicators used in the assessments (see 4.3.3). Since many indicators are value laden, certifiers may be unable to evaluate a site if the non-value-laden social and/or natural science constraints are not integrated into an analysis of system health. Satisfying value-based indicators does not guarantee that a landowner's forest ecosystem will not degrade. It is recommended that the reader see Chapter 3 for a discussion of the values that are currently integral to certification protocols.

This chapter will analyze the approaches for indicator selection that will move the analysis from being predominantly value based to identifying other indicators that reflect social and natural science constraints which control the functioning of the ecosystem. To effectively evaluate management impacts, there is a need to determine the minimum information necessary for any site. This can be accomplished by moving away from using just the value-based indicators. However, determining the relevance of value-based vs. non-value-based indicators is not easy if there is no good framework to facilitate this process. Simply combining value- and non-value-based indicators as part of the analysis will not allow for a better analysis of the system. Including all indicators will only encumber the analysis process if there is no mechanism for identifying which indicators are relevant to each site. The next section presents our framework for determining the minimum information that is needed in any site to evaluate its management and also how to select those indicators which are most sensitive for that site.

The indicator selection section will be followed by a presentation of other legacies that can affect how a system functions. These legacies are social and natural science-based. It is important that an evaluator determines whether these legacies exist in the site. If these legacies are relevant for a site, indicators reflecting these conditions must to be included in any evaluation. The legacies presented in the second part of this chapter have not been integrated into most of the certification protocols.

Since the idea behind certification is to assess the impact of management, it is important to know what the impacts are of different management activities on different parts of the ecosystem. Understanding where the impacts of management activities can be detected in an ecosystem will help to identify indicators for specific management activities. In Chapter 6, the indicators that have been assessed as part of specific management activities will be presented. Since the purpose of certification is to assess the impacts of management in forest ecosystems, indicators identified in certification should be similar to those identified as part of disciplinary-based research studying management impacts.

5.1   Indicators selection criteria

Kristiina A. Vogt, Daniel J. Vogt, Anna Fanzeres, Bruce C. Larson, Peter A. Palmiotto

5.1.1   Rationales for indicator selections

Much of the effort by the certification community has been put towards the better definition, development, and application of the criteria and indicators used to assess a management unit (Wijewardana et al. 1997). This has resulted in a drive to better define criteria and indicators, and to come up with a sufficiently large number of variables that can be used to assess any site. Wijewardana et al. (1997) summarized many of the regional and international activities that have occurred to define criteria and indicators for sustainable forest management in tropical, temperate, and boreal forests. Over one-hundred countries are involved in the development and application of criteria and indicators. The philosophy driving this work has been to develop tools to assess national trends in forest conditions and their management, which integrate their environmental and socio-economic benefits to society (see 2.1, 2.2). The goal of these activities is to use indicators to provide a common framework for improving national-level policies on sustainable forest management. Therefore, indicators reflect national trends in managing forests (Wijewardana et al. 1997). Much of this discussion has also implicitly defined what indicators should mean at the ground level. This has occurred, although Wijewardana et al. (1997) warned that they "cannot be applied wholesale to other levels of assessment" and be assumed to make sense at the level of the management unit.

The format of many of the national and international criteria and indicator developmental efforts have been at the level of identifying and listing all indicators considered relevant to assess a site.

> This type of approach is extremely useful at one level, but creates other problems because no consistent framework has been developed to identify which indicators are relevant to use at the management-unit level. Similar problems have plagued ecosystem management (Vogt et al. 1997a) where "laundry lists of variables" have been identified, but without a mechanism for weighting the relevance of the information to specific management units.

In ecosystem management, it is frequently heard was that it is impossible to implement ecosystem management because the data needs are too high (Vogt et al. 1997a). Many of the protocols analyzed in the Toumey case study (see Chapter 4) had such high data needs that they were judged to be cumbersome by the students. The students also concluded that most forestland owners would be discouraged from pursuing certification because of the significant need for different information.

Collecting large quantities of data, however, does not mean that a land manager is doing a commendable job of managing their forest in a highly sustainable manner. Most likely, a smaller and more focused database would be sufficient to show that land is being managed in a certifiable manner (Vogt et al. 1999ab) even though the certification assessment might not reach the same conclusion. In fact, by eliminating many indicators upon further testing of their protocols, the certifying organizations have recognized the inadequacies or weaknesses of some indicators. The Forest Stewardship Council has dropped about a third of the indicators originally identified as relevant in their analysis system (Ervin 1998). LaPointe (1998) suggested that several indicators could not be evaluated sufficiently, and that their use in assessments should be questioned. Jack (1998ab) stated that the AF&PA Sustainable Forestry Initiative was continuously improving the indicators which they had identified as weak.

Another problem with identifying all indicators which could potentially impact an ecosystem has been the desire by some to develop a global or similar set of indicators that could be used in any forested ecosystem. This approach towards identifying indicators ignores the changing driving variables controlling the functioning of a forest in different locations and the legacies of human impact or natural constraints that may exist in that same system (see 5.2). The similarity of criteria and indicators given by the different certification protocols (see 4.3) reflects this drive to identify a common set of indicators for all systems.

> This similarity in indicators for different assessment protocols should be expected when one realizes that they capture the human values that are desired out of the system (see 3.2.2). Since the social and ecologically based values desired by humans from natural systems do not vary much, similar indicators will be identified for use in most systems.

This inability to use a uniform set of indicators to sensitively assess a site is further discussed by Vogt et al. (1997a, 1999ab).

Most certifiers are presented with a cookbook approach to their evaluation — again, the idea being that this is the only way to have a credible system and to deal with the different technical backgrounds of the evaluators. However, it is time to recognize that the cookbook approach was a necessary step during the initial stages of developing the basic concepts of forest certification, but there is now a need to move beyond it. A cookbook approach to certification will not effectively assess whether good forest management is occurring. In addition, the cookbook approach drives an evaluator to use many indicators, even when they may not be relevant for the evaluation of a particular system. The approach of identifying generic indicators also results in the indicators not being very useful for different sizes of forest-management projects (see 7.4, 7.5).

The concept behind the development of a uniform and standardized set of indicators is fundamentally good. It seeks to provide a common reference point or threshold to which any given site can be compared in order to assess how far a system is from the selected condition. However, there has been great difficulty in identifying what a comparison site or forest condition should be for it to be classified as sustainable. Many consider a sustainable site to be one that has minimal human activity. In the United States, the precolonial forest has been identified as the forest condition to aim for by the FSC-type assessment (see 4.1.3.2, 4.3.5.3).

If it is not clear what forest condition should be managed for, an assessment becomes very subjective because there is no consensus on specific guidelines and indicators. Other certification protocols call for comparing the site to the natural variability existing within that forest type. This approach raises the question of what the natural variability is for that system. No matter what approach is used to identify what the "model" system should be, the standards must indicate whether a system is degrading or not (Vogt et al. 1999ab).

> It will be extremely difficult to assess sustainability if the evaluator does not understand how the choice of particular indicators will determine the outcome of an assessment.

There is a need for an evaluator to be able to weight the relevance of all of the information and utilize only those indicators pertinent for an assessment. This requires an ability to weight the importance of the all variables typically identified for a generic ecosystem and make them site specific.

Drescher (1998) suggested that simple indicators are needed for certification and there must be practical methods for their application at the site. He stated that indicators could be found: 1) through observations; 2) through controlled experiments; and 3) as common sense (the old foresters' approach). Because Drescher (1998) believed the focus should be on assessing ecosystem health, these indicators should reflect that ecological condition. He further stated that we needed "to lose the human filter that constrains us" in managing a forest and "do good forestry on the ground" (Drescher 1998). According to Drescher (1998), it is appropriate to consider the socio-economic factors and take care of the people closest to home once ecosystem health has been achieved. In addition, Drescher (1998) suggested that all the arguments for balancing the social and natural sciences in certification would disappear if the focus were on ecosystem health.

Any discussion of indicators has to determine what is considered "bad" forestry and what the indications are of system imbalance due to some management activity (see 3.1.3, 5.1). It would be an interesting comparison to determine how the information needs would vary for assessing ecosystems for their social and natural science sustainability and for good forest management. The types of information required by both approaches should vary, since satisfying the sustainability equation requires measuring more values that are socially based.

There is a need to identify which indicators should be used to measure social systems and how we can measure the resilience of these systems beyond satisfying values. If one examines the certification protocols (see 4.2.1), the types of information required to be collected for assessing social sustainability is very similar in all of them. Some of the protocols have a greater requirement for public participation in the process and for utilizing the public to generate the local values expected from the forestland. Much of the social-science information revolves around data which addresses our social values and public participation in that process. There is no question that local communities and stakeholders have to be involved in how resources are used (Merino 1996, Grove and Burch 1997). However, it is important to ask how much other stakeholders should drive the certification process and at what point they should participate in certification. Also, there is a need to ask whether the right type of social science information is being collected to assess social sustainability in certification. For example, certification assessments include only a small part of the information needed to understand the human ecosystem and the role of humans in their environment, as summarized by Burch (1992) and Machlis et al. (1997).

An examination of the types of indicators used to assess whether natural systems are degrading shows how: 1) the selected indicators are not mechanistically based; and 2) they do not reflect past indicators of ecosystem degradation. For example, measuring natural system sustainability emphasizes collecting information on species and using biodiversity indicators as one of the principle objectives of an evaluation. However, diversity is only one of the indicators which reflects the health or sustainability of an ecosystem. For example, the following indicators are a few that have been used to identify when an unmanaged system is degrading (Vogt et al. 1997a):

1. Increase in disease, parasitism, and insect/pest outbreaks

2. Decrease in symbiotic microbes, and increased dominance of symbionts less beneficial to plants

3. Decrease in species diversity

4. Decrease in net primary production or net ecosystem production

5. Increase in annual transfer of organic matter to the decomposition system

6. Increase in plant or community respiration
7. Increase in losses of nutrients limiting plant growth
8. Bottlenecks of nutrients in long-term storage pools

Arbitrarily using one of the eight indicators of ecosystem degradation to assess all ecosystems should be used with caution. For example, species diversity and ecosystem function or productivity is not a generality that can be applied to all systems. In fact, species diversity and ecosystem health are frequently not directly correlated to one another (Johnson et al. 1996, see 3.2.1). Therefore, the appropriate indicators of ecosystem sustainability may be missed as part of an assessment if the focus is predominantly on biodiversity. Since biodiversity is a value that we desire from a system, indicators reflecting this must be included in any assessment, but should not be the primary focus of data collection in most ecosystems. Interestingly, most of the criteria listed above are not included in certification protocols (see 4.2.1). Most of the certification protocols do mention insect and pest outbreaks, but do not articulate guidelines for determining whether the outbreak is the result of management or is a natural occurrence that should be allowed to run its course (Vogt et al. 1999ab).

It is also important to realize that the eight indicators of ecosystem degradation may also reflect secondary responses by the system and not be the primary causes of ecosystem degradation. For example, by the time pest and pathogens are observed in forest ecosystems, the systems have already been stressed by other factors (e.g., acid deposition) that have caused the system to degrade (see 5.2.1.4, Vogt et al. 1999ab). If an indicator reflects secondary effects occurring in the ecosystem, it will probably not be a parameter that can be monitored to indicate the real ecosystem condition.

Inclusion of any of the eight indicators of ecosystem degradation in certification will add an element of complexity to an assessment and will require different indicators to be monitored. Since many certification protocols are already cumbersome in their data needs, there will be resistance to including more variables for any data analysis. However, if there were a more selective process for weighting which data were collected during any assessment, there would be no need to collect additional data for all sites. Instead, initial analysis would identify whether existing indicators in the protocols and/or any of the eight signs of ecosystem degradation were relevant for a site followed by a focus on selecting which indicators best reflected those signs. Currently, the choice of indicators in certification protocols is not based on factors that are clearly linked to system degradation for a particular management unit. Instead, indicator selection is based on the types of information that it is believed can be easily collected. This working model results in systems being evaluated using indirect determinants of the condition of the system. For example, long-term inventories of tree growth are very useful in showing that the system is functioning as expected, but not in what might be causing changes in the measured inventory growth data.

However, much of the ecological data may require sophisticated equipment for monitoring indicators of system degradation that will not be available to most forestland owners. For example, much of the clearcutting issue would require measuring leaching losses of nutrients from the site and not just soil loss with erosion (see Chapter 6). It is unrealistic to ask landowners to collect nutrient leaching data for their forests. This is where the existence of regional databases on the natural system could be useful to broadly state whether nutrient losses may be a problem with particular management activities for that region (see 7.5.4). It is then possible to determine which indicators are relevant for addressing whether management practices can potentially cause degradation at the site.

5.1.2   Approach for determination of the minimum information needs of a certification protocol

Part of the rationale for having a large, uniform, standard set of minimum indicators is to systematize the assessments by certifiers who are coming into these evaluations with very different knowledge bases. At present, two groups of certifiers can assess one site and produce very different conclusions on the certifiability of that forestland (this still happens because of the subjectivity that exists in implementing indicators, see 4.2.3, 4.3.5). In addition, because certification evaluates sustainability and needs to be conducted at a holistic or ecosystem level, the amount of data needed to conduct an evaluation increases dramatically. A smaller number of indicators cannot be currently selected "a priori" which would effectively evaluate system sustainability. This happens because the tools to measure social and natural science sustainability are still being developed and consensus does not exist on which are currently the best indicators.

At present, it is easier to require consistent and uniform data collection so that evaluators with very different knowledge bases are less able to be subjective in their analyses. The potential fear of flexible protocols is understandable. The skill of the certifiers would then become even more important in controlling the success of an evaluation process. Therefore, there is probably a close link between the number of indicators required to be assessed and trying to maintain credibility in the final assessment. This creates problems, because when the data needs are high, the ability to weight the quality and importance of the different information becomes more difficult. In this case, an individual would need to have an ecosystem-based background so they could adjust the weighting of the information to determine what should be collected. If the evaluators did not have an ecosystem background, there would be a need to have an oversight review panel to justify the ranking. Therefore, when less information is required as part of an assessment, greater pressure is placed on having competent system-based certifiers.

Some of the certification protocols incorporate flexibility into the system so the certifiers in the field can make decisions on adjusting or modifying the standards. For example, the SFI system is quite concise, but is also quite flexible in allowing the certifier to adapt the system to the local conditions. However, other aspects of the SFI system were perceived to be weak since there were insufficient guidelines of the evaluation process (i.e., needing firmer words related to sensitive areas). The SFF system is also flexible in allowing the evaluators to modify the standards. By pursuing the development of regional standards, the FSC also recognized that different indicators would be needed in different regions. In the FSC approach, they recommend two levels of criteria — one that is general and applied to all sites, and another that is regionally specific. In the FSC, the evaluator should be able to override some of the standards by presenting the information to a committee. This flexibility, however, is not very useful unless the evaluators have effective systems for filtering through the myriad information they need to deal with in an assessment.

The large data need of certification also gives the impression that a forestland owner would not be certifiable. This is similar to court cases where the assumption is that the defendant is guilty. In court cases, more information is needed to prove that one is not guilty. This happens because the defendant needs more information to prove their innocence than their guilt.

A framework must be developed which allows an evaluator to determine what indicators are relevant for assessing the impacts of management on any particular forestland. The real condition of a site cannot be obtained by retaining all of the indicators given in the different protocols (see 4.2.1). A framework has to be developed for weighting and selecting appropriate indicators. The inability to select if few, sensitive indicators will result in the collection of a large amount of data where the critical driving variables reflecting

TABLE 5.1

Non-value-based legacies and driving variables potentially maintaining both the natural- and social-sciences functions in a wet tropical forest — in this case, the Brazilian Amazon.

Natural Science	Social Science
Nutrient-poor soils (i.e., low availability of P, K, N, Ca) diminishing options for plant growth on a site	Institutional structures managing or contributing to how natural resources are used (e.g., church, community, labor unions, government, international organizations)
Toxic trace elements in soil (e.g., Al)	Poor people with limited livelihood options
Fauna — role in plant regeneration	Poor human-waste management
Low diversity of crops able to grow under the biological limitations of the growth environment	Conflicting human values for natural resource management and extraction of resources; subsistence to large-scale, biodiversity, alleviation of poverty among indigenous families, communities, and national, international organizations
Low water quality	Past periods of colonization
Breeding grounds for diseases	People living integral with the natural resource; forests
High rainfall	Products being managed: agriculture, timber, cattle, non-timber forest products; palm fronds for roofing, medicines, pharmaceuticals, etc.
El Nino droughts or excess rainfall	Land-use practices: shifting agriculture, fire, timber harvesting, etc.
Loss of land surface due to building ditches; water-buffalo grazing causing formation of large rivers washing land blocks to the ocean (Raffles 1998)	Sustainable harvest of natural resources from the forest by local to international organizations
Shifting agriculture effects on soils	Laws and regulations controlling resource uses and management
Landscape mosaic and fragmentation patterns due to timber cutting, agriculture, river tributaries, road systems, settlements	

the impact of a management activity can be entirely missed because of the need to deal with too much data. Most frameworks have not been designed to weight the importance of different indicators (Hart 1984, EPA 1992, Machlis et al. 1997, Vogt et al. 1999ab).

The Legacy Framework developed by Vogt et al. (1999ab) has been designed to explicitly identify the critical variables that drive or modify how the system functions from the myriad information that potentially exists for a site. The usefulness of this framework for identifying relevant social and natural science indicators is documented using wet tropical forests as an example (Tables 5.1–5.4). Initially, the framework requires identifying the legacies and driving variables particular to that system (Table 5.1). This is followed by identifying the desired end points (e.g., values) desired for that system (Table 5.2). This is finally followed by the indicators that would be identified using only the value-based, desired end-points approach for that system and the indicators that would be identified using the non-value-based approach using legacies and driving variables (Tables 5.3, 5.4).

The different data needs (e.g., indicators to monitor or analyze) generated by using the desired end points and those identified using the legacies and driving-variables approach are quite apparent from examining Tables 5.1 through 5.4. Using the value-laden, desired-end points approach to define the data needs ends up defining large data needs for an assessment, but not data needs specific for the management unit (Table 5.4). The legacies and driving-variables approach require several other variables to be measured which focus specifically on the biological, ecological, and social constraints that limited the options available for managing a particular site (Table 5.3). Some of the indicators were the same when using both approaches (i.e., satisfying laws and regulations that limit how a site can be managed or resources extracted). The legacies and driving-variable approach tend

TABLE 5.2

Value-based desired end-points for the social, ecological, and management systems in a wet, humid tropical forest (e.g., Brazilian Amazon).

Ecological	Social*	Management*
Healthy and functional systems	Alleviation of poverty	Sufficient supply of quality tree species to be harvested in the long term
No loss of species important in maintaining ecosystem functions and resilience	Life quality high (i.e., also low disease incidence or physical injuries)	Availability of non-timber forest products — wildlife, plants, fruits and berries, decorations, housing structural materials, etc.
Natural diversity of ecosystems and species maintained within the landscape	An economically viable economy at all levels (local, national, international)	Sufficient management options for harvesting, silviculture, regeneration
Mosaic landscape of forest developmental stages and diversity of functional groups similar to non-human modified systems	Employment options with an ability to acquire hard cash for clothing and medicinals	Forests stands within all developmental stages
No chemical contamination disrupting system function. Clean air and water.	Maintain human values in ecosystems (e.g., rare species and habitats, biodiversity, wetlands, unique landscape features, aesthetic features)	Sufficient economic return on management and harvesting resources
Mammals, reptiles, amphibians, insects, microorganisms at levels comparable to healthy systems	Continuous supply of forest products to trade, for personal use, to sell in markets	Control consumers stealing crops and other natural resource products
Minimal loss of soil and nutrients from forests	Organize labor (e.g., unions) to negotiate job security, salary commensurate with job, safety on job	Sufficient floodplain land area for agriculture and to manage floodplain tree species
Fragmented landscape similar to natural	Eliminate predators and animals harmful to humans	Productivity levels optimal for the site
	Low fragmentation value in the landscape	Minimize invasive species
		Low levels of insect/pest outbreaks
		Maintain wildlife levels for hunting

* Scaling from individuals to families, to local, national, international organizations, to governments.

to focus on identifying the biological and ecological constraints which limit the possible social options and uses of a landscape unit. The desired end-points approach tends to have the social part of the system (as part of human values) driving the type and quantity of information that is required. Many of the certification protocols (see 4.2.1) have identified indicators that strongly reflect the social values (e.g., alleviation of poverty, right to organize, etc.).

There appear to be similarities between the two approaches being used to identify indicators, and some of the data needs appear to be the same (see Tables 5.1 to 5.4). For example, tree-growth-rate data are needed as part of assessing the site's capacity to grow trees. However, yield data are needed to know when trees can be harvested and as part of assessing the health of the system.

Both approaches require similar types of information as part of the management plan. For example, the desired end points in the management system require a sufficient supply of quality tree species to be harvested in the long term and the availability of non-timber

TABLE 5.3

Non-value-based possible indicators identified using legacies and driving variables for the wet, humid tropical forests (e.g., Brazilian Amazon).

Document and incorporate into management plan those disturbances occurring at short- and long-term temporal frequencies (e.g., drought, hurricanes, monsoons) that will affect feasibility of management plans

Document ecosystem health by measures of recovery rates of natural and social systems strongly affected by different disturbances (human and non-human generated)

Document the constraints and options of the soil to maintain plants and animals (low inherent nutrients, chemical toxicity due to Al), and to manage the ecosystems

Types and amount of products harvested from the forest by local communities, national and international organizations

Impact of land-use activities on ecosystem health and sustainable extraction of natural resources and under conditions of changing intensity of particular land uses

Maps of the land uses within the landscape (e.g., fragmentation patterns; types and extents of land uses; location of rivers, roads, settlements, reserves, etc.) and the management unit in relationship to them. The impact of these activities on management-unit resilience after disturbances.

Surveys and maps of species and resources (plant, wildlife) within management area

Landscape matrix and how ecosystems outside of the management unit affect the resilience of the management unit

Document and identify institutional mechanisms that are impinging on natural resource use — from local to international, historical legacy of institutional structures

Document the constraints and options existing for local communities for their livelihood within their particular environment

Identification of all human values (from local to international) that drive the management and extraction of natural resources and which are dominant in controlling resource uses

Laws, regulations (local, national, international) controlling resource uses and management

Land-use practices and options particular to that area because of biological and ecological constraints

TABLE 5.4

Value-based possible indicators identified using desired end points for the wet, humid tropical forest (e.g., Brazilian Amazon). Most criteria require documentation to show that these indicators have been satisfied.

Plant and animal biodiversity surveys at landscape levels

Indigenous people's ownership (land tenure), rights, and security

Tree-growth rates inventory and mapping of trees

Organizational ability of workers for safety on job, equitable pay scales, etc.

Economics of natural resources — availability of resource, price, profitability of company or organization

Set aside areas (reserves) for human values — riparian zones, biodiversity, rare species, etc.

Wildlife abundance surveyed

Management plan that enhances habitats; harvesting rules to minimize system degradation; regeneration planned for; set aside protected areas for human values that are endangered

Maps of vegetative communities and ecosystem types

Document plans to protect water and soil — mainly with respect to harvesting (erosion, riparian protection, no chemicals)

Satisfy laws and regulations

Document resource utilization to all interested groups

forest products. However, it is not always clear what a management plan should incorporate — most protocols require a plan, but do not specify what the structure of such a plan should be. The desired end-point approach further requires strong public participation in formulating a management plan. With such a scenario, human values external to the management unit may drive the design of — and the elements considered relevant to — the management plan.

Part of the requirement for maps of vegetative community types and ecosystem types is also part of both approaches. The legacies and driving-variables approach use these maps to identify the links to the natural system. However, the desired end-point approach does not use that information to identify the links between the social and natural systems. In the latter approach, the emphasis is mainly on having documentation that the information exists and identifying the location of all resources.

The desired end-point approach does not specifically consider the spatial matrix within which the management unit is embedded, nor whether the proximity to different systems changes the thresholds at which the systems may degrade. This type of spatial analysis requires the evaluator to be an experienced individual who is capable of interpreting the landscape and understanding how the different parts of the landscape interconnect. This type of analysis requires the certifier to understand the local landscape. It also requires the certifier to be able to adjust their interpretation of an assessment based on silviculture; economics of forest growth; plant and wildlife ecology; and human impacts on these forests. Since most companies seeking certification face a multitude of problems at any given time, there is a need to evaluate and balance the potential impact of each problem holistically. For example, a landowner may have to deal with the loss of softwood cover types because of the spruce budworm epidemics; the need to reduce the susceptibility of these forests to the budworm in the future; the loss of quality timber because of the epidemics; and the potential impact of the epidemics on changing the quality and quantity of wildlife habitats.

Identification of the many values which can be obtained from a forest is similar to the philosophy of multiple use that was the paradigm for public land management in the 1970s to 1980s (see 3.2.2). The feasibility of maintaining all resource products out of a given piece of land is unrealistic and is one of the reasons that the shift away from multiple use to ecosystem management has occurred (Vogt et al. 1997a). The importance of being able to weight the relevance of different extractions and activities on a given piece of land in a manner which recognizes that if one product output is optimized others will not be optimal. Who should decide what should be optimized? Should it be the public or the landowner? Where do public property rights and private rights begin?

Acknowledging how many of the indicators reflect our value for biodiversity begins to show us the importance of human values in driving indicator selection. Much of the emphasis on biodiversity in the desired end-point approach is driven by our desire to maintain species and to not allow any species to become extinct. In this case, emphasis on measurements that focus on species may result in species becoming extinct, because species do not always reflect how sustainable or healthy a site is (Johnson et al. 1996, see 3.2.1). There is a perception that it is sufficient to focus on managing species if there is the desire to maintain their habitat (Johnson et al. 1996). Some data has suggested that there is a relationship between species and ecosystem function (see 3.2.1). However, most of these relationships have been recorded in grassland ecosystems and not in forests, so there is a question as to whether they are transferable to forests (Johnson et al. 1996). It is unfortunate that this relationship does not exist for woody tree species, because it would definitely simplify an assessment.

An important point made above is that the general paradigms developed in ecology do not transfer automatically to all natural systems — they are relevant for some systems and should be used only in those cases. Therefore, it is a fallacy to assume that the development of a protocol of biodiversity assessment will lead to a good assessment of the site. When biodiversity assessments drive the data needs of assessment protocols (see 4.2, 4.3), these protocols become quite site specific and are not useful tools for sites in general. The fact that information needs do change with locality has been recognized by most certifiers and can be seen in their desire to structure protocols at two levels: national and regional.

Perhaps the national and regional approach should be further adapted to having one set of general indicators that summarizes all the potential criteria and indicators that would be useful in examining any ecosystem. This general list of data needs could be used to stimulate the assessor during the initial analysis to make sure that relevant information is not being missed during the assessment. The regional criteria and indicators should be modified to use a framework that allows site-specific indicators to be selected, using a tool similar to the Legacy Framework of Vogt et al. (1999ab). This would require, even at a regional level, that there is no single, consistent set of indicators that would be automatically analyzed to assess a site. For each certification, there should be an exercise conducted to identify those legacies and indicators that are particular to a given site. If this type of approach is not taken, information will be averaged in a manner that will potentially dilute and negate the importance of those indicators reflecting how the site is functioning.

5.2   Non-value-based indicators relevant for incorporation into forest certification

Kristiina A. Vogt

The discussion in Chapter 3 was an analysis of the issues that currently drive the structure and approaches used in most certification protocols. However, other constraints may determine how a particular ecosystem responds to disturbances (natural or anthropogenic). These constraints have been classified as legacies (see Vogt et al. 1999ab) that can be affected by a combination of socio-economic and ecological factors that control how a system responds. For forestry certification protocols, these factors need to be better identified in order to make certain that system functions are maintained, and to prevent ecosystem degradation.

A good analogy for these constraints is a play being acted on a theater stage. The constraints can be considered as the stage and the surrounding environment that will determine how effectively the players will be able to present their monologues. Even the best actor will not be able to give their best performance if the lighting is too strong, if the orchestra is playing too loud, or if there are frequent loud noises occurring backstage.

Legacies in natural systems, which have been modified or formed by a past disturbance, will affect the way an ecosystem functions, and may modify the response of a site to future disturbances. Frequently in the natural system, the constraints are not readily visible (i.e., legacies of the soil) so it may not be immediately apparent to the observer that a particular system is not similar to an adjacent system (Palmiotto 1998). These hidden legacies may also be produced due to land use activities that occurred temporally disjunct (non-synchronized) from the current assessment. Therefore, the casual observer may not realize that some previous land use activity has changed how that system functions even though it currently controls the system's resilience and resistance (i.e., agriculture and the decreased organic carbon in the soil, see 5.2.1.1). We are just beginning to understand some of these legacies, such as those associated with invasive species (see 4.3.4), edges (5.2.1.2), insects, and pathogens (see 5.2.1.4).

Examples and a brief discussion of some of the legacies which can exist in forest ecosystems follows in Sections 5.2 and 5.3. These legacies can be natural or social science based and it is important to ensure that they do not control the functioning of the forest being certified.

5.2.1 Natural science legacies constraining natural resource uses

5.2.1.1 *Land uses and soils*

Ragnhildur Sigur∂ardóttir, Daniel J. Vogt, Kristiina A. Vogt

Soil quality and soil conservation may be viewed as the backbone of any land management program and is of fundamental importance to sustainable forestry projects. Soils provide the support and nutrient medium for plant growth, control the availability of water in the ecosystem, and are the habitat for myriad living organisms, from microbes to small mammals and reptiles. Properties of the soil often determine the nature of the vegetation present and thus, indirectly, the number and types of animals (including humans) that the vegetation can support (Brady and Weil 1996). The multitude of functions occurring within the soils can make them more susceptible to disturbances that may even lead to ecosystem degradation.

The following have been identified as some of the signs of soil degradation (Brady 1990, Stott and Kennedy 1999):

- Soil erosion
- Nutrient runoff
- Acidification
- Compaction
- Crusting
- Organic matter loss
- Salinization
- Nutrient depletion
- Accumulation of toxins

The term "soil quality" has been defined as the capacity of a soil to function within ecosystem boundaries to sustain biological productivity, maintain environmental quality, and promote plant growth (Soil Science Society of America 1999). Soil-quality assessment should be a useful tool in forest certification protocols. These assessments not only look at the system in terms of its own sustainability, but they also examine the system in the context of the landscape. This landscape approach is relevant for soils because soils protect watersheds by regulating infiltration and partitioning of water and they prevent water and air pollution by buffering potential pollutants (NRC 1993). Soil-quality assessment may therefore satisfy the multiple value-based goals of forest certification, whether they are ecological, social, or end points desired by management objectives (see 3.1.2).

For soil-quality assessment, it is important to identify the most important indicators of soil quality. This should be done by identifying the site's key soil legacies which control the function of the soil system and its resilience and resistance to disturbances. These legacies can be modified by either natural or human-induced disturbances.

Human activities can create legacies in the soil environment, many of which lead to the degradation of the system. The major human-induced perturbations can be separated into the following categories: agricultural, urban, or industrial. These anthropogenic forms of soil degradation are of concern because they can accelerate the rate at which soil processes occur, especially when compared to the slow process of soil formation and recovery (Brady 1990).

The degrading effects of industry on soil processes can be examined at two scales:

1. The small geographical scale where the impacts are localized (e.g., soil compaction; severe chemical contamination adjacent to the release; or spill of substances)
2. The larger scale where the impacts occur over a large geographic area (e.g., atmospheric deposition of chemicals)

If the impact on the ecosystem is localized, it will be easier to determine which indicators to monitor. However, human activities that impact at larger spatial scales (i.e., across the landscape) are more difficult to monitor because visible signs of soil degradation are difficult to detect. In some cases, chemical analyses may have to be conducted on soils to determine whether soil chemical or physical properties are changing (e.g., acid deposition and decreased soil-base saturation).

Human activities that impact soils are quite varied making it difficult to articulate that certain variables should always be measured in soils. There is a need to determine how soils have been affected for each area. This can be done by initially documenting human land-use activities that have occurred at that location using a historical perspective. For example, human settlements in urban areas can cause soil degradation because of pollution, such as urban and sewage waste, or due to soil compaction or erosion. Knowing that these activities have occurred will help to focus the types of analyses that need to be obtained. The diversity of indicators reflecting a particular human land-use activity is high. For example, the following degrading effects have been recorded to be the result of agricultural practices:

- Loss of species
- Changes in species dominance in ecosystems
- Introduction of weedy species
- Changes in soil chemical and physical properties
- Chemical toxicity
- Decreased nutrient storage in ecosystems
- Soil compaction and crusting
- Accelerated erosion

Agricultural processes can also be an important non-point source of chemical pollution due to the widespread practice of using fertilizers and pesticides. All of these changes affect the resistance and resilience characteristics of ecosystems and therefore change how the system responds to and recovers from other disturbances.

Harvesting activities may have a large impact on the soil quality and often form legacies that determine the future productivity of a site (see Chapter 6). Many harvesting activities reduce surface cover and cause soil compaction that can result in a loss of organic matter and increased runoff and erosion. Erosion generally decreases productivity of forests by decreasing the available soil water for forest growth and through loss of nutrients in eroded sediments (Elliot et al. 1999). The effect of a given disturbance to a site is highly site specific, and its severity depends on both soil characteristics and climate.

Since these legacies vary across and within the landscape, ecosystems appearing to be similar in structure may actually have very different functions and recovery rates after disturbances. Therefore, they should not be evaluated in the same manner. It is imperative for any assessment to identify the different legacies existing in the system. This should be followed by analysis of how these factors affect assessments of sustainability or good forest management. There is a pressing need to develop standardized and simple tools that can help to evaluate and assess whether soil degradation is a potential problem. These assessments may have to be site- or regional-specific and be determined on soil characteristics and abiotic factors, such as climate.

Steps have been taken to identify the most sensitive indicators of soil quality (Doran and Parkin 1996, Stott and Kennedy 1999). These indicators need to be sensitive to both management and climate. These changes must also be detectable immediately after and several years following a disturbance (Stott and Kennedy 1999). Doran and Parkin (1996) classified soil-quality indicators as being physical, chemical, or biological with each indicator being highly dependent on the other. Doran and Parkin (1996) identified the following physical, chemical, and biological indicators for soils:

Physical indicators:

- Soil texture
- Depths of soil, topsoil, and rooting
- Bulk density
- Infiltration rate
- Water retention characteristics

Chemical indicators:

- Total organic C and N
- pH
- Electrical conductivity
- Extractable N, P, and K

Biological indicators:

- Microbial biomass C and N content
- Potentially mineralizable N
- Soil respiration
- Water content
- Soil temperature

It is obvious that the indicators listed above will not to be monitored by a certifier. However, it is relevant for a certifier to determine what the historical land uses have been for a given area. Based on that assessment, it may be relevant to conduct some type of soil analysis or to consult with state or federal organizations, or with soil experts. It is important to ensure that a forest does not have the potential to become degraded by the addition of management activities on top of old or new soil legacies. Presently, soil erosion is the only soil-related indicator mentioned by most certification protocols. Elliot et al.

(1999), however, showed that erosion alone is seldom the cause of greatly reduced site productivity. They found that erosion in combination with other factors (e.g., loss of nutrients, mycorrhizas, and organic matter) might cause severe ecosystem degradation on the scale of decades and centuries.

Soil organic matter is a crucial component of the soil and a prerequisite for ecosystem health and productivity. It is the primary sink and source of plant nutrients in terrestrial ecosystems. Higher soil organic matter levels increase the ion exchange capacity of soils and their water holding and infiltration capacity; it ameliorates soil structure and is the main energy source for the soil biota (Paul and Collins 1998). Soil organic matter is the major bio-available source of several of the most limiting macronutrients (e.g., nitrogen, sulfur, and phosphorous), and micronutrients (e.g., iron, manganese, copper, boron, molybdate, and zinc).

Cultivation of agricultural land leads to dramatic losses of soil organic matter (Schlesinger 1986). These losses are due to decreased annual production of plant residues and an increase in decomposition rates due to increased soil temperature, aeration, moisture, and the use of fertilizers. Schlesinger (1986) calculated that 36 Pg C were lost from soils due to agriculture between 1860 and 1960, with a current rate of about 0.8 Pg C/yr.

After agriculture abandonment, there is a short-term selection for plant species adapted to higher nutrient levels because high residual nutrient levels can exist after crop fertilizations. This should result in initially higher plant growth rates. However, once residual nutrients have been depleted, these species will be outcompeted by plants adapted to lower nutrient levels.

Changes in soil organic-matter levels have dramatic effects on the plant and microbial communities found inhabiting this environment. Since soil organic matter is an important source of nutrients to plants, a decreased pool of soil organic matter results in a decreased pool of soil-nutrient reserves. These factors will reduce plant net primary production and select for plant and microbial species adapted to acquiring abiotic resources at lower levels. Furthermore, loss of soil organic matter reduces the water-holding capacity of the soil which reduces plant growth and selects for plant species with lower moisture requirements. This, in turn, feeds back to further reducing the annual production of plant residues and production of soil organic matter.

Agricultural and chemical pollution often leads to decreases in the number and diversity of microbes and symbiotic associations with plants. These changes in the microbial community will feedback to decrease ecosystem production and decomposition rates (and, indirectly, plant available nutrient levels with lower decomposition rates and less-efficient mycorrhizal associations).

After chemical pollution, excess inputs of one nutrient (especially nitrogen) can cause plants to become deficient in other nutrients so that plant growth rate is lowered. Some species experience increased mortality rates due to their physiological inability to adapt to the changing nutrient environment of the soil. Where nitrogen levels have been dramatically elevated, it is common to find an increased incidence of pests and pathogen problems (Vogt et al. 1997a).

High inputs of nitrates and sulfates cause excess leaching of bases from the soil. The decreased base saturation will decrease plant growth, eliminate species not adapted to acquiring low Ca levels from the soil, and eliminate animal species dependent on acquiring Ca or other bases indirectly from the soil.

Agriculture and forest management can produce several distinct legacies in the soil environment which can be summarized as follows:

- Decreased nutrient reserves in the soil
- Decreased soil organic matter
- Lower diversity and different populations of plants, microbes, and soil animals

The legacies which can exist in a soil environment are demonstrated by examining the development of the shifting agricultural systems in wet tropical areas. For hundreds of years, these ecosystems have been exposed to high rainfall which has resulted in leaching losses from the soil of several critical plant-required nutrients. Shifting agriculture has developed in response to the soil legacy of low nutrients where the thresholds for these elements are repeatedly reached during agricultural use of these lands. When the nutrient-deficiency threshold is reached, the inherent capacity of the site to support plant growth will be dramatically diminished, since the soils are unable to provide the abiotic resources needed for growth. This threshold is a legacy of the site and must be considered when managing these ecosystems.

Many legacies can be detected in the soil after agricultural lands have converted back to their pre-agricultural ecosystem (Johnson 1992, Richter et al. 1994). For example, significant decreases in soil organic matter have been detected on agricultural lands after 10 to 20 years of conventional agriculture (Brady 1990). Much of agricultural management has focused on managing soil organic matter and on the development of tools to increase soil carbon levels. These legacies of soil changes resulting from agriculture can be detected in the soil for decades after agricultural production has ceased. It has been suggested that agricultural use of previously forested areas can be still detected close to one-hundred years after the land is allowed to return to forest cover (Motzkin et al. 1996, Marsh and Siccama 1997). Interestingly, these soil legacy changes recorded with agriculture have not been recorded when forestlands are managed for timber production and harvesting (Johnson 1992). Synthesizing many studies which determined changes in soil carbon after harvesting trees, Johnson (1992) reported that only a 10% decrease or increase in soil organic matter was generally recorded.

Another legacy created in the soil when forests are cleared for agriculture is the change in diversity and composition of the microbial community — especially the symbiotic associations. The dominant symbiotic associations present in agricultural systems are not those which are found to colonize trees (Schenck and Kinlock 1980, Boerner et al. 1996). This can become a problem when agricultural lands are allowed to revert to forests without management intervention. The problem arises because the innoculum of the symbiont needed by trees is present at low levels or not present at all. This means that a matrix of forestlands has to exist in close proximity to these agricultural lands in order for the spores to be reintroduced into that environment (Perry et al. 1990). Since many tree species are highly dependent on symbionts to grow, this may require the reinoculation of symbionts into a newly forested landscape in order for a forest to be able to regrow on that site (Mikola 1973).

Chemical pollution also causes significant changes in the legacies (at the levels of the plants, animals, and soil) in our forested landscapes. Pollution has changed the species that are able to survive within a given landscape — mainly by increasing the amount of a nutrient to excess levels. This imbalance in nutrient availability can result in the site becoming dominated by weedy, frequently invasive species capable of outcompeting native flora in the new legacy environment (see 5.2.1.5). Atmospheric pollution can also cause dramatic changes in the capacity of the soil to retain base cations because pollution causes increased nitrate and sulfate levels that effectively leach soil bases (Lawrence et al. 1999). Those tree species (e.g., spruce, sugar maple) that are inefficient at accessing the required bases from the soil under conditions of low availability experience greater mortality rates and reduced growth rates (Lawrence et al. 1999). This results in changing the composition of the forest which will be responding to new disturbances and creates new legacies. Similarly, a chemical legacy in the soil due to pollution resulted in reducing the reproduction of passerines in Holland (Graveland et al. 1994). In this case, acid depositional losses of Ca from the soil created a legacy of lower Ca levels. Snail population levels declined as Ca levels decreased (i.e., shells are high in Ca), which eventually affected passerines who needed to eat snails to obtain their Ca for egg production (see 3.2.1.4).

Many ecosystems characterized by naturally low N levels are also not adapted to using the excess N that is added with pollution. In this case, the new legacy is higher soil N levels. Waring (1987) suggested that boreal and subalpine coniferous forests were particular endangered by nitrates since they are not adapted physiologically to these higher inputs and are unable to utilize them. This means that enrichment of a site by N will require other species to be able to dominate a site because of the soil changes that reduce the ability of native species to grow.

The effect of atmospheric N additions is similar to what has been recorded with N-fixing tree species. Again, a soil legacy results from having higher levels of one nutrient, resulting in an imbalanced nutrient availability in the ecosystem. N-fixing trees have been planted for erosion control, to stop the spread of deserts, or have begun to dominate after selective harvesting of other tree species (van Miegroet and Cole 1984, Vitousek and Walker 1989). Nitrogen-fixing tree species change the soil-nutrient status and, therefore, create a new soil legacy which must be considered when assessing management impacts on a site. A well-documented example was recorded in Hawaii where the introduction of an N-fixing shrub to control erosion resulted in soil legacies and an inability of native vegetation to grow in the new, soil-chemical environment (Vitousek and Walker 1989). Excess N inputs into an ecosystem, whether from a natural source such as N-fixing trees or from atmospheric pollution, result in similar soil legacy effects. The following soil legacies occur with high atmospheric N inputs into ecosystems (van Miegroet and Cole 1984, Lawrence et al. 1999):

- Soil acidification
- Increased mobilization of Al
- High levels of nitrates and sulfates
- Losses of base cations

One of the most common approaches used by humans to create soil legacies is the addition of human waste to environments. Sometimes this occurs at small spatial scales, but the end result is the same. This is typically observed in ecosystems characterized by naturally low levels of N where weed complexes have begun to dominate after the addition of high N. An example of this is the dung heaps that formed around human habitation in the tundra (Chernov 1985). There are many examples in the literature of humans having to eliminate and dispose of their waste which has resulted in the enrichment of adjacent natural systems.

Human activities frequently produce chemical legacies in the soil environment that can cause ecosystems to degrade or to not be as healthy. The classic example is when lead was previously added to gasoline — gasoline exhaust added lead to the environment and eventually resulted in dramatically increasing lead retention in forest ecosystems (Friedland et al. 1992). Proactive measures were taken to eliminate lead from gasoline and lead levels have started to decrease in the northeastern United States. Lead becomes a legacy in the environment because it complexes with organic molecules within the surface organic layers on top of the soil or within the soil itself (Friedland et al. 1992). This means that higher lead levels will continue to be detected for several decades as an imprint within the soil, since the decomposition rate of organic matter in the soil is slow.

A legacy of unintentional inputs of base cations by human activities also exists. These inputs have decreased as atmospheric pollution inputs have been regulated. For example, Hedin et al. (1994) reported steep declines in atmospheric inputs of bases in several regions in Europe and North America. In this case, changing human land-use activities decreased the nutrient inputs into forests already experiencing decreased soil bases levels due to acid deposition. These earlier atmospheric inputs of bases were the result of road building with asphalt, production of cement, and burning of forested landscapes — common in the United States prior to the 1700s (Thomas Siccama, Steve Hamburg, personal communications).

5.2.1.2   Landscape spatial patterns and edges
Brooke A. Parry, Kristiina A. Vogt, Karen H. Beard

Past land uses have a significant effect on creating legacies within a landscape that cause individual management units to respond differently to management activities within that landscape. This means that there is not going to be a uniform response to a particular management activity in similar forest types because the legacies modify how the management unit responds. The legacies in the landscape will determine how that ecosystem responds to disturbances; how susceptible it will be to being degraded after disturbances; and its recovery rate after disturbances. The landscape matrix with its different legacies will, therefore, have a major influence on how a system responds and maintains its functions. However, the landscape matrix and its legacies within and outside the management unit have typically not been considered in most certification approaches (see 4.2.1). Spatial-scale considerations in certification have focused on:

1. Determining the minimum size of forestland where the financial returns from resource extraction are sufficient to support the cost of certification itself (see 7.6)

2. The costs of tracking individual trees from their source to the market (e.g., chain of custody; see 7.7)

An important management tool for understanding the forest matrix landscape can be obtained by studying the amount of edge environment created as part of management. Edges appear to be particularly sensitive locations within ecosystems to assess whether ecosystem function is changing (Gosz 1993). Different land uses create legacies (which are not frequently visible) which change the resistant and resilient characteristics of edge environments and how far edges extend into the interior forest. The type and quality of edge environment existing around a management unit will significantly control its functioning and structure, and therefore its sustainability.

Managing landscapes and the specific management unit of interest should not primarily focus on mapping the location of reserves or other sensitive areas, nor should it be assumed that it is sufficient to take information collected at smaller spatial scales and automatically transfer them to larger scales. Transferring the same bits of information from small to large scales does not work because a greater number of ecosystem components need to be assessed at small scales and the information needed to predict ecosystem processes varies by scale (Vogt et al. 1997a). This suggests that the collection of uniform spatial data for all sites being assessed for certification is not wise (see 5.1). Instead, the management site should be characterized by the legacies that exist within the landscape and by focusing on environments which sensitively reflect the system (e.g., edges).

Obtaining information on the amount and type of edge environments particular to a management unit is a useful initial approach to understanding the management unit within the matrix landscape. To show why it is important to understand them, a brief discussion of edges follows. For example, edges have been typically viewed as negative to system health and sustainability. However, edges also exist naturally in less human-impacted systems and can be important locations of higher resource and species diversity. In fact, almost all intact old-growth Douglas fir forests in the Pacific Northwest region of the United States can be classified as edge environments. This high incidence of edge environments is produced because of the disturbance cycles that cause single-tree falls to occur at a rate of 1% annually (Vogt et al. 1999ab). It is then critical to identify the thresholds at which the types and amounts of edges have changed sufficiently enough to cause the system to degrade.

5.2.1.2.1 Edges

Studies of edge ecosystems do not allow generalizations to be made about the role of edges and boundaries in landscapes. An understanding of natural edges and their importance in the landscape that is distinct from those edges that are perceived to be negative has not been developed. Most documentation presents edges as always being negative, without discussing the important natural roles they play in ecosystems. For example, existing documentation of edge effects suggests that edges are zones that can impact ecosystem sustainability and species maintenance in negative ways (Forman and Godron 1986, Hunter 1990, Saunders and Hobbs 1991, Murcia 1995).

Edges have been defined in several ways, including the borderline between a clearing and a forest; the depth to which an edge effect penetrates; and a transition zone where two dissimilar, adjoining ecosystems interact (Chen et al. 1992, 1995). Edge effects have been defined as gradients of energy, nutrients, and species occurring adjacent to a boundary (Forman and Godron 1986) and as the outcome of interactions between two adjoining ecosystems divided by a boundary (Murcia 1995).

Leopold (1933) originally defined edge effects as an increase in "species richness" at the intersection of a clearing and a forest. At the time, this conclusion generated interest in creating edges wherever possible. It is now understood that Leopold's observation of species richness at edges may be the result of exotic plant invasion, pioneer herbs, or generalist animal predators (Saunders and Hobbs 1991). This type of species richness can have a negative impact on forest sustainability and therefore is not desirable in management. Groups of "edge" and "interior" species are now recognized (Lagerlof and Wallin 1988). In some cases, edge species are highly valued and deserve protection. In Sweden, field margins serve as habitat for 75% of endangered vascular plants. As use of land for agriculture has declined and fields are reclaimed by forest, there has been a marked loss of species associated with Swedish field margins (Lagerlof and Wallin 1988).

Edges can be created by natural disturbances (e.g., hurricanes, single-tree fall gaps), as well as through human land-use activities (i.e., agricultural production next to forests, etc.). They are also zones that develop because of changing vegetation, soils, bedrock, or topography within the landscape (Vogt et al. 1996, Palmiotto 1998). "Induced" edges are those created by forest harvesting or burning, as well as by other disturbances such as blowdowns. "Inherent" edges are found in areas of ecosystem transition, such as the ecotone between two forests or soil types (Hunter 1990). Some tropical forests have transition zones in close proximity to one another because of small-scale changes in soil types within a stand. Changes in soil types across scales of meters were found to have significant effects on which tree species dominated and the diversity of tree species in the landscape in tropical forests on the island of Borneo (Palmiotto 1998). This stratification of tree species by soils eventually affected the resilience of these tropical ecosystems to other disturbances (e.g., drought) because of the different efficiencies that these species had for utilizing abiotic resources at each microsite (Palmiotto 1998).

In most cases, human-induced formation of edges has been viewed as negative, as evidenced by the change in terminology from "edge species richness" (Leopold 1933) to the "edge creep" (Soulé 1986). The term "edge creep" evokes vivid negative images of species loss and decreased ecosystem health that would result in reduced ecosystem sustainability through physical and biotic changes occurring near edges. Changes in edge environments have been described in several different ways. Some examples of ecosystem changes occurring in edge environments follow:

- Wind can be responsible for introducing seeds of weeds and exotic species to edge environments (Hobbs and Atkins 1988). In a tropical forest in Queensland, Australia, germination of plants adapted to disturbance was noted up to 500 m from clearcut edges (Laurence 1991).

- Wind penetration at edges can also cause plant death, another form of increasing the depth of the edge environment (e.g., edge creep). Blowdowns and tip-ups can create new edges. When these susceptible edges are subsequently blown down, the edge creeps further into the forest. Mature trees at newly created edges have developed in an undisturbed environment and, therefore, are less likely to have formed the required supporting root structures and canopy shape that is capable of withstanding increased wind (Saunders and Hobbs 1991).

- Increased tree mortality at edges is caused by physical damage due to wind, increased evapotranspiration, and soil desiccation (Lovejoy et al. 1986, Saunders and Hobbs 1991). These can also be natural processes of mortality and regeneration associated with certain species that have been called "fir waves" (Sprugel 1984).

- Edge creep can take the form of weeds creeping toward the forest interior and changing the floral dynamics of the system.

- Disease can also creep from the edge through the forest when susceptible edge trees are infected by a pathogen that then spreads through the root systems killing other trees (see 5.2.1.4). Wind can be responsible for introducing airborne pathogens, soil and waterborne fungi, and other diseases to physically damaged or susceptible edge-plant species. Incidence of pathogen and fungal infections has been positively correlated with physical damage, a trait common to edge trees and plants, particularly in areas where there has been logging, road construction, or mining (Packham et al. 1992, Weste 1994). In Australia, Neyland and Brown (1994) described an area where harvesting eucalyptus to a rainforest edge resulted in most of the rainforest becoming a stand of standing dead trees. In that study, living trees adjacent to the dead stand are currently dying of Myrtle Wilt, the presumed cause of death of a majority of trees at that site (Neyland and Brown 1994).

Abiotic edge effects (i.e., changes in light, temperature, wind, and water regimes) are dynamic forces that influence the type of edge creep that occurs. Changes in the physical environment affect the distribution and abundance of plant and animal species living at or near an edge. The magnitude and extent of abiotic edge effects can be influenced by forest architecture, edge aspect, slope, time of day, and time of year (Chen et al. 1995, Murcia 1995). For example, in the Northern Hemisphere, radiation penetration at south-facing edges can extend 60 m into the edge environment while radiation changes at north-facing edges can penetrate about 20 m from the clearing edge (Chen et al. 1995). As expected, the reverse occurs in the Southern Hemisphere where north-facing edges can experience four times the radiation typically observed at south-facing edges (Young and Mitchell 1994). Likewise, changes in air-flow dynamics following edge creation depend on the architecture and biomass of the vegetation adjacent to the edge, edge aspect, the topographic position of the edge, and the prevailing wind direction (Oliver and Larson 1996). Wind profiles change with changes in vegetation structure and with increased distance traveled by wind into the edge ecosystem (Raynor 1971, Chen et al. 1995). Most studies estimate that depth of wind penetration is two to three times tree height (Reifsnyder 1955, Chen et al. 1995).

Changes in forest ecosystems due to the presence of an edge can have a significant impact on sustainability. As edge creep increases, it indicates that an edge has the potential to be the zone of influence in the landscape — an ecosystem that can change the state of or engulf the ecosystems it adjoins. Because of the feedback associated with certain kinds of edge creep, edge-related changes can even spread throughout an extensive reserve or landscape (Soulé 1986). Our traditional definition of an edge, as a zone in which adjoining ecosystems interact (Chen et al. 1992), does not provide certifiers with an explicit basis for incorporating the impact of edges on forest sustainability.

It is clear that edges and their impacts should be incorporated into certification assessments. However, edges need to be understood as being inimical to natural ecosystems and not always something that is negative. There have been discussions about managing forests to maintain the types of edges that naturally occur as part of disturbances that are particular to any ecosystem. But it is also important to understand that many human activities do change the type of edge environments that exist and result in less-resistant and resilient ecosystems and landscapes. Therefore, edges must be included in ecological assessments of a site since they can result in the failure of conservation efforts and produce systems that are not sustainable. Again, it is important to understand that edges are also a natural part of the system disturbance cycle and it is important to be able to distinguish these natural processes (e.g., fir waves) from those being generated by human activities.

5.2.1.2.2   Forest fragments

Depending on the extent of edge creation and ecosystem modification, forest fragmentation and patch-isolation issues must be considered in certification. The property size, as well as where the property is situated in the matrix landscape, need to be assessed because of the relationships that exist between patch size and ecosystem sustainability. As suggested above, seed rain can be of particular concern in small woodlots as species composition may gradually change if edge species spread toward the interior of the lot. Disturbance- or edge-adapted plants are often found to dominate small patches or forest patches with irregular shapes (Ranney et al. 1981).

As our awareness of the potential effects of forest fragmentation has increased, so has our understanding that the impact of an edge is difficult to separate from the influence of patch size and the surrounding landscape. For example, humidity and topography influence fire regimes; factors controlled both by edges and the size of the patch, as well as the matrix in which the patch is embedded (Janzen 1986). In Costa Rica, Africa, and Australia, forests existing within fire-prone grasslands can eventually be replaced by grasslands as each burning episode burns trees at the edge, replacing them with more flammable species to be burned during the next fire. The encroachment of the grass species can change the water regime of the area, which further contributes to making the system susceptible to hotter fire. In a small patch, the risk of flammability increases because of decreased humidity and the increased influence of the flammable perimeter. Janzen (1986) suggested that although a 100 ha patch may contain a core uninfluenced by edge, it is unlikely that the core will be sustained in the face of biological interaction among dynamic landscape zones.

Many of the arguments concerning the sustainability of different-sized patches developed from Island Biogeography Theory (MacArthur and Wilson 1967). MacArthur's and Wilson's research in the Pacific archipelagoes indicated that larger islands contain more species than smaller islands and that islands closer to the mainland have more species than islands located farther into the ocean. The Minimum Critical Size of Ecosystems Project was designed to test theories between forest patch size and ecosystem change.

The Minimum Critical Size of Ecosystems Project (Lovejoy et al. 1986) is an excellent example of the importance of having indicators change with the size of the land parcel undergoing the certification process. Although this study is limited, in that it described different fragmentation effects in isolation from one another (e.g., bird distribution, plant mortality rates, etc.), it showed that ecosystem change can depend on patch size and the matrix surrounding the patch. For example, bird behavior differed in one- and ten-ha forest fragments. In the one-ha fragments, birds migrated to the edge of the patch to feed on aggregated insects, which were in turn feeding on the abundant new growth at the forest edge. In contrast, birds in the ten-ha patches remained in the forest interior, despite the fact that similar changes in plant and insect populations were occurring at the patch edges. Crowding effects following patch isolation were also dependent on patch size. More birds were found in the one-ha patch than in the ten-ha patches, presumably because there was relatively less space for refugees in the one-ha patches. Only the ten-ha patch showed no increase in refugee birds following isolation of the patch as part of the experiment. The suggested reason for this anomaly was that the patch was separated from the forestlandscape on three sides by only 100 m. Therefore, it was likely that the clearing dislodged fewer birds (fewer than if thousands of square meters had separated the patch from the nearby forest) and that there was ample space for refugees in both the patch and the nearby forest. Thus, the matrix surrounding the patch can be equally important, if not more so, than patch size.

These studies support the need for certification assessments to consider ecological processes that occur at the edge of the property and how management changes the type and extent of edge environment. In addition, certification must consider the matrix landscape as much as the processes occurring within a property, particularly when certifying smaller lots. To quote Sample and Aplet (1993), "We are urged to look beyond the borders of public forestlands, which usually do not correspond with ecological boundaries, and consider all lands within the ecosystem as important to its overall functioning and sustainability."

5.2.1.3   *Species diversity and spatial scales*
Karen H. Beard

5.2.1.3.1   *Maintaining species diversity in isolated areas*

The diversity of species in an area is a function of two general processes. First, local interactions in an area determine the number of species, such as local abiotic processes and interactions between species. Second, the surrounding matrix plays an important role in determining the number of species in a region. Therefore, both pieces of information are crucial to determining whether the diversity of species in an area is likely to change as a result of different management practices.

Conservation biologists have acknowledged that isolated, small areas will not likely be able to preserve the functioning of communities and ecosystems. Instead, they realize that networks of reserves should be designed to increase reserve area, allow migration of species, and buffer preserves from human impacts (Noss 1983). Since species are likely to migrate between core preserves, facilitation of migration and the habitat quality of the matrix between reserves are extremely important.

Landscape corridors might offer one of the best solutions for conserving biological diversity (Noss 1987). The idea behind landscape corridors is that they can link reserves and make them effectively larger. Although corridors are intuitively appealing, there might be some unforeseen consequences of using corridors. For example, they may facilitate the movement of pest species, disease, and fire, increase the risk of predation, and have detrimental edge effects (Simberloff et al. 1992). In addition, the evidence of successful corridors is limited (Rosenberg et al. 1997), and they are extremely expensive considering that their ecological benefit has yet to be shown. However, in general, corridors could be invaluable in addressing the concerns of habitat fragmentation.

The uses in the surrounding matrix are also crucial for conserving species (Primack 1993). The general conservation principle of surrounding core conservation areas with core buffers (traditional human activities, monitoring, nondestructive research) and transition zones (sustainable development and experimental research) could prove important for minimizing the effects of habitat fragmentation on species diversity (Harris 1984)).

5.2.1.3.2 *Species-area relationships, Island Biogeography Theory, and Metapopulation Theory*

Ecological theory can be used to predict the number of expected species in reduced habitat. For example, two ecological theories — the species-area relationship and the Island Biogeography Theory — when used together, can estimate the fraction of the number of species expected in an area of a particular size. The species-area relationship states that species richness increases with area (Gotelli 1995). The Island Biogeography Theory (MacArthur and Wilson 1967) describes the relationship between immigration and extinction rates onto islands using area and distance effects. The implication from these theories is that larger and less-isolated areas contain more species. Unfortunately, these relationships are not as simple as an area relationship between species and their habitat.

One of the drawbacks of using these theories to understand the impact of habitat fragmentation on species is that they assume we know the sizes of the areas into which species must migrate and/or the distances species must migrate. In addition, these theories treat all species as equal. The Island Biogeography Theory predicts the equilibrium number of species on the island from a balance of migration to the island and extinction on the island, but does not specify which species will remain on the island. Furthermore, the theories are only suggestive. There are undoubtedly other environmental factors besides size and area that are important determinants of the number of species in an area (i.e., barriers to migration). This discussion also ignores the complexity and diversity of habitats which can exist in any given area and how this influences the number of species that can occur in that area.

Another ecological theory that can be used to understand the effects of habitat fragmentation on species persistence is the metapopulation concept (Hanski and Gilpin 1991). This concept is used to describe populations of species living as collections of subpopulations, each occupying a suitable habitat patch in a landscape of unsuitable habitat. A certain number of subpopulations are necessary to ensure the survivorship of a population and suitable, unoccupied habitat is important for species persistence. If a species has a metapopulation-type structure, the arrangement of suitable habitats will be important to the persistence of the species. As fragmentation continues, the spatial arrangement of available habitats changes. For species exhibiting a metapopulation-type structure, these changes will be important.

The metapopulation concept might be difficult to apply in practice because the species must exhibit a metapopulation-type structure, which has only been shown for certain species. In addition, the concept would be difficult to use without knowing the life-history traits of the species, such as the dispersal rates, and the locations of the subpopulations. However, it can be useful in considering the implications of reduced habitat size on the status of a species. For example, the metapopulation theory used with Island Biogeography Theory suggests the probability of local population extinction as the area for local populations and the number of surrounding local population diminishes and the distance among local populations increases. Although these theories might be used to understand and predict the number of species in an area, they may not be used to understand how the functioning of the system has changed with species diversity changes.

5.2.1.3.3 *Spatial scale and ecosystem stability*

In ecology, it is now recognized that processes in nature are sensitive to the space and time scales at which they are considered (Wiens 1981, 1986). For example, one might expect the relationship between species diversity and ecosystem stability and resilience to be different for different spatial scales (sized areas). Island theory may be used to understand how spatial scale might affect the relationship between diversity and stability. If islands can be viewed as habitat fragments, this theory may have great implications for the effect on ecosystems of reducing areas. The following explains why the relationship between species diversity and ecosystem stability and resilience may be expected to be different on continents and islands and how this might be useful for understanding the implications of habitat fragmentation (see 3.2.1.3).

First, islands often lack the full complement of taxonomic groups found in comparable non-insular regions because of their isolation and the variable dispersal abilities of potential colonists (Cushman 1995, MacArthur and Wilson 1963, 1967). Island communities also often appear to have fewer species in each taxonomic group; for example, compared to their continental counterparts, forest systems are often dominated by one plant species. (Mueller-Dombois 1981, 1995). Therefore, islands, in general might be described as inherently species-poor ecosystems.

Second, island ecosystems are thought to have more open "niche space" or less "species packing" and less functionally redundant species than continents simply because of the inability of species to colonize islands (D'Antonio and Dudley 1995). Competition among species is therefore not likely to be a major driving force in island communities. This might explain why there is a great amount of adaptive radiation observed on islands (niches grow wider as opposed to narrower).

Finally, islands have been found to have fewer species performing any particular ecosystem function than on comparable mainlands (there are likely more "keystone" species on islands) (Fownes 1995, see 3.2.1.4). Comparable continental communities have a much greater variety of species playing trophically equivalent roles (Reagan and Waide 1996, Stewart and Woolbright 1996). Also, new species often do not replace species, but rather fill new functional roles.

Thus, islands have fewer species and fewer species per functional group, and they have less functional roles and functional redundancy than their continental counterparts. This suggests that islands are unstable or susceptible to disturbance because fewer species are available to fill functional roles when the system is disturbed (Cushman 1995). Island ecosystems have long been considered unstable — or more susceptible to disturbance — than continental counterparts because they are particularly vulnerable to exotic invasions (Loope and Mueller-Dombois 1989, see 5.2.1.5). Again, certification should not use the model relationships developed for biodiversity and their spatial patterns of persistence in the landscape from studies conducted on islands. Unfortunately, much of the informa-tion has been produced on islands because of their ease of experimental manipulation.

The idea that islands are easier to invade than mainlands is a pervasive hypothesis in ecology and is widely stated throughout the ecological literature (starting with Elton 1958). It is often argued that islands are more subject to invasions for the exact reason stated above — their native biota is unsaturated and lacks mechanisms to buffer against change (Carlquist 1965, D'Antonio and Dudley 1995, MacDonald and Cooper 1995). The theory derived from Elton (1958) in combination with theoretical modeling (Case 1990) may be used to understand how alien invasions on island communities are due to inadequate or incompletely interacting species in the island ecosystem. Although the hypothesis that islands are easier to invade has not been well tested empirically, there is great natural history evidence for the idea (Elton 1958).

Again, much of the information relating spatial scales and ecosystem stability has been collected on islands because fewer constraints exist to manipulation. However, the appli-cability of these theories to continental areas that are reduced in size has not yet been determined. This must be determined prior to its use in describing these systems. It is likely that these two systems will respond differently because of their different evolution-ary histories. Therefore, certification should not use the model relationships developed for biodiversity and their spatial patterns of persistence in the landscape from studies conducted on island. Although this information may be a useful starting point for thinking about these systems. In general, the theory states that areas reduced in size will have fewer species and will have less-stable ecosystems.

5.2.1.3.4 *Spatial scale and ecosystem functioning*

Another issue relating to spatial scale and ecosystems is the relationship between spatial scale and ecosystem functioning. For example, how much area is required to maintain some level of primary productivity? Island theory may again provide some insight. Although communities on islands and mainlands are different according to the Island Biogeography Theory, ecosystem functioning is often not different. In other words, even though islands are smaller and have few species, they do not necessarily function any differently than mainland communities (Mueller–Dombois 1995). Therefore, based on island systems, it appears that there is *not* an overriding principle, like the species-area relationship for species richness and spatial scale, for ecosystem functioning and spatial scale, or for ecosystem functioning and species richness.

This pattern can be explained by evolutionary adaptation on islands and the level in the biological hierarchy at which the maintenance of ecosystem functioning is important on islands (i.e., genetic differences may be more important than on mainlands). Small, isolated continental areas that have not evolutionarily adapted to their size probably do not have ecosystem function being maintained at the same level in the biological hierarchy as island communities. Therefore, as a more general rule, if the area of study is small enough and enough diversity is lost at the levels in the biological hierarchy determining ecosystem function, there will be an effect on ecosystem functioning. In conclusion, the spatial scale necessary to maintain ecosystem function and the level in the biological hierarchy indicative of ecosystem functioning will be dependent on the system, but the maintenance of ecosystem funtioning is likely to be affected by changes in spatial scale.

Because of the use of the management unit to define scale in certification, it is important to know that no single scale can be automatically selected for use at each site. The Island Biogeography Theory is probably not useful for developing species conservation strate-gies. But what is apparent from the previous discussion is that there is no general rela-tionship between species and ecosystem function. It is also clear that the scale appropriate for each management unit must be determined for each site. Since conserving biodiversity is such an important part of many certification protocols, it is critical that a certifier realizes that no general paradigm explains biodiversity levels and spatial scale. It is important to assess the quality of the landscape for each site and how landscape fragmentation affects the function of each management unit (see 5.2.1.2).

5.2.1.4   Pest/pathogens

5.2.1.4.1   Introduction

Kristiina A. Vogt, William H. Smith

Pests and pathogens have traditionally been viewed as something "undesirable" in the environment that should be controlled using chemical or biological management. That view is changing and there is a greater realization that insects and microbial pathogens (e.g., root decay fungi) have important roles in natural systems. These organisms have important roles in determining the diversity of plant species found in forests and in determining the cycling rates of nutrients. Frequently, pests and pathogens become more dominant in ecosystems that are already degrading. Their increased presence appears to be a mechanism for accelerating the manner in which a degrading system recovers and moves on a trajectory toward a "healthy" system. In natural systems, there is increasing evidence that insects which feed on plants and microbes that infect plants play a number of vital roles in determining the character, organization, and processes occurring in these systems.

These natural roles of pests and pathogens must be distinguished from the many examples of pests and pathogens that are having unusually significant and devastating impacts on ecosystems. Most of these negative examples are due to exotic and invasive species that have been inadvertently — or on purpose — introduced into areas outside of their normal ranges (see 5.2.1.5). In other cases, cultural activities of human managers have elevated to damaging levels the status of native microbial pests and pathogens. In both cases, these pests and pathogens produce a legacy in the ecosystem because they drive the expression of ecosystem structures and functions. Therefore, it is important to be aware that insects and fungal diseases have important roles in ecosystems; they are not just problems that should be eradicated. In most cases, when epidemic outbreaks of insects or pathogens occur, they are noted by the public because they are unusual events and result in many dead trees. Dead trees are not valued by the public even though they are beginning to realize the importance of dead wood in carbon and nutrient cycles in forest ecosystems.

In many cases, by the time the insect or fungal pathogen outbreaks have occurred, the primary factor causing system degradation is unclear. This uncertainty is created because pests/pathogens are frequently the secondary agents taking advantage of other changes that have already occurred in the ecosystem (e.g., air pollution or climate change predisposing trees to be weakened, etc.). Insect/microbial activity may be an excellent indicator of the health status of the system, but it does not establish the initial cause of the problem.

Pests and pathogens are mentioned by almost all of the certification protocols, although what data to collect and how to use that information is not articulated (see 4.2.1). No indicators are specifically given on how to assess their role in ecosystems and to determine whether the impacts are negative or positive (i.e., increasing long-term productivity, Romme et al. 1986). Many of these pest situations create considerable difficulties for small landowners since these pest/pathogen outbreaks are frequently not confined to just their lands. The potential for outbreaks to spread from neighboring lands may be very strong and a single landowner may be unable to even consider managing for the pests or pathogens by themselves.

5.2.1.4.2 *Forest ecosystem structures and functions controlled by phytophagous insects and microbial pathogens*

William H. Smith

Phytophagous insects and pathogens perform multiple roles in natural forest ecosystems by:

- Controlling the structure and function of ecosystems
- Influencing the distribution of plant species in the landscape
- Influencing plant species composition and therefore herbivore composition, by controlling the food sources
- Modifying species diversity within the landscape
- Significantly affecting ecosystem productivity
- Changing the resistance and resilience characteristics of ecosystems
- Influencing the biogeochemical cycles by changing the rate of nutrient cycling

Their ecosystem roles are important and their loss from the ecosystem would be detrimental to ecosystem functioning — and therefore its sustainability — so it is important to consider in certification protocols. A brief discussion follows on the roles they play in natural forest ecosystems.

The best examples of insect-regulated structural impacts on forest ecosystems are derived from studies of succession. Examples of insect-driven succession can be found in both coniferous and deciduous forests. Good examples of insects controlling species richness and density can be found with the mountain pine beetle, the spruce budworm, and the gypsy moth, to name a few (Smith 1986).

For example, the mountain pine beetle (*Dendroctonus ponderosae*) plays a critical role in lodgepole pine succession in both the absence and presence of fire (Amman 1976). Depending on the site and fire history, the mountain pine beetle can both favor and restrict lodgepole pine success. In the absence of fire, beetle infestations will kill the large, dominant lodgepole pines. Residual and more shade-tolerant species will grow more vigorously. This happens since fewer living trees will compete for limiting abiotic resources until the next cycle (20 to 40 years) of beetle outbreak. These cycles are repeated until lodgepole pine is eliminated from the stand. This contrasts with what happens when fire is present. With fire, the large accumulations of beetle-killed lodgepole pine results in hot fires which eliminate other tree species that compete with lodgepole pine. Since lodgepole pine is a serotinous species, these fires help it to regenerate on these sites.

Species richness and density characteristics of natural ecosystems are also influenced by microbial pathogens. A good example of this can be found in the high-elevation, mixed-conifer forests in the Cascade Range in Oregon. In the forests that are dominated by mountain hemlock, root infection by *Phellinus weirii* causes hemlock mortality to occur in circular disease centers that expand approximately 34 cm in radius per year (Smith 1986). In time, these "disease pockets" are invaded by less-susceptible lodgepole pine; true firs, including Pacific silver fir and subalpine fir; and mountain hemlock (Nelson and Hartman 1975). A net effect of *P. weirii* infection is an increase in forest diversification.

Microbial pathogens can regulate the distribution of trees by specifically infecting some plants, but not others. This results when disease eliminates some species from growing in particular environments. For example, white pine blister rust disease caused by *Cronartium ribicola* limits the reproduction of eastern and western white pine by killing young trees. In Australia, *Phytophthora cinnamoni* restricts eucalyptus species to ridge-top sites because root infections of eucalyptus are quite severe in the lower valleys (Burdon and Shattock 1980, Shea et al. 1983). Another exotic fungus (*Ophiostoma novvulmi*) is the causal agent for the Dutch elm disease and, along with elm phloem necrosis caused by a mycoplasmalike organism, has severely impacted American elm. Both diseases severely reduced the distribution of American elm between 1930 and 1980 in mesic and lowland forests of the eastern United States (Smith 1986). Annosus root disease of pine caused by *Heterobasidion annosum* can eliminate pine on managed sites with sandy or sand-loam textured soils with no high seasonal water table (Anderson and Mistretta 1982). A microbial pathogen was responsible for the elimination of the American chestnut from the canopy of eastern hardwood forests. The result of this microbial pathogen was the conversion of oak-chestnut type forests to oak-hickory or oak-type forest (Shugart and West 1977). Prior to its elimination from the landscape, the American chestnut was one of the most important hardwood tree species in the eastern United States. It also had considerable commercial value for a variety of wood-based products, provided food for wildlife, and was a keystone species throughout much of its natural range (Smith 1986).

> Phytophagous insects and microbial pathogens have significant roles in two of the most critical processes in regulating ecosystem function: nutrient cycling and productivity.

It is largely through their influence on nutrient cycling that these biotic agents impact forest productivity (Smith 1986). Phytophagous insects and a large variety of saprophytic soil arthropods play numerous essential roles in ecosystem-level biogeochemical cycles. The former group, especially foliage feeders, has received much attention because of their obvious and very dramatic effects. In temperate zone forests, foliage insects generally consume 3% to 8% of the annual leaf production. However, when outbreaks occur, this percentage can rise quite dramatically. Short-term decreases in productivity have been recorded with moderate (above 40%) and heavy (above 75%) defoliation. These decreases in tree growth are mainly the result of reallocation of carbon resources from wood production to refoliation to reestablish foliar biomass. Long-term increases in forest productivity, however, may follow severe defoliation events. This effect is largely due to the increased availability of both nutrient resources and other limiting factors, such as light and moisture. Surviving plants are able to access greater amounts of abiotic resources. Similar responses have been recorded with bark beetles and lodgepole pine where long-term productivity increased dramatically after a bark-beetle attack (Romme et al. 1986). In addition, Mattson and Addy (1975) showed that fifteen years after an eastern spruce budworm outbreak, wood production in the defoliated forest was higher than in an undefoliated forest. In another study, Wickman (1980) was able to show increased radial growth of defoliated trees being maintained for 36 years following defoliation.

The impact of insect defoliators on ecosystem carbon and nutrient cycles was demonstrated by Swank et al. (1981) using the watershed approach. These investigators showed that with chronic defoliation (approximately 33% maximum defoliation) by the fall cankerworm (*Alsophila pometaria*), a substantial increase in stream export of nitrate nitrogen followed. This study showed that the defoliation caused a temporary shift of carbon allocation from wood to leaf production, increased rates of nutrient uptake, and increased the amount of litter being added to decomposition.

The infection of roots by fungi results in root disease that ultimately leads to reduced nutrient and water uptake, reduced wind-stress resistance and may eventually result in tree death. Several basidiomycetes (e.g., *Phellinus weirii, Heterobasidion annosum, Armillaria mellea*), which are poor competitors in the soils, have evolved a specialized capacity to parasitize woody roots of trees. Many of these fungi are ubiquitous and can cause significant tree loss in managed forests (Smith 1986). Many pathogens strongly influence tree productivity by altering the flow of carbon through a tree and its fixation and storage. Pathogens infect trees at several different locations and can disturb carbon fixation, translocation, and storage in foliar tissues, branches, stems and roots.

Phytophagous insects, and microbial and other pathogens play varied and important roles in forest ecosystems. The natural roles for these organisms must be identified, understood, and separately assessed from events that can be classified as exceptional or adverse. It is important that indicators are able to distinguish those events which, if ignored, will result in system degradation. It is important to determine what management activities will result in insect and microbial pathogen being elevated to pest status. Management must ensure that the important ecosystem roles of these organisms are not hampered.

5.2.1.5 *Invasive alien species*

Bronson W. Griscom, Laura A. Meyerson, Kristiina A. Vogt

Globally invasive species are now viewed by ecologists as a threat to the structure and function of ecosystems (D'Antonio and Dudley 1995, Vitousek 1990) and should be recognized as such in the design of forest-certification protocols. Humans are responsible for virtually all alien species invasions occurring today through our role as dispersal vectors and agents of novel forms of disturbance in ecosystems. The introduction of an invasive species into an ecosystem differs from other types of anthropogenic impacts, because invasive species may continue to spread and modify the system after human activities (i.e., introductions or disturbances) have stopped (Cronk and Fuller 1995). This can create a pervasive and lasting threat that leaves an ecosystem legacy that may affect future management efforts. This discussion shows that the ecological paradigms typically presented for invasive species are not supported by the literature and that our present understanding of invasives and forestry is elementary.

Invasive species can hamper or limit the management options in forests and can therefore constrain a forestland owner's ability to manage their lands sustainably. Some management activities (see Chapter 6) temporarily create environments that invasive species can take advantage of and some of the tree species used in management have been shown to become invasive (see 5.2.1.5.3). The biology of invasive species must be understood in order to respond to their increased occurrence and to refine forest practices so that the spread and detrimental impacts of invasive species can be minimized. This understanding should also provide the tools which allow forestland owners to reduce the invasibility of an ecosystem.

Invasive species (i.e., those that spread beyond the site of initial introduction) can have severe biological and economic costs. In fact, biological invasions follow land conversion as a leading cause of the loss of biological diversity (Meyerson et al., 1998). It was estimated by the congressional Office of Technology Assessment (U.S. Congress 1993) that the economic losses due to 79 of the estimated 300 invasive plants in the continental United States averaged $97 billion between 1906 and 1991 (Marinelli 1996). This congressional study concluded that alien invasive species are having major impacts on natural areas, agriculture, industry, and human health. The scale of this issue is also suggested by the dramatic number of alien invasive species comprising present day floras. It has been estimated that one-fifth to one-third of all plant species in the continental United States and Canada are alien (Marinelli 1996). Nearly half of all plant species in Hawaii are alien (Randall 1996).

A small percentage of the alien species present in an ecosystem can have a disproportionate impact. It is estimated that only 1% of the species introduced to an ecosystem cause native species displacement or alteration of ecosystem function (Williamson 1996). Despite an extensive literature on invasive species, it has been difficult to generate consistent supportive rationales on why alien species begin to dominate in some habitats (Higgins and Richardson 1996) and the role disturbance plays in this process (McIntyre and Lavorel 1994). Humans are intimately involved in the phenomenon of alien species invasion:

- As the primary instigators introducing species to different habitats
- As the generators of novel disturbance regimes which facilitate alien species spread
- As stockholders or stakeholders in the resulting environmental changes and the economic losses that result

5.2.1.5.1   *Factors controlling invasive species success*
Bronson W. Griscom, Laura A. Meyerson

Despite the extensive descriptive literature on species introductions around the world, it may come as a surprise that only weak statistical generalizations can be produced on the attributes of invasive organisms. As mentioned above, predictions on the precise outcome of invasive-species introductions are relatively unreliable due to the creation of ecosystem legacies (Higgins and Richardson 1996, Williamson 1996, Lodge 1993). Even the most commonly known attributes of alien species such as high r (intrinsic rate of natural increase), high abundance and range in native habitat, and good climate and habitat matching between donor and target ecosystems are often unrelated to the success of an introduced alien species in becoming established. Both Higgins and Richardson (1996) and Lodge (1993) argued that the extensive evidence from case studies around the world (MacDonald et al. 1986, Mooney and Drake 1986, Drake et al. 1989, di Castri et al. 1990, Richardson et al. 1992, Williamson 1996) supported few consistent generalizations on invasive species success.

In attempting to understand and predict alien species invasions, the authors have focused on four issues:

1. The introduction phase of alien species invasion (Williamson 1996)
2. Attributes of the alien species (Baker 1986, Bazzaz 1986, Heywood 1989, Reichard and Hamilton 1997)
3. Attributes of the target ecosystem (Lawton 1984, Price 1984, Walker and Valentine 1984)
4. The role of disturbance (Hobbs and Huenneke 1992)

The complexity and lack of generalizations on invasive species' success is due to the fact that all four factors (e.g., introduction, alien species attributes, target ecosystem attributes, disturbance) may be involved (Lodge 1993). Furthermore, each factor is relevant during some stages of alien species invasion (e.g., introduction, establishment, spread, and equilibrium), but not in others (Williamson 1996). Given the variety of factors involved in the outcome of alien species invasion and the shifting relevance of each factor as an invasion proceeds, it is understandable that the outcome of alien species invasions is not yet predictable. Nevertheless, we can reduce the probability of detrimental alien species invasions in managed forests by understanding the biology of alien species invasions.

Many of the mechanisms responsible for successful invasions by animals (and other motile organisms) are different from those involved with plants. Two differences stand out when comparing animals and plant invasions: 1) animals often alter the trophic structure of the target ecosystem (Vitousek 1990) and 2) animals tend to be less dependent upon disturbance for successful invasion (Ehrlich 1989). The most transformative impacts on species demographics and easily measured impacts on ecosystem-level characteristics have resulted from alien animal invasions (Vitousek 1990). The exploits of *Homo sapiens* provide an example that need only be mentioned. Animal introductions are particularly disruptive on oceanic islands that lacked any large generalist herbivores or predators prior to human arrival. The dramatic influence of the introduction of pigs, goats, and cattle on soil erosion, nutrient cycling, and other ecosystem-level attributes has been documented (Stone 1985, Vitousek 1990).

The four factors contributing to explaining the success of alien species are briefly discussed below. The role of disturbance will be discussed in more detail because of its potentially important role in facilitating invasions by alien species. It is important to obtain an understanding of the potential for management activities to facilitate invasions of exotic species within the landscape.

5.2.1.5.1.1. *Introduction of alien species*

To understand the invasion process, it is helpful to conceptualize the biotic and abiotic environment as a series of "filters" through which the alien species has to survive (Keddy 1992, Higgins and Richardson 1996). Williamson (1996) separated alien species invasion into four different stages (i.e., arrival or introduction, establishment, spread, equilibrium) where each stage can function as a filter for successful establishment. Other researchers have presented more elaborate sets of stages (Kornas 1990, Wade 1997). These filters influence demographic processes and thereby prevent unsuited plants from establishing, growing, reproducing, or dispersing. The filters become more refined as alien species approach later stages of invasiveness accounting for the low probability that introduced species become invasive.

In modern times, the initial stage of a successful invasion often has little to do with attributes of the target ecosystem and more to do with human activities (Roy 1990) and attributes of invasive species (Williamson 1996). A surprisingly large number of introductions were intentional efforts by agriculturalists, foresters, and hunters to introduce species as part of management. About half of the alien plant species in North America were brought in deliberately for horticulture alone (Marinelli 1996). In some cases, whim was the only incentive for introduction. For example, the North American population of the common starling (*Sturnus vulgaris*) derived from 60 birds released in Central Park in 1890 and 40 more in 1891 — by a club wishing to establish every bird named in the works of Shakespeare (Long 1981, Williamson 1996). In addition to human interest, ease of transit seems to play an important role (Williamson 1996). In the case of unintentional introductions, such characteristics as clonal growth in plants and the ability to survive without live food during transport in animals are important.

It is with the shift from introduction to establishment that attributes of the alien species and the target ecosystem become increasingly important, although the interaction between the two is often still strongly buffered by disturbance and/or human activities (Wade 1997). Two important phenomena are involved with the establishment phase: 1) lack of "enemies" and 2) propagule pressure (Williamson 1996). The term "enemies" refers here to pathogens, herbivores, predators, competitors, and parasites that may have a detrimental impact on individual fitness. Often these enemies are left behind when an alien species is removed from its native ecosystem and introduced to a new one (Roy 1990, Williamson 1996). This phenomenon confers a competitive advantage on the invader against native species. For this reason, plants often tend to reproduce more vigorously and grow taller in alien environments than in their native ecosystems. For example, both *Chrysanthemoides monilfera* (introduced to Australia from South Africa) and *Acacia longifolia* (introduced to South Africa from Australia) have seed production rates that are an order-of-magnitude higher than in their native habitats (Crawley 1987, Noble 1989, Blossey and Notzold 1995). This "lack of enemies" phenomenon may also help to explain why *Impatiens noli-tangere*, a native herb of Britain, is declining in its native range and expanding in the United States, while *I. capensis*, a native of the United States, is expanding in Britain (Williamson 1996).

5.2.1.5.1.2. *Attributes of alien species*

Some authors focus on attributes of invasive species to explain why certain alien species dominate (Baker 1986, Bazzaz 1986, Heywood 1989, Reichard and Hamilton 1997). Bazzaz (1986) discussed autecological attributes which confer invasive success. In models of woody plant invasions, Reichard and Hamilton (1997) found the most reliable predictor of invasive success was whether or not a species was known to invade elsewhere in the world. Panetta (1993) suggested that attributes of alien species themselves will provide the primary explanation for the success of certain alien species and that attributes of the target habitat are secondary. However, this conclusion assumed that the invaded ecosystems did not share important attributes which may have interacted with the species in question to confer invasibility. While attributes of alien species are certainly critical and will be discussed below, we believe that they must always be considered with reference to the equally important attributes of the target ecosystem.

Although lack of enemies confers to invasive species an inherent advantage, they are generally at a disadvantage due to a limited supply of propagules needed to establish a stable population (di Castri 1990). Even the most invasive species have had numerous failed introductions. Limited initial population size creates vulnerability to chance events (Williamson 1996). This is one of the primary reasons species with a high intrinsic rate of increase are generally considered to have an advantage as alien invasive species (Baker 1986, Ewel 1986, Crawley 1989, Bastl et al. 1997). Certain tree species have received considerable research attention because they are economically important. In many cases, these tree species are introduced as part of industrial forestry into habitats where they were previously not found, and therefore are aliens. These species tend to be fast growing, highly productive, commercial species with characteristics that make them potentially invasive if not properly managed. Surprisingly, the evidence that a high intrinsic rate of increase confers a strong advantage in invasive ability is limited and contradictory (Lawton and Brown 1986, Williamson 1996).

5.2.1.5.1.3. *Attributes of target ecosystems*

By developing a concept of "vacant niches," other authors invoked attributes of the target ecosystem to explain why alien invasive species dominated in some habitats. (Lawton 1984, Price 1984, Walker and Valentine 1984). This argument stated that ecosystems with a reduced array of organism functional groups (relative to other similar ecosystems) have a high number of vacant niches and, therefore, an introduced alien species is more likely to establish successfully. This theory has been attacked on the grounds that vacant niches do not exist because: 1) introduced species usually reduce resource availability to native species either through competition or altered ecosystem processes; and 2) a niche is not a property of an ecosystem, but is an emergent property of each organism and its environment (Whittaker et al. 1973, Herbold and Moyle 1986, Williamson 1996). In fact,whether or not invasive species reduce resource availability to native species depends on the relative importance of competition vs. mutualism in community structure (Perry 1994). While the notion of vacant niches may not be appropriate, the importance of the addition of novel plant functional mechanisms is an important underlying concept that should not be discarded.

5.2.1.5.1.4. Disturbance

Disturbance has also been invoked as the critical phenomenon facilitating alien invasions. Most of the research studying the role of invasive species and disturbance has been conducted by ecologists studying natural systems or their edges with different land uses. Less research has been conducted in "working forests" specifically focusing on management activities as disturbance agents.

Disturbance plays an important and distinct role in the establishment of sessile species that compete for space (e.g., plants and barnacles) (Wade 1997). Motile species, on the other hand, often do not depend upon disturbance for establishment (Williamson 1996). It is at this stage in alien invasion that different mechanisms drive the invasive success of plants, animals, and other organisms. There are many examples of animal invasions where disturbance plays a limited, if any, role in ecosystem invasibility (e.g., wild boar [*Sus scrofa*]; brown tree snake [*B. irregularis*]; predatory snail [*Euglandina rosea*]; cats [*Felis catus*]; and minks [*Mustela vison*]) (Lever 1985, Williamson 1996). In a number of other cases, disturbance is associated with animal invasiveness (Elton 1958, Leidy and Fiedler 1985, Isenmann 1990).

Elton (1958), drawing heavily on insect examples, argued that disturbed areas are particularly vulnerable to invasion (Simberloff 1989). Higher incidence of insects in disturbed habitats may be due to inherent aspects of invasive insects, sampling error (humans inhabit disturbed habitats), or the far higher number of deliberate attempts to introduce insects into agricultural systems (e.g., disturbed systems). Only until recently has there been any quantitative data to support the assumption that alien insects are more abundant in disturbed habitats (Simberloff 1989). Surveys by Medeiros et al. (1986) and Reimer (1994) have found that few alien ant species had proliferated into undisturbed cooler montane ecosystems above 1000 meters in Hawaii (where the majority of intact ecosystems remain). However, where alien ant species have invaded these systems, they are having a dramatic impact. Reimer attributed the limited entry above 1000 meters to the fact that species that are adapted to these cooler environments must first survive at their point of entry in the lowlands, which acts as a filter.

It is apparent from this review of the literature that it is important to differentiate between two potential roles of disturbance in influencing invasive success. Disturbance can affect target ecosystems by:

1. Opening growing space
2. Altering historic (endogenous) disturbance regimes

The opening of growing space previously occupied by native species increases the likelihood of invasion simply due to chance and opportunity. Alterations in the disturbance regime of an ecosystem increase the likelihood of invasion because they create new contexts within which species can compete for resources. Alterations in the disturbance regime can either reduce the likelihood that native species will have pre-adapted mechanisms for surviving and being competitive or increase the likelihood that alien species will have pre-adapted mechanisms for surviving and being competitive. The first scenario suggests that any disturbance event will benefit alien species. However, the second scenario suggests that only disturbances that represent a shift from the historical frequency and pattern will benefit alien species. Considering that these two phenomena may occur simultaneously helps to explain the difficulty which many authors have had in explaining the relationship between disturbance and alien species' success (Lodge 1993, Higgins and Richardson 1996, Williamson 1996).

A number of studies on plants have demonstrated that disturbance is correlated with higher diversity or abundance of alien species (Baker 1986, Ramakrishnan and Vitousek 1989). Hester and Hobbs (1992) censused species' presence along a gradient from the edge to the center of a native shrub and woodland nature reserve surrounded by agricultural and pasture land in the western wheat belt of Australia. They found that density and cover of non-native species was consistently greatest at the reserve edges. DeFerrari and Naiman (1994) studied the relationship between disturbance and the abundance (measured as cover) and richness of alien plant species by comparing census data from upland and riparian sites with different ages since disturbance in forests of the Olympic peninsula in Washington state. The alien plant species they encountered tended to be annual and perennial ruderals with small, wind-borne seeds. They found that both the abundance and richness of alien species showed the strongest correlation with intensity of disturbance. Riparian areas disturbed by recent flooding had the highest alien species abundance and were subjected to the most severe (in terms of percent of leaf-area removal) disturbance. Upland clearcut sites came second in terms of alien species abundance and richness. The timber harvesting removed the overstory, but much of the understory cover remained, unlike flooded sites. On both upland and riparian sites, alien species abundance and richness declined with time since disturbance. The relationship between alien species and native species was not studied. However, in at least some sites (i.e., beneath some alder stands and on some riparian cobble bars), alien species appeared to occupy a niche that was not otherwise occupied by native species.

These studies demonstrate that higher abundance and richness of alien species are often found in more disturbed sites, whether it be natural or anthropogenic disturbance. This correlation is likely due to some combination of: 1) factors that determine which species are introduced to a target ecosystem; 2) factors that determine how many propagules of a given species reach different sites within a target ecosystem; and 3) opportunities of open growing space created by disturbances. However, in order to understand how different types of disturbances may have different roles in determining which alien species *dominate* in a given ecosystem, it is important to consider the relationship between the richness and abundance of both alien and native species. In addition, it is important to consider the extent to which disturbance has changed through time.

In order to understand the relationships between native and invasive species, it is necessary to look beyond disturbance intensity. The literature reviewed here suggested that the historic disturbance regime relevant to the evolution of the species should be conceptualized as an attribute of an ecosystem. This is similar to climate, geology, soils, and species that are considered attributes of an ecosystem. Denslow (1980) hypothesized that any natural community would be richest in species adapted to respond to the type of patch most commonly created by the historic disturbance regime (referenced in Hobbs and Huenneke 1992). The alteration of this regime is likely to create contexts for which native species have not had the opportunity to evolve mechanisms (Hobbs and Huenneke

1992). This is the rationale used to explain Hobbs' and Huenneke's statement, "Total diversity of native species at the landscape level will be greatest when disturbance occurs at its historical frequency and in the historical pattern" (Hobbs and Huenneke 1992).

What will happen when historic disturbance regimes are altered? The alteration of disturbance regimes has not been a random process, especially within the last four centuries due to the prevalence of "European-model" disturbance as the source of novel-disturbance regimes within many ecosystems outside of the Mediterranean Basin region (di Castri 1990). At the beginning of the Neolithic period (i.e., ca. 10,000 years ago), changes in the distribution of plants in Europe became de-coupled from climate change and human activities became the primary controlling factors (Vernet 1990). For example, a high diversity of species have adapted to soil disturbances and fire associated with agricultural activities (Trabaud 1990, Hobbs and Huenneke 1992). This diversity of species adapted to human disturbance is apparently due to the long history of human activities and the associated migration and evolution of plant species in the regions surrounding the Mediterranean Basin (Kornas 1990). Outside of the Mediterranean Basin, five other regions are considered centers of domestication and may have been regions where a similar occurrence of plants adapted to human disturbance could be found (Simpson and Ogorzaly 1986).

It is informative to examine the responses of native and alien species to both historic and altered forms of disturbance to determine whether human disturbances that differ from natural disturbance regimes may increase the presence of invasive species. Some examples are summarized below:

- McIntyre and Lavorel (1994) studied the influence of human-disturbance variables on relative native vs. non-native plant species richness in the temperate grasslands of New South Wales. They found that in the presence of alien plants and various types of human disturbance, the intermediate disturbance hypothesis (Connell 1978) did not apply to this site. Alien species increased while native species decreased with increasing levels of soil disturbance and water enrichment. Alien species did not vary significantly with different grazing intensities (considered an "endogenous" disturbance), but native species declined at the higher levels of grazing. Green and Baker (1997) found an increasing abundance (density and canopy volume) and richness of alien species along a gradient from rural to urban sites along riparian corridors in the lower Sonoran Desert, Arizona. In the extreme disturbance case of lawns and golf courses in the United States, where a large influence of human decision is involved, 73% of species have been found to be Eurasian and only 12% native (Baker 1986).

- The presence of native species is particularly low in much of California, where lowland terrestrial summer-green communities are rare, and few native species are adapted to these conditions. However, rice fields in wet alkali soils in California are a form of human disturbance that more closely mimics tule swamps, and here 70% of plant species are North American and about 20% are Eurasian (Baker 1986).

- In the French Mediterranean region, Trabaud (1990) studied the influence of fire on the invasion of sites by alien plants in the eight most frequently encountered plant communities (e.g., grasses, forests, etc.) in the region of Montpellier, France. While Trabaud (1990) found that some annual species entered the sites briefly, all were gone within the first few years following the fire. All native species regenerated from surviving underground organs or by seed. These results were attributed to the fact that these communities have been subject to fire disturbance for millennia (Trabaud 1990).

- Hester and Hobbs (1992) studied the influence of fire on the relative abundance of regenerating native vs. alien species on a nutrient-poor site of *Allocasuarina* shrubland and a more fertile site of *Acacia-Eucalyptus* woodland in the western Australian wheat belt. Fires were reported to be a natural part of the vegetation dynamics for both vegetation types, and the sites were last burned 60 years ago. They found that in the shrubland, fire (burning leaves and small branches, but leaving large, woody stems standing) did not affect the abundance (stem density) of alien plant species in the post-fire regenerating stand, but increased the abundance of native species — potentially creating a competitive advantage for native species. In the woodland, fire (only in the understory) did not affect the relative abundance of native and non-native species (Hester and Hobbs 1992).

- During and Willems (1986) argued that the loss of most lichens and reduced bryophyte diversity in Dutch chalk grasslands was due to the cessation of grazing (Hobbs and Huenneke 1992). Cowling et al. (1986) found that fire suppression has been responsible for the conversion of a South African open, grassy veld to a vegetation now dominated by non-native shrubs (Hobbs and Huenneke 1992).

- One of the most dramatic and large-scale "natural experiments" on the differential role of disturbance in alien plant species invasions is the response of temperate grasslands to the arrival of Europeans (cattle and their humans) (Mack 1989). The temperate grasslands in Australia, South America, and western North America (to be referenced as the "vulnerable" grasslands) all experienced massive changes in species composition because of the introduction of European cattle. The temperate grasslands of Eurasia, southern Africa, and central North America (to be referenced as the "resistant" grasslands) experienced comparatively little loss of native species and alteration of community structure. According to Mack, the critical differences which conferred greater resistance to the "resistant" grasslands were 1) the presence of large, hooved, congregating mammals in the Holocene (or longer), and 2) dominance by rhizomatous grasses (as opposed to tussock grasses). Tussock grasses originally dominated the vulnerable grasslands. Tussock grasses have a tightly packed cluster of erect tillers and an erect apical meristem that is more exposed to grazing and trampling than the low-sprawling form of rhizomatous grasses in the other group of grasslands. Furthermore, the tussock grasses reproduce via seed and the inflorescences are vulnerable to grazing, while the rhizomatous species have many axillary buds and can regenerate vigorously after destruction by trampling hooves. Consequently, the "vulnerable" grasslands experienced a rapid invasion and dominance by alien rhizomatous species such as *Poa pratensis*. The "resistant" grasslands also received a number of alien plant species, but native rhizomatous species remained relatively abundant (Mack 1989). This is a clear example where only the combination of a novel form of disturbance and the addition of a novel functional group lead to dominance by alien species.

In some instances, disturbance can have positive effects on both alien and native species richness within a given community. In these situations, the timing of disturbance can have an important influence. For example:

- Gopher excavations in California can create islands of diversity for rare native plant species on the one hand (Martensen et al. 1990) and facilitate invasion by alien invasive species on the other hand (Hobbs and Mooney 1985, Hobbs and Huenneke 1992).

- Grazing on upland British grassland can have both positive and negative effects on the competitive success of native species relative to alien species (Rawes and Welch 1972). In this case, the timing of grazing to avoid the sensitive phases in the lifecycles of plant species vulnerable to grazing was found to be important for maintaining native diversity (Wells 1969, Hobbs and Huenneke 1992).

There can be an important feedback link between an alien invasive species and disturbance or resource availability (Vitousek 1990, Higgins and Richardson 1996). In some cases, this feedback loop can result in significant modification of the target ecosystem and allow an alien species to dominate (Hughes et al. 1991, Ramakrishnan and Vitousek 1989, Vitousek 1986, Vitousek and Walker 1989, Walker and Vitousek 1991).

Not all feedback loops increase alien species abundance. The alien crucifer *Lepidium latifolium* L. established monoculture stands in wetlands along the Pacific Coast of the United States. Blank and Young (1997) found that *L. latifolium* caused reduced soil fertility. Eventually, the aggressive grass, *Bromus tectorum*, becomes more competitive under reduced levels of soil nutrient availability and replaces *L. latifolium*. Ramakrishnan and Vitousek (1989) reviewed other examples of the effects of alien species invasion on ecosystem-level characteristics.

The literature review previously discussed suggests four concepts relevant to invasive species:

1. The functional mechanisms of alien invasive species, and the functional mechanisms present within the native species in the target ecosystem tells us something about the likelihood of dominance by the alien species.

2. Shifts from the endogenous (historic) disturbance regime in an ecosystem can alter the types of functional mechanisms that are most successful.

3. Alien species can have dramatic effects on ecosystem-level processes, in terms of both energy and material cycles, and disturbance regimes. These effects can create positive feedback loops with alien species abundance and species richness.

4. Invasive species are frequently found to be animal species or herbaceous, grassy, or shrub plant species. Woody perennial tree species are not as commonly found to be invasive.

5.2.1.5.2   *Ecosystem-level effects: legacies from invasive species*
Laura A. Meyerson, Bronson W. Griscom

All of the issues discussed above (introduction, attributes of alien species, attributes of target ecosystems, and disturbance) play a role in the eventual impact of alien species on ecosystem function. These factors do not operate in isolation, but interact to create emergent processes that may determine the eventual impact of an alien species introduction. Thus, it is both the variety of factors involved, and the resulting emergent processes that explain the complexity of alien species invasions. Many of these emergent processes are a result of positive feedback loops between alien invasive species and disturbance or resource availability which create legacies (Higgins and Richardson 1996).

Disturbance forces can interact with certain species to create positive feedback loops when disturbance-tolerant invasive species also increase the susceptibility of an ecosystem to that disturbance (Hobbs and Huenneke 1992). For example, Hughes et al. (1991) found a positive feedback loop between fire disturbance and invasion of fire-tolerant alien C_4 grasses into seasonally dry shrubland habitats in Hawaii. Their results demonstrated that grass invasion took place in two distinct phases involving at least two alien grass species: *Schizachyrium condensatum* and *Melinis minutiflora*. *Schizachyrium condensatum* is effective at invading never-burned shrubland habits and providing fine fuels necessary to carry fire. *Melinis minutiflora* depends on fire disturbance for invasion, and, following fire, replaces *S. condensatum* as the dominant alien grass. Furthermore, *M. minutiflora* increases the intensity of fire, and the damage of fire to native species. *Melinis minutiflora* also (unlike *S. condensatum*) expands vegetatively with a rhizome mat and, once established, is capable of smothering native vegetation. These alien grasses have also be shown to alter the water regime (Mueller-Dombois 1973). Hughes et al. (1991) predict that the conversion of native shrubland to exotic grassland is permanent.

It is generally easier to detect the ecosystem-level changes wrought by invasions of animals than those caused by plants, perhaps because modification of the uppermost trophic level results in more rapid, large-scale modifications (D'Antonio and Dudley 1995). Feral pigs (*Sus scrofa*) invaded the Great Smoky Mountains National Park in North Carolina and Tennessee during the 1940s. This species roots around the forest floor and has a significant impact on the structure and composition of forest understory communities. Comparisons of soils in areas disturbed by the pigs to non-disturbed areas revealed thinner forest floors, mixing of organic and mineral horizons, and a greater proportion of bare ground. These structural changes were accompanied by increased cycling of nutrients in these ecosystems (i.e., higher concentrations of nitrogen and potassium in the soils to 100 cm depths, higher concentrations of nitrate-N in the streams) (Vitousek 1986).

The alteration of resource availability by alien species can also have dramatic ecosystem-level effects. *Myrica faya*, introduced into Hawaii for erosion control along highways, presents one of the best examples. This plant is a nitrogen-fixing species native to the Canary Islands that has invaded primary successional sites (e.g., lava flows > 20 years old) and large areas of the mesic and dry woodlands of Hawaii (Vitousek et al. 1987). It is able to colonize young lava sites and spread rapidly, changing the rate of species change on primary successional sequences. The main effect of *Myrica* occurs because it alters nutrient cycles by adding three to four times the amount of N to soils than what would occur with native species (Vitousek et al. 1987). Nitrogen-fixing plants often dominate the early stages of primary successional sequences around the world. Unlike the introduced *Myrica* species, however, native N-fixing plant species in Hawaii (Acacia and Sophora) do not have effective seed dispersal mechanisms (Vitousek and Walker 1989). Lack of colonization by native N-fixers is not due to a low presence of the required N-fixing symbiosis. Because of the resulting high soil N levels, Vitousek and Walker (1989) speculated that *Myrica* facilitates the invasions by other invasive plant species, such as strawberry guava (*Psidium cattleianum*) which, in turn, attract another introduced invader, the feral pig.

On the other hand, some invasions appeared to have little ecosystem-level impact even though they resulted in the loss of native species, although further research is necessary to confirm these findings (D'Antonio and Dudley 1995). For example, the American chestnut tree (*Castanea denatata*) has been almost completely extirpated from the eastern forests of the United States by the introduced fungus *Endothia parasitica* (D'Antonio and Dudley 1995). However, chestnut has been replaced by oak (*Quercus* spp.) so that species composition has changed but no detectable ecosystem level changes have been recorded (D'Antonio and Dudley 1995). Introduced species have been found to cause extinctions of native species on islands, but few studies have been done to document the effects of species extinction on ecosystem-level processes. This lack of ecosystem-level research done simultaneously as the documentation of population-level changes makes it difficult to establish the link between introduced species and alteration of ecosystem-level processes (D'Antonio and Dudley 1995).

Given that some invasive species will have ecosystem-level impacts, the rates of change caused by the invasion will depend on several factors. Rates of change will depend on how quickly the species becomes established and disperses; its competitive ability and resource-use efficiency; which species it displaces (e.g., a keystone species); as well as the type of ecosystem invaded and its resistance and resilience. Whether the invasive species becomes dominant or simply "naturalized" will impact the rate of change that occurs. A naturalized species may be one that integrates into the ecosystem but does not significantly alter the community. However, some species cause dramatic changes in the ecology of ecosystems and accelerate or decelerate processes such as succession.

5.2.1.5.3   *Implications of invasives for management*

Kristiina A. Vogt, Laura A. Meyerson, Bronson W. Griscom

Most forest management activities are not automatically contributing to increasing the presence of alien invasive species in the landscape. A few cases have been documented where industrial forestry has been responsible (directly or indirectly) for the spread of alien species around the world. Active forest management has the potential to contribute directly to the invasive species problem because exotic tree species with fast growth rates are desirable in industrial forestry and they have been introduced into areas where they were not previously present. This only becomes a problem, however, when and if the species become naturalized. Since most forest operations harvest trees prior to them reaching the stage where they are reproducing, the movement of trees from planted forests where they can become invasive is minor. Many of the invasive tree species have been introduced into the landscape from ornamental tree plantings or trees planted for erosion control and not from "working forests." Humans are the primary disturbance agents involved in the introduction and spread of nearly all cases of invasive species (Williamson 1996).

Indirectly, some forest management activities may provide opportunities for other invasive species to establish. Some forestry practices, such as road building or harvesting, have the potential for changing the resistance and resilience characteristics of landscapes and can make them more prone to being invaded. These activities are not, however, unique to active forest management. For example, many highways and development areas have extremely high numbers of exotic invasive species. It can be generalized that when a forest practice highly disturbs the environment or changes how disturbances are expressed across the landscape (e.g., fragmentation), the greatest potential exists for invasive species to become established.

Some of the invasive species that are found to be pests worldwide are woody, perennial species of early-successional, not late-successional stages. Most of the early-successional tree species are fast growers and have the general characteristics identified with many alien species which make them so successful in invading other habitats. It is important to recognize that many woody plants that have become invasive alien species are not part of forestry operations. For example, many alien woody species — intentionally introduced for erosion control (*Myrica* example from Hawaii), to restore degraded lands, or as part of agroforestry — have become invasive in their new habitats (Richardson 1998). Many exotic trees (e.g., Christmas trees, city road-side plantings of Norway maple) and shrubs have contributed to increasing the presence of other non-tree alien invasive species. This occurs because they harbor insects, plant pathogens, and other weed species that can invade ecosystems and devastate native flora (Pimentel 1986).

It has been noted in forestry that the most invasive trees are the are those that have been the most widely planted for the longest periods (Richardson 1998). Furthermore, the ecosystems that are most greatly affected are those that have the longest history of intensive planting (Richardson 1998). For example, *Pinus* spp. is an especially problematic genus in the Southern Hemisphere where at least 19 species have become invasive (Richardson 1998). The invasiveness of *Pinus* has been attributed to three variables in particular: short juvenile period, small seed mass, and short intervals between large seed crops — which leads to early and consistent reproduction and rapid population growth (Rejmánek and Richardson 1996).

In light of the previous discussion, managers need to ask the following questions when it comes to invasive species:

1. What disturbance regimes will enhance the dominance of invasive species?
2. What management activities, other than the introduction of trees species, will enhance the dominance of invasive species?
3. What management activities change the landscape-level interactions and cause invasive species to spread?
4. What legacies exist in the ecosystem from past forest management activities that are conducive for the spread of invasive species?
5. What management activities can be incorporated into management that will minimize the introduction of invasive species into forests?

Disturbance regimes often occur at a variety of spatial and temporal scales, and also have different vectors of disturbance (drought, fire, and flood), which influence microsites differently within the ecosystem. There is an unavoidable process of trial and error in identifying disturbance regimes under which alien invasive species do or do not exclude (or minimally exclude) native species. This necessitates having a temporal framework of analysis when studying the impact of an invasive species in an ecosystem.

Due to the role of disturbance in alien species invasion, managers may be faced with having to use particular management activities to maintain native diversity, even though these management tools create pathways for alien species to invade. The challenge is to find practical means of managing forests without allowing management practices to function as vectors for alien species spread. The central challenge to managing alien species invasions will not be maintaining alien-free ecosystems, which is not practical, but identifying alien species that are likely to exclude native species and change ecosystem processes before they become invasive.

Identifying potentially problematic alien species at an early stage is critical to minimizing the detrimental impact of invasives on ecosystem health. Because of the present inability to predict the impact of invasive alien species, the common approach for dealing with alien species is eradication (Meyerson et al. 1998). Even following successful eradication, invasive species may leave a legacy in the environment that impedes recolonization by native species. Many of the eradication procedures also employ chemicals that may augment potential legacy effects. However, most of the certification protocols frown on the use of chemicals in management operations and penalize forestland owners when they use them. This results in a conflict between the forestland owner and the certifier when the dominant mechanism for controlling invasives is chemical control.

For a long time, forest management has had to manage invasive species and the problems they create (Smith et al. 1997). Since silviculturalists are concerned with controlling species composition to obtain their management objectives, they are frequently faced with controlling invasion by undesirable species. Controlling species composition means regulating disturbance to favor the species being managed for, and sometimes directly removing unwanted species through "cutting, poisoning, controlled burning, or regulated feeding by various forms of animal life…" (Smith et al. 1997). There are several examples where the more disruptive land management regimes associated with agriculture have led to poor regeneration in forests found in the same landscape. In Southeast Asia, agricultural practices are causing natural forests to be deforested at a rate of 16,000 km^2 per year (Kuusipalo et al. 1995). After cutting these forests, shifting cultivation with short fallow periods causes land to be invaded by *Imperata cylindrica*, a rhizomatous perennial pan-tropical grass. This invasive species has a high growth rate and allelopathic effects on tree seeds and other plants, thus preventing natural regeneration of trees (Kuusipalo et al. 1995). Since *I. cylindrica* is also a pest species for agriculture, farmers prefer slash-and-burn cultivation to reclaim their lands. Since this grass makes poor fodder for domesticated animals, those regions invaded by this grass have essentially become wastelands. Reforestation efforts have met with limited success because of annual fires, lack of standing trees, and a poor seed bank (Kuusipalo et al. 1995).

This aggressive Asian grass (e.g., *I. Cylindrica*) has become one of the critical forest management issues in the humid tropics. It is invading highly disturbed and degraded agricultural soils, resulting in the arrested succession of old fields in both the old and new world tropics (Kuusipalo et al. 1995). Woods (1988) compared the density of invasive *I. cylindrica* grass to native forest understory plants in patches of forest in Sabah, Malaysia that had been subjected to different degrees and types of disturbance: 1) logging; 2) fire; 3) logging and fire; and 4) no disturbance. On the "logging and fire" sites, the otherwise natural fire disturbance was apparently enhanced as a result of the logging. Woods found that only on patches subjected to the combined effect of human disturbance through logging and fire (because of the 1983 drought) was *Imperata* able to invade and dominate the understory.

Ramakrishnan (1992) found that in northeast India, the presence of native competitors prevented *I. cylindrica* from arresting succession. In addition, Kuusipalo et al. (1995) found that where *I. cylindrica* is not native, the introduction of a fast-growing alien pioneer-tree species (*Acacia mangium*) successfully overcomes the arrested succession of old fields. However, this approach should be used with caution, since it is using one exotic to control another invasive species.

How disturbances contribute to increasing the susceptibility of a management unit to being invaded by alien species is presently not known. This is an important topic of research and must consider how management affects the types of edges that are created and, therefore, the health of edge environments and their resistance to invasive species (see 5.2.1.2). It becomes important to understand why alien species are dominant in some habitats and what role disturbance plays in this process so that management activities can be regulated to hinder rather than facilitate the increased presence of invasive species. This type of information is important for certification where the sustainability of forest management is a goal and invasives can decrease the health of an ecosystem.

Invasive species do not appear to be able to invade intact forest ecosystems. Overall, fewer introduced species have invaded intact ecosystems, primarily because the lack of disturbance has provided fewer opportunities. A study in central Indiana of relatively undisturbed forest "islands" surrounded by alien agro-ecosystems found that introduced species moved only one or two meters into the forest (Brothers and Spingarn 1992). In this study, the introduced species did not enter the forest interior and were not abundant relative to native plants (Brothers and Spingarn 1992). The authors attributed this to low light levels in the forest interior, the low levels of disturbance in the fragments themselves, and potentially to the inability of these species to disperse into the forest without human assistance.

Forest management often introduces exotic tree species in plantations in order to attain attributes valuable for timber production (e.g., high productivity, stem straightness, and abbreviated rotations) (Smith et al. 1997). In contrast to agricultural crops, most commercial forests in the United States are managed native species (Pimentel 1986). Silviculturally, foresters ideally select species composition based on the best economic and biologic alternatives (Smith et al. 1997), and not based on their ecosystem function. Intensive management may artificially maintain many ecosystem functions through fertilization, irrigation, and control of "weed" species. These systems are created to produce products of economic value; therefore, maintaining productivity is a more important goal than maintaining ecosystem function. At times, these systems are artificially maintained in a state that closely resembles early succession — lower diversity and simpler structure (Huston 1994).

Examples of some tree species and the problems that have occurred with them are summarized below:

- In other parts of the world, exotics have been used to supplement native species. Exotic species have several advantages; these include: rapid growth and an adaptability to produce a wide range of wood products (Richardson 1998). A good example of this is *Pinus* in the Southern Hemisphere. Trees from the genera *Pinus* and *Acacia* are some of the most widely planted and are featured prominently on weed lists and in reviews of invasive species (Richardson 1998).

- Monterey pine (*Pinus radiata*) is the most widely exported tree species in the world and forms the major softwood resource in Australia (Chilvers and Burdon 1983, Smith et al. 1997). As an introduced species, *P. radiata* has been highly successful, having been planted to cover nearly one million hectares of land in different parts of the world. This sharply contrasts with its limited natural range that is estimated to cover only 4,000 ha (Cronk and Fuller 1995). Interestingly, *P. radiata* does not exhibit invasive characteristics in its native range in California or when introduced to the Mediterranean basin, Atlantic France, or Chile. However, it has become a very aggressive invader in South Africa, Australia, and New Zealand (di Castri 1989, Richardson 1998). This is a good example of how an introduced species can sometimes have greater growth in a new region rather than in its native habitat because of more favorable climatic conditions (Smith et al. 1997). In Australia, *P. radiata* that invaded a native eucalyptus forest had a growth rate ten times greater than that of native trees (Chilvers and Burdon 1983).

- Eucalyptus is another example of a genus introduced for many purposes around the world (e.g., ornamental landscape tree, reclamation of degraded lands, etc.). Some eucalyptus species were considered desirable to plant when using them to reclaim dry woodlands because of their capacity for deep rooting (average ~ 35 meters) (Stone and Kalisz 1991). This deep rooting has allowed it to thrive in droughty regions because of its ability to access deeper water sources; this results in it having higher productivity rates than native species. However, dense plantations of this genus are now recognized as a problem in some regions because they can cause water tables to decrease at rates higher than is natural (Vogt et al. 1997a).

- Australian pine (*Casuarina* spp.) was introduced into south Florida in the late nineteenth century for use as windbreaks, shade, and lumber (Langeland 1990). Since then, it has become one of the worst pest plants in Florida. Australian pine has three mechanisms by which it appears to dominate (Langeland 1990). It forms dense stands that out-compete native species. It accumulates litter that inhibits the growth of other plants. It is capable of fixing nitrogen, which may give it an additional competitive advantage. It has become so abundant along some sandy shores that it interferes with the nesting of sea turtles and American crocodiles (Cronk and Fuller 1995). Cutting, in combination with herbicide, is currently the only means of control (Langeland 1990).

- Research on tropical forest restoration has begun to focus on biological barriers to natural regeneration and on jump-starting succession by utilizing "nurse trees" on degraded and impoverished sites. This method is useful for restoring soil nutrients and structure and improving microclimatic conditions to favor regeneration. While exotic species are sometimes used because of beneficial capabilities (e.g., legumes), research is underway to utilize native-tree species in Costa Rica. This ongoing research by Guariguata et al. (1995) is investigating the impacts of single and mixed-species plantings on the invasion success of the plant understory. While invasion by forest trees is enhanced under one species (*Vochysia*), shrub invasions are more abundant under all other treatments. There was no woody recruitment after one year in the unplanted controls. This type of research with native species is important for the regeneration of native tropical forests that are often degraded by invasive species due to poor forest practices and because it does not require the introduction of another exotic.

Currently, it is impossible to determine whether a more simplified ecosystem (e.g., plantations) can be managed to minimize the spread of invasive species or whether they are inherently invasible to other species. This question is not being asked by certifiers who are generally not interested in having plantations as a dominant type of forest management activity. Most certification protocols prefer not to have plantations as part of sustainable forest management strategies, although some have developed criteria and indicators for plantations (e.g., Smartwood of Rainforest Alliance). At this stage, it is probably more relevant to ask whether the introduction of exotic species in plantations have irreversible ecological impacts in the ecosystem and not whether they increase the dominance of other invasives in the landscape.

When searching for evidence for negative ecological impacts of introducing an exotic species, eucalyptus is typically mentioned. The eucalyptus tree provides a very good example of problems that can arise with plantation monocultures with high productivity maintained at short rotations. Eucalyptus planted in semi-dry regions uses excessive amounts of water resulting in soil erosion problems, impoverishment of soil nutrients, and the creation of poor habitat for wildlife (Smith et al. 1997). However, eucalyptus is not responsible for these problems, but rather the practice of using short rotation plantation monocultures with high growth rates (native or exotic). In cases like these, it would not be possible to manage the forest sustainably because rooting depth and nutrient uptake are difficult to control. Under these conditions, eucalyptus should not be used as part of the forest-management strategy. In fact, many of the eucalyptus examples are derived from land reclamation efforts and not from forest operations. What this states is the importance of adapting active forest management for the particular forest that is being managed. This uses the characteristics of the species (i.e., this is silviculture) to determine where and how forest tree species should be used in management. This type of information feeds into certification at the level of the management plan where the decisions are being made on the silvicultural system which are most relevant for each forest.

5.3   Social legacies constraining natural resource uses

5.3.1   Human values driving natural resource conditions

Kristiina A. Vogt, Joyce K. Berry

Use of natural resources is strongly controlled by several socioeconomic variables that are mainly driven by human values. Many of the conflicts in natural resource use have been produced by the existence of institutional structures and regulations reflecting public values. Much of the current conflict has revolved around issues of:

- Public and private property rights and economic growth within a matrix of mixed ownership.
- State or federal laws controlling land development and its uses (Loomis 1993; Gunderson et al. 1995).
- People's preference and value for familiar species which they import into their current habitats (see earlier section on invasive species).
- People's preference for landscapes that fit their aesthetic view of what a natural landscape should look like.

The human desire for particular types of commodities out of a natural system base has contributed significantly to changing forest function and its health (Rapport 1997, Vogt et al. 1999ab). Within the last 100 years, there has been an unprecedented change in the conditions of natural systems due to human activities. Most of these changes can be attributed to heavy grazing, selective timber extraction, and fire exclusion (Agee 1993, Sackett et al. 1993, Vogt et al. 1997a). The desire of managers for particular economic products has driven many of these management approaches. For example, the desire to extract valuable tree species has resulted in the remaining tree species becoming more dominant within the landscape. This means that the characteristics of the remaining trees become important determinants of the health of the ecosystem and its response to other disturbances. For example, when Douglas fir was selectively harvested in the Pacific Northwest, white pine remained and became more dominant. However, white pine is not very tolerant of fire and does not survive well without fire control. Because of fuel loads that had been allowed to accumulate during fire control, many fires that would have been small scale in the past have now become large scale. In addition, the legacy forest also experiences atypical amounts of insect pest problems which reflect its degraded condition due to human land uses (Rapport 1997). The insect problems result in more dead trees and help to fuel the spread of fire in these forests.

Human values have left a disturbance legacy that has made the landscape in the Pacific Northwest and southwest United States more prone to devastating fires (Agee 1993, Sackett et al. 1993, Rapport 1997). This has happened mainly because people do not want fires to occur around their homes or in nearby forests. Controlling fires in forests early in the century (Williams 1989) created forest conditions where species not sensitive to fire began to increasingly dominate the landscape. Recent realization of the natural role of fire in "cleansing" forests has validated the need to allow fires to occur within the landscape (Knight 1991). However, the legacy landscape resulting from earlier fire controls has produced a forest that is not adapted to the current fire regimes. Human values of land uses have resulted in changing the natural frequency and intensity of fires in many parts of the world, while the fires of human origin have dramatically increased within the last hundred years (Specht 1991). Many of the places where the fire frequency has increased are in areas that have been classified as degraded. These same areas have been reported to have increased insect pest problems, root pathogen problems, and weed complexes outcompeting native species (Specht 1991).

Human values for particular plant or animal species and a desire to plant or introduce them into newly occupied areas have created many new, well-documented problems (Mooney and Drake 1986, see 5.2.1.5). Many times, the introduced species do not have natural predators and, therefore, are able to spread rapidly within the landscape. Some of the introduced species have changed the disturbance cycles to those which favor their continued dominance. For example, some of the grasses introduced into Hawaii have increased fire frequency and thus, are eliminating native non-fire-adapted species (Loope et al. 1988). In addition, these introduced grasses are not active during the rainy season, so erosion has become a significant problem (Loope et al. 1988).

Human values also change with time, so that what was acceptable in the past may no longer be acceptable today. A good example of this is Yellowstone, where economic values for resource extraction used to be the dominant value for how the Greater Yellowstone Ecosystem should be managed (Clark and Minta 1994). Today, environmental values and species preservation have become the overriding values guiding what types of goods and services should be produced (Clark and Minta 1994). Concurrent with the environmental values driving the management of this ecosystem are other resource-use values associated with fire control, disease control on bison, tourism, private property rights, and mainte-nance of viable ranches outside of the park's perimeter. Prior to humans establishing artificial park boundaries, fire was an important disturbance controlling the composition and structure of these systems. It was not until after the great fire in 1988 that the role of natural fires and the effects of excluding fires became better understood (Knight 1991).

In addition to changing values within the Greater Yellowstone Ecosystem (GYE), its legacy landscape is very different from what existed more than a hundred years ago. Isolating the GYE within defined boundaries has changed how the ecosystem responds to current disturbances and how easily it can sustain the mix of vegetative communities needed as habitat and food by park animals. Many of the large mammals (especially bear and bison) living within the boundaries of the park are unable to obtain their habitat and food requirements within the confines of the park (Clark and Minta 1994). In this example, different social values often conflict with one another and produce different legacies within this landscape that will impact future park management.

5.3.2   Regulations and laws as constraints

Joyce K. Berry, Kristiina A. Vogt

Regulations create legacies within a system because they are rules that cannot be elimi-nated by democratic process. They define, or can limit, the options or types of activities that can occur on public and private lands. Thus, they should be understood as constraints that will strongly control what can be potentially applied on a given piece of land. Frequently, laws reflect public values for forestlands and, therefore, are a good barometer of past and current public values. However, they may not reflect ecological constraints inherent in many systems. They may also not be targeted toward those factors causing system degradation.

On public lands, regulation of the management practices and the goods and services to be provided from these lands has been a long-term process in the United States. For example, the nineteenth and twentieth century statutory laws established a system of public land management by professionally trained foresters for the multiple benefits of all Americans. These laws can be summarized as follows: The 1891 Forest Reserve Act, 1897 Organic Act, 1905 Transfer Act, 1960 Multiple-Use Sustained Yield Act, 1974 Forest and Rangeland Renewable Resources Planning Act, and the 1976 National Forest Management Act.

The benefits of these acts were to (Vogt et al. 1997a):

- Improve and protect the forest
- Secure favorable conditions of water flows
- Furnish a continuous supply of timber for the use and necessities of citizens

However, the public became increasingly concerned about the public lands' capacity to provide multiple goods and services without habitat degradation and species loss. During the 1980s and 1990s, this concern was reflected in changing public values that often conflicted with congressional mandates (such as timber production targets), sharp differences among the public, and forest-policy gridlock. With decreasing satisfaction with land management, and declining timber production, more public attention focused on private forests and their ability to provide goods and services.

Regulation of private landowners is not a recent phenomenon. Pinchot believed that private forests needed to be regulated because of over-exploitation of these lands (Ellefson and Cheng 1994). Concern over management of private lands can be traced back to 1934 when Article X of the Lumber Code was enacted to control logging on private lands. (This code was struck down as unconstitutional by the Supreme Court less than one year later.) However, regulation of private landowners has blossomed in the last few decades and has fomented conflict between proponents of common and private property rights.

The statutory laws enacted for public lands, of course, do not directly transfer onto private lands, but there are many other laws that are relevant for private forestland owners. Laws exist at the federal, state, and local levels that are specifically geared to controlling landowners interested in harvesting trees or any other natural resource from their lands. Ellefson and Cheng (1994) concluded that the legal limits to forest practices have become a way of life for landowners in every state in the United States. They describe a common regulation pathway that starts with passive announcement of voluntary guidelines, moves to laws imposing penalties if guidelines are not followed, and ends with a law requiring a plan, notification, reviews by the public and agencies, and inspections. Satisfying all of the laws can be a formidable undertaking for a small landowner just wanting to clear or harvest a few acres of their land. Not all these regulations, however, drive small landowners from choosing to manage their lands for timber and the government supports various incentives and cost-share programs to encourage good forest practices. The majority of private landowners also share the public's desire to protect the environmental attributes of the land and to improve the condition of the natural resources (Ellefson and Cheng 1994).

There has been strong opposition to federal involvement in private land uses (Cubbage et al. 1993), although several federal laws exist that affect private property rights. Some of the federal laws that directly affect private landowners follow:

1947 Forest Pest Control Act
Protects all lands, regardless of ownership, from destructive forest insects and diseases. First legislation to direct management across ownership and political boundaries.

1947 Insecticide, Fungicide and Rodenticide Act
Required registration of pesticides.

1970 Clean Air Act
Established standards to control pollutants, including documentation of adverse effects and identification of control techniques and costs.

1972 Federal Environmental Pesticide Control Act
Was an amendment to 1947 Insecticide, Fungicide and Rodenticide Act. This gave the EPA more regulatory control and required registration of all pesticides and certification of applicators on private and commercial lands.

1972 Federal Water Pollution Control Act
Addressed industrial point and non-point sources of pollution. Forest activities and agriculture were considered non-point sources of pollution. This act required permits from the Corps of Engineers to discharge dredged or fill materials. Silviculture was exempt, except when wetlands were altered or area was brought into forestry. Relevant for protecting wetlands and riparian areas. Road building required a permit.

1973 Endangered Species Act
The purpose of this act is to save species and their habitats. It bans taking species on endangered list and allows for preservation of critical habitats.

1977 Clean Air Act

1986 Safe Drinking Water Act

1987 Clean Water Act
Revisions to Federal Water Pollution Control Act resulted in this act. Each state was required to prepare a water-quality management plan that included identifying non-point sources of pollution that violate water quality, describing control mechanisms, and Best Management Practices (voluntary).

1990 Clean Air Act
Amended the 1970 Clean Air Act. This provided acid rain compliance and listed 189 chemicals to be regulated.

The two acts that have had the greatest impacts on private landowners are the 1987 Water Quality Act and the 1973 Endangered Species Act (ESA). The Water Quality Act exempts "normal" silviculture from requirements permits unless a wetland is being altered or when an area is being converted into forestry. The definition of "normal' is a continual concern for landowners since it is not clear what kinds of activities this word encompasses. In order to obtain a permit under this act, the following factors need to be included in any evaluation: "conservation, aesthetics, economics, general environmental concerns, fish and wildlife values, water quality, energy needs, property rights, and the needs and welfare of people" (Cubbage et al. 1993).

The 1973 Endangered Species Act has probably had the most impact on private landowners. The ESA contains two main processes to ensure the continued existence of species:

1. Designation of species and their critical habitat through "listing"
2. Protection from illegal trade, "taking," or agency actions that jeopardize species or their habitats (O'Laughlin 1992)

The term "take" is defined as harass, harm, pursue, hunt, shoot, kill, trap, capture or collect or engaging in any conduct that attempts to do any of the previously listed activities (Environmental Law Institute 1977). "Harm" is defined to include the modification or degradation of habitat (Code of Federal Regulation, 50 C.F.R. Section 17, 1994). Under the ESA Section 10, non-federal parties (including landowners) may apply for an "incidental take permit" by developing a habitat conservation plan.

From Pinchot's time to the present, no federal regulation has *directly* restricted timber harvesting on private lands in the United States (Cubbage et al. 1993). (Federal regulations do *indirectly* influence forestlands through the ESA and wetlands regulation.) Most of the initial attempts to regulate private forestlands occurred at the state level in response to the threat of federal regulation (Dana and Fairfax 1980). Many of these attempts go back to the 1920s, but most were not very successful in developing regulation for private lands. Some of the regulations that do exist can be summarized as follows (Cubbage et al. 1993):

- In 1903, the first seed tree law was enacted.
- By 1940, eleven states had some type of forest practices act.
- By 1950, one-fifth of commercial forestlands were under some type of legislation (mainly prescribed cutting restrictions, regeneration requirements).
- In the 1970s, state forestry acts, especially in California, Washington, and Oregon, included protection provisions for runoff, erosion, water quality, and fish and wildlife species and habitat.
- By the 1990s, 22 states had some type of state forest-practices legislation:
  1. Not all were comprehensive forest-practices acts, but each included some control of timber harvest, regeneration, or voluntary Best Management Practices.
  2. Regional differences exist in private forest legislation. The Northeast is mainly concerned with logging and its effects on water quality, wildlife, and aesthetics. The South is mainly regulating logging and setting regeneration standards. The western laws limit the role of local governments in the regulatory process.

Many regulations related to forests also exist at the local-government level (i.e., one-hundred in the state of New Jersey), with the type of ordinances varying regionally throughout the United States (Martus et al. 1995). In the northern states, ordinances are more concerned with protecting environmental quality and aesthetics. In the southern states, regulations center more on safeguarding local government investments in roads, bridges, and highway structures. Hickman and Martus (1991) described five general ordinance categories enacted by local governments: 1) public property and safety; 2) urban/suburban environmental protection; 3) general environmental protection; 4) special-feature habitat protection; and 5) forestland preservation. Martus et al. (1995) noted that these local ordinances have increased ten-fold in the past ten years and they predict that they will continue to increase.

5.3.3   Examples of tax laws that affect private forestland owners

Frederick A.B. Meyerson

Virtually every law affecting property ownership has some direct or indirect effect on private forestland owners. These effects range from direct impacts on land use, such as those of the Endangered Species Act, to the much subtler — but more widespread and pervasive — effects of income tax, property tax, and inheritance laws.

5.3.3.1   *Federal income tax*

The federal income tax was adopted in 1913 and changed both the face of American wealth and the land-preservation movement. In the late nineteenth and early twentieth centuries, private conservationists such as the Rockefellers began to purchase, protect, and donate significant forested lands, including Acadia National Park, Redwoods, and Muir Woods. Prior to 1913, the year the income tax was established, there was essentially no financial incentive for this action — it was more a function of what Alexis de Tocqueville once described as the American tradition of fulfilling public needs through private action.

The creation of the income tax and the charitable deduction resulted in an additional motivation for land transfers from private to public hands. This became a significant factor when income tax rates began to rise in the 1930s and especially before and during World War II. Organizations (e.g., Nature Conservancy, Trust for Public Land) have recognized that the tax benefits of donations effectively lower the cost of parkland conservation. They have used this fact to facilitate and manage land transfers or bargain sales to non-profit conservation groups and government entities within and outside the United States.

However, the income tax (and other taxes) cuts both ways for the private landowner. While the income tax provides some incentive for transfer of land for conservation purposes, tax burdens in general create the necessity for landowners to find ways to produce income from their land. This fact — in combination with the equity that an investment in forested land represents — tends to drive exploitation of the forest resources and/or sale to realize profits.

5.3.3.2   *Property taxes*

Property taxes are an annual cost of land ownership that create an incentive or necessity for exploitation of forest resources to meet tax obligations. Rising land values increase the tax burden which, in turn, increases the necessary level of exploitation.

Many states have enacted laws that reduce the property tax burden on forested land developed, maintained, and/or managed in accordance with certain rules. A typical law might require the creation and implementation of a forest management plan in exchange for a substantial reduction in the property tax rate. While these laws may slow the sale of forestland for development purposes, they also have other effects. They may require the commercial exploitation of forest resources and therefore be counterproductive to efforts to maintain old-growth stands. The fact that these laws vary substantially from state to state is relevant to any efforts for a coordinated timber certification effort.

5.3.3.3   *Inheritance Laws*

Laws that affect the intergenerational transfer of property rights have a substantial impact on long-term land management by families. Until recently, all estates in the United States with a total value greater than $600,000 were subject to inheritance taxes at rates which rose rapidly to 55% for estates greater than $3,000,000. A change in the tax law will gradually increase the $600,000 minimum to $1,000,000 over the next several years, but the essential effect remains the same.

For example, the estate of a person who dies owning several million dollars worth of forested land will be subject to a federal estate tax of a million dollars or more. This can often be satisfied only through the sale of some or all of the land, unless the decedent has other substantial assets. Therefore, family members or other beneficiaries of the estate may be unlikely to maintain the existing parcel in its entirety or as forest.

In some states, relief from the effects of the inheritance tax (and property tax) may be found through the sale or donation of development rights, which reduces the assessed value of the land and, therefore, both the property and the inheritance tax. In certain cases, special provisions for the inter-generational transfer of family farms may also be applicable and operate to reduce estate taxes. Finally, a few additional methods of inter-generational transfer prior to death are available which can operate to substantially lower the tax burden and facilitate keeping forested lands intact and within a particular family. Due to their complexity, many of these methods are essentially available only to those with access to substantial legal resources.

The combined effects of income, property, and inheritance taxes on the ownership, use, and future of private forested land cannot be underestimated. The size of the yearly tax burden affects the way in which private owners view their land. Either they view it as an inexpensive asset that can be developed over the long run or left unexploited, or as an expensive asset which must be sold or exploited to offset or eliminate the tax burden. The inheritance tax affects the longer-term perspective of families towards forested land. The anticipation of a large, future estate tax bill may induce present exploitative behavior towards the forest resources. On the other hand, it may propel a family towards one of the tax-reduction or avoidance mechanisms, such as the sale of development rights. In any event, these long-term solutions obviously have the potential to affect or interact with certification regimes.

6

Direct and Indirect Impacts of Natural Resource Management Practices on the Ecosystem

Bruce C. Larson, Daniel J. Vogt, Michael Booth, Kristiina A. Vogt, Laurie E. Koteen, Peter A. Palmiotto, Jennifer L. O'Hara

6.1 Introduction

In this chapter we address the fundamental question, "What is good and bad forestry?" We approach this topic by examining various impacts of management practices on different forest ecosystems. It is intuitive that not all management practices will produce the same effects in a given forest and that particular treatments will vary in their impacts across a variety of forest types. An understanding of these consequences lends itself to the selection of the most appropriate indicators of sustainable forest management. Presently, small land owners face the problem of having too many indicators to monitor management impacts on ecosystems. With the informed selection of a few site-specific indicators, monitoring the effects of forest management becomes more tractable and economically viable.

If one examines all the issues related to the condition of a forest, many different human values exist that impinge on this analysis (NRC 1998) and that are not encompassed by the values incorporated into certification. These encompassing values were grouped by the 1998 NRC report into the following categories: biodiversity; forest fragmentation and habitat isolation; rare and endangered species habitat; forest management intensity; forest fires; air pollution; carbon sequestration; forest insects and disease; alien invasive species; and watershed integrity. This report dealt with institutional relationships and policy, and governmental initiatives as important issues that crossed the landscape, but not as directly affecting the forest condition (NRC 1998).

Some of the values listed above are also those factors which have been recorded as impacts in the field, especially in tropical forests. For example, selective logging conducted in mixed species stands often acts as a thinning operation that releases the less valuable tree species This occurs in many forests around the world, both temperate and tropical (Daniel 1980, McGee 1982, Oliver and Kenady 1982, De Leo and Levin 1997). Thus, after multiple rotations, a long-term sustained yield of the same commercial species is unlikely from the forest stand. Also, a decrease in total species diversity may result since the valuable timber species decrease significantly in numbers. Other ecological effects of selective logging practices include an increased inbreeding among extracted species (Murawski et al. 1994); significant alteration to forest structure (Cannon et al. 1994);

0-8493-1585-9/00/$0.00+$.50

destruction of forest undergrowth; decrease in bird species richness and abundance (Thiollay 1997); and soil compaction on tractor tracks which significantly delays natural regeneration (Guariguata and Dupuy 1997).

Management decisions used in the practice of silviculture and harvesting of forest resources typically consider several different types of information that have their own sets of choices that must be made as part of management. Depending on what type of activity is implemented, consideration may be given to many of the points articulated here (Smith et al. 1997):

1. Regeneration system to be used (matched to species and site)
2. Rotation lengths, availability of different developmental stages, and stand growth rates
3. Product markets
4. Rate of stand development
5. Natural disturbances characteristic of the location (ice storm, fire, etc.)
6. Protection of soil; water quality; unique natural features in the landscape; and wetlands
7. Setting the stage for efficient harvesting that considers the ability to control harvesting operations and the aesthetics of the resulting operations: existing road networks for access; road construction; equipment availability and costs; preventing landslides
8. Capacity of the site to continue to grow a managed forest — sustained management
9. Wildlife species, habitat requirements of wildlife species
10. Habitat utilization by domesticated animals (cattle, sheep grazing in forests)
11. Economic consideration on market availability and financial returns
12. Invasive alien species
13. Pests and pathogens particular to a region
14. Control of waste and damage, and protecting remaining trees during harvesting

The dominant tools used for site management and harvesting include:

1. Mapping of resources: trees, soils, aquatic systems, area in different stages of development
2. Computer simulations of growth models, landscape interactions between forest units
3. Silvicultural systems and site treatments: broadcast burning, woody-residue management, thinning, fertilization, herbicide applications, etc.
4. Methods of regeneration (natural or artificial) (Smith et al. 1997):
   a. High-forest method: regeneration of stands mainly from seed
   b. Clearcutting method: removal of the entire stand at one cutting followed by natural or artificial regeneration
   c. Seed-tree method: cut entire stand at one time except for a small number of seed trees
   d. Shelterwood method: cutting an old stand over a short period of time to produce a cohort of regenerating plants

e. Selection method: developing an uneven age distribution of trees in a stand by periodic cutting of individual or groups of trees

f. Coppice-forest methods: production of stands mainly from vegetative regrowth of individuals

For greater detail on these different regeneration methods, the reader is referred to Smith et al. (1997). The first two tools have minimum impact on the environment from the activity itself. The next two tools are the ones that have generated the most controversy. Today, it is apparent that past fire control policies have resulted in forested ecosystems that are not considered healthy and are susceptible to more pests and pathogens (Rapport 1997). Other controversies center on our desire to use forests as sites of high carbon sequestration to reduce atmospheric carbon dioxide levels. This desire to sequester carbon conflicts with management designed to minimize the presence of coarse woody debris in order to either increase efficiency of wood utilization or to reduce negative impact of woody debris on regeneration of some species. We also desire to preserve old-growth forests to maintain them as habitats for the diversity of those species dependent on them (Vogt et al. 1997a). This, of course, conflicts with the desire to harvest forests in a way that maximizes mean annual growth rates. The use of chemicals to control pests and pathogens is also quite controversial because there is disagreement on how to measure the life of these complex carbon-based compounds and the multiplicative effect of several toxic chemicals is not known (Wargo 1996).

Issues driving the negative reactions to forest management by the public have centered on topics that have been implicated in both species loss and degradation of ecosystem and/or landscape functions (see 3.2):

- Clearcutting (e.g., leaching of nutrients, species loss in the tropics, decreased ecosystem health, increased invasive species, fragmentation in the landscape) (see 5.2.1.1–5.2.1.5)

- Use of single species in monocultures (e.g., decreased diversity of plant species, loss of habitat for associated plant or animal species, decreased ecosystem health) (see 5.2.1.1–5.2.1.5)

- Road construction to harvest forests (e.g., migration of landless people into previously impenetrable forests, erosion potentials and associated nutrient losses, impacts on survival and reproduction of species — anadromous fish utilizing streams located in forested areas) (see 5.2.1.1–5.2.1.5, 5.3)

A useful way to identify which indicators most sensitively reflect the impact of management methods on the forested ecosystem is to compare them to natural system processes and functions (temporal, spatial, and disturbances; Vogt et al. 1998). When the degree of variability of response to an activity extends beyond the natural thresholds, the variables changing should be useful to monitor as an indication of system health and sustainability. The literature review of the impacts of different management activities will be synthesized using this approach to show when management is mimicking the variability recorded naturally (not a good indicator of system change) and when it is not.

Society in some way affects the growth and development of all forests in the world. Forest land owners, however, can affect (either permit or exclude) activities that *directly* impact the forest. This control is what we call forest management and it is this management that certification hopes to influence. Forest management modifies the disturbance pattern of the forest by either preventing or causing disturbances. Either way, the forest structure is changed, as is the susceptibility of the forest stands to other disturbances. The link between the values that humans desire to maintain and the management activities has not been made to show whether they match one another or are clearly distinct from one another. This chapter will summarize some of the literature that exists on the different management activities and how they are translated across the landscape.

Management activities fall into two broad categories:

- Those that are intended to regenerate forest stands, or at least create new age classes
- Those that impact stand structure or growth, but are not intended to regenerate a new cohort of trees

The establishment of regeneration means that growing space has been made available and a regeneration mechanism has been enabled. A regeneration mechanism can consist of the following: exposure of bare mineral soil for regeneration by seed, or preservation of the understory for the release of advance reproduction. Management activities that are not intended to lead to regeneration are usually called intermediate operations. These operations focus on either shifting the growing space from certain trees or other plants by thinning or competition release, or ameliorating the site through activities such as fertilization or drainage. It is important for certification programs to consider how each specific aspect of forest management operations affects the ecosystem at immediate and long-term time scales.

All forest management operations also require the construction of an infrastructure. This may be as simple as making skid trails and landings or as extensive as creating a major road system. On many occasions, the infrastructure causes more degradation of the ecosystem than the operation itself (e.g., road building). Many certification systems endorse or condemn a particular forest management activity based on an assumption that certain infrastructures will be created or used. It would be better to investigate and evaluate these structures separately.

6.2   Forest management activities leading to regeneration

6.2.1   Processing of logs for transport (from stump to landing)

Processing of logs before their movement from the stump to the landing can be classified as one of three types. The logs can be processed and transported as:

1. Tree length — the very top and all the limbs are removed
2. Cut-to-length — similar to "tree length," but the stem is cut into shorter lengths from 1.2 to 12.2 m long (4 to 40 ft)
3. Whole tree — the treetop and limbs are left attached

The type of "log processing" should be matched with the "log transport" system as described later (see 6.2.2). Certain of these combinations are unworkable from a machinery standpoint (i.e., whole trees moved with a forwarder), whereas others are impractical from a cost standpoint (i.e., cut-to-length logs moved with a grapple skidder). Forest management activities must be planned with a specific "log transport" and "log processing" combination in mind.

The effects of different "log processing" types on the forest are largely related to the increased difficulty of moving large objects through the woods. Long logs are inflexible and sweep broad swaths as they are skidded around corners. To minimize the disturbances of this action either a straight-line pathway should be created from the stump to the landing (often undesirable) or the skid trails should be cut wider than normal. Failure to plan for one of these two options results in great damage along the skid trails to seedlings and saplings, as well as bark wounds to residual, uncut trees.

Moving logs with the limbs attached has several implications for the assessment of logging impacts on the ecosystem. The most direct effect is that the limbs add greatly to the width and weight of the material being moved to the landing. As described later, this larger size results in not only greater potential for soil compaction, but in wider skid trails, as well as more associated damage to the vegetation along the skid trails. As the limbs are being dragged through the forest, they "sweep" the ground removing the forest floor, killing small vegetation, and exposing large quantities of bare mineral soil (see 6.2.2.1).

The choice of which "log processing" method and which utilization standards are used for the logging operation will determine how limbs will impact the forest ecosystem. The method and standard used will determine where the limbs will be left and how much will be left as part of the operation. For example, the limbs may be removed from the log and left near the stump. Or the limbs may be transported on the log/tree to the landing and then removed. Or the tops of the cut trees are sometimes chipped and the residue completely removed from the site to be used as fuel for energy. More often, the limbs are removed from the logs at or near the landing. However, even if the limbs are returned to the forest, their controlled placement is typically not what occurs naturally in the forest (e.g., slash piles). Thus, the influence of slash piles on the ecosystem would be spatially concentrated.

Logging will affect the placement as well as the amount of foliage and woody debris that is left in the forest stand. The amount of woody debris remaining in the forest stand is directly affected by logging. When the trees are felled, debris is left on the forest floor. The amount of debris will be determined by the size of the trees being cut; for example, if some small-diameter trees are cut and the whole tree is removed.

The amount of woody debris that accumulates in a stand can also be determined by natural processes occurring after the harvesting operation. Most of these inputs are indirect. Natural processes of mortality can increase or decrease the amount of woody debris that is added to a stand as the stand develops. For example, by harvesting some of the trees, self-thinning in the stand will be reduced which will decrease the amount of woody debris inputs. In this case, the reductions in the natural inputs of woody debris to the forest floor are related to the density-dependent mortality of trees. If even small-diameter trees are removed during logging (i.e., high utilization), the onset of self-thinning of the trees will be delayed even more.

The effects of residual branches and foliage from the logging operation will affect the forest stand physically and chemically. Their physical effects include influences on the moisture and temperature conditions in the soil, as well as potential modification of the habitats for insect pests and pathogens (see 6.2.2.1). Another physical effect can result from possible soil compaction due to the increased weight of the machinery used in the operations or even a decrease in surface soil compaction due to organic matter being worked into the soil.

The chemical effects are more complex and are either direct or indirect effects in the forest stand. The direct effects would occur as a release of the nutrients because of the decomposition of the branches and foliage. If the branches and foliage are moved from the stump area to some other location (e.g., the landing), nutrient export from the stand will occur as with the removal of the branches and foliage. Over time, this would decrease nutrient levels around the stump (but increase them at the landing). This practice results in a significant export of nutrients from the stump area if logging is conducted in a low site-quality forest. Since nutrient availabilities are more limiting in a low site-quality forest compared to a high site-quality forest, tree removal from low site-quality stands will more quickly result in increasing the deficiency of nutrients for that forest.

Therefore, an important consideration when disposing of tree tops would be the nutrient status of the foliage. It becomes important to consider the timing of harvesting because this determines the magnitude of nutrients removed from a site during a forest operation. Nutrients do translocate from the foliage to the branches, stems, and/or roots depending on the tree species and the time of year. If the harvest is conducted before nutrients in the foliage have been translocated (i.e., prior to foliage senescence) and the foliage is left to decompose near the stump, then the nutrients will become available to nearby uncut trees or regenerating seedlings in that area. However, if the foliage is removed to some other location (e.g., the landing), then over time the spatial variability of the soil nutrient concentrations could increase even more across the landscape creating a unique soil nutrient legacy. If this occurs, forest management would then need to consider fertilization plans that vary spatially.

Indirect chemical effects of the slash residue (especially the woody residue) on the ecosystem are mostly related to the availability of the soil nutrients, on both a short- and a long-term basis (see 5.2.1.1). The short-term and immediate effect on soil nutrient availability is related to the fact that woody debris typically has a high carbon-to-nitrogen ratio (much more C than N). For the microorganisms to decompose woody debris, they have to immobilize some of the available nutrients from the soil (especially N). So when large amounts of fresh woody debris are applied to the forest floor, the available pool of soil nitrogen would probably decrease (microbes competing with the trees for nutrients). However, as the woody debris begins to decompose (or mineralize), these immobilized nutrients are slowly released back to the soil. Thus, the woody debris may initially immobilize the available nutrients, but later act as a slow-release fertilizer.

In a forest stand of high site quality, the effects of soil nutrient immobilization by the decomposing woody debris may not be noticeable. But the effects may be quite dramatic in a low site-quality stand where some nutrients already limit tree growth. Even a temporary decrease in productivity and/or chlorosis of the foliage may be noticed in these stands, especially for the understory plants. However, in an open area (e.g., scarified soil, clearcuts) where there are no plants, this would not be a problem. In fact, this temporary immobilization of nutrients by the decomposition of woody debris may even benefit some ecosystems and act as a nutrient conservation mechanism. For example, during a logging operation the foliage from cut trees could quickly decompose and mineralize nutrients (relative to woody debris). The mineralized nutrients, however, could leach from the soil if there is no root uptake and/or if the soil has a low nutrient exchange capability (i.e., low cation exchange capacity) as found in a sandy soil or a highly weathered tropical soil. But if woody debris is also decomposing, its nutrient immobilization could include nutrients released from the foliage, thus potentially conserving the nutrients from leaching.

6.2.2 Methods of log transport (from stump to landing)

Forest management activities that receive the most scrutiny by certifiers include some form of timber harvesting. Harvesting trees requires the movement of logs from the stump to a road network where they can be loaded onto trucks. The logs can be skidded (one end elevated and dragged on the ground), cut into small lengths and carried on wheeled or tracked vehicles (e.g., forwarders), or elevated and "flown" (e.g., skyline cable or helicopter systems). These three logging systems have very different effects on the forest ecosystem.

6.2.2.1   Ground skidding

Skidding logs can be accomplished by either pulling the logs with a tractor (e.g., a rubber-tired skidder or a tracked dozer) or a cable system (e.g., high-lead system). Ground skidding with a tractor is usually accomplished by utilizing a series of skid trails. The skidder will get from the stump to the trail by the shortest distance possible and then repeatedly use the same trail to the landing. Logs can be skidded either with the branches attached or after delimbing. In addition, logs can be skidded singly (e.g., a high-lead cable system) or in groups (e.g., bunched behind a skidder with a cable or a grapple).

Disturbances to the forest ecosystem created by skidding will vary in the amount of area disturbed and the intensity of the disturbance depending upon which method is used. These disturbances can initially be separated into two categories:

1. Those disturbances created by the specific equipment used to remove the logs
2. Those disturbances related to the movement of the logs across the soil as they are being moved from the stumps to the landing

Skidders are usually large, heavy machines so soil compaction could be a problem in forests where these machines are used. The type of wheels or tracks used on the skidders can greatly influence the weight exhibited per soil surface area (g/cm^2 or lb/in^2). Extra-wide, high-flotation tires or wide tracks can be used on the skidders. The tradeoff of the extra-wide tires is that a wider skid trail is required. Wider skid trails, of course, mean that there is less soil compaction but more soil disturbance.

Today most rubber-tired skidders distribute their weight across a greater soil-contact area than the older skidders, so the rutting and compaction to the soil are much less. Tracked dozers generally distribute their weight across an even greater soil-contact area, so soil compaction is even less (per vehicle weight). However, the soil is disturbed even more with tracked dozers, relative to rubber-tired skidders, when making turns, especially sharp ones.

Skidding disturbances are relatively minor compared to the total forest ecosystem area according to Weetman (1998) (at least on soils derived from glacial parent material, e.g., coarser soil texture). However, Malmer and Grip (1990) in Sabah, Malaysia found 25% of the logged area was degraded due to skidding disturbance. Abdulhadi et al. (1981) reported an even higher area disturbed by skidding (e.g., 30% disturbance) on highly weathered, fine-textured soils in East Kalimantan.

The intensity of soil disturbance may be influenced not only by the frequency of trips over the same tracks, but by the soil wetness, texture, and organic matter. If the soil is too wet (e.g., operations should be halted if the water table is at or above a depth of 38cm [15in.]), rutting and compaction by vehicles and logs are at a maximum (Soil Survey Staff 1993). Skidding on wet soils could result in ruts up to a meter deep (or to a depth that is equal to the height of the bottom of the skidder's chassis above the ground). These ruts can be avoided by limiting skidding to dry seasons, to sandy soils in wet weather or to frozen soils. "Rut" disturbances could leave their legacies for years affecting tree growth, seedling regeneration, and water infiltration rates (Malmer and Grip 1990), especially in areas with thin forest floors, lack of rocks, or lack of a freeze-thaw period.

If the soil texture is fine (e.g., clayey) or the soil is organic (e.g., peaty or mucky), soil moisture conditions will further exacerbate the disturbance impact of skidding. If the soil is saturated (or under water) when skidding, ruts may not form; however, the soil structure will be destroyed. Loss of soil structure will reduce soil aeration and water infiltration, again potentially reducing plant growth. Krag et al. (1986) in British Columbia, Canada found that slope steepness also influenced the degree of disturbance. They determined that skid road disturbances were significantly greater on slopes > 20%. Obviously, the intensity of the disturbances are site-related, as well as seasonally related. For example, surface compaction can be partially offset by rocks in the soil, which help dissipate the weight of the machinery. Or, skidding on frozen ground can prevent soil compaction. This information is commonly used to schedule tree harvesting in the wintertime instead of the summer months in northern latitude forests.

The same disturbance effects in the soil that result from the logging equipment can also be ascribed to logs being dragged across the soil, either singly or by groups and whether they are dragged by tractor or cable. When logs are skidded in bunches or hitches, the impact of such an operation is spread out over a larger soil area. Branches left on the log may distribute the weight of the log over a larger area, but they also add more total weight to the "hitch." The redistribution of weight across a greater surface soil area may create less soil compaction, but at the same time there may be more soil surface that is scarified. This scarification may be beneficial for seedling regeneration since bare mineral soil favors the regeneration of light-seeded species. It is not uncommon, many years after the harvest, to be able to detect the location of skid trails by the tree species found regenerating there.

However, increased mineral soil exposure can create other problems. For example, exposed mineral soil is exposed to raindrop compaction and erosion (especially on steep slopes) (Krag et al. 1986). Since fine roots tend to be located close to the forest floor (surface organic layer) and surface soil (Krag et al. 1986), injury to these roots can be high. However, if the soil scarification is only a few centimeters deep, and not on steep slopes, it is widely thought that it could function as a silvicultural prescription for seedbeds. Conversely, if the soil surface (e.g., the "A" soil horizon) erodes (or is dozed) to a deeper depth, the "B" soil horizon may be exposed. The exposure of this "B" horizon soil could physically be analogous to a compacted soil surface. It generally has a higher bulk density, and chemically may be more nutrient poor relative to the "A" horizon (i.e., the "A" typically has more organic matter and nutrients present). For example, Oswald and Brown (1993) found that Engelmann spruce (*Picea engelmannii* Parry) seedlings planted on cutbank soils ("B" and "C" soil horizons exposed) had chlorotic needles and other seedlings were found to be more susceptible to insect damage. Also, soil scarification may result in decreased nutrient availability when the surface organic layer and/or "A" horizon is removed (Uhl et al. 1982, Weber et al. 1985). This decrease in soil nutrients, however, may not significantly affect tree growth until years later, and its effect on growth may diminish through time depending upon the development of forest humus (Munson et al. 1993).

Studies of soil disturbances have not determined if there are advantages or disadvantages of limbing logs in the forest. Certainly leaving the limbs and foliage in the forest to decompose and release nutrients for root uptake is an advantage, especially in poorer site-quality forests. Also, if the site is drier or wetter, the slash may decrease soil water evaporation or erosion, respectively. Some studies have attributed insect infestations and pathogen infections to the presence of slash in the logged sites (Furniss and Carolin 1977). This is one reason the slash is typically burned. Slash is also burned for seedbed preparation, to reduce plant competition, and to produce an immediate release of available nutrients for seedlings planted into the area, analogous to the slash and burn technique commonly used in the tropics.

Not only are the limbs and foliage potentially available to be decomposed, but the forest floor has typically been mixed into the mineral soil by the equipment dragging the logs. This incorporation of organic matter into the soil will decrease the soil density, enhance soil aeration and water infiltration, increase the daytime surface soil temperature, and increase the decomposition rates of the organic matter. However, on the negative side, soil water evaporation may increase. Furthermore, nutrients released by the quick decomposition of the forest floor may be leached before plants are able to assimilate them (relative to the slower release of the nutrients when the forest floor is decomposing undisturbed).

If the log removal methods must involve skid trails, insisting on prior skid trail planning and supervision of operations may reduce soil disturbance. Froehlich et al. (1981) found that planning reduced soil disturbance 45%–65% compared to non-planned operations.

6.2.2.2 Forwarding

Forwarding requires that the trees be "processed" before movement. In order to fit on the forwarder, the limbs must be removed and the logs cut into short lengths. This operation is only affordable if highly mechanized. This would require that large harvesting machinery be used in the forest stand. Forwarders put the weight directly on the wheels or tracks of not only the large machine, but the weight of all the logs as well. The total weight on the ground is no more than with skidders, but the weight is distributed differently. Often, the surface compaction from forwarding is less than from skidding. However, deep-soil compaction is similar since this is a function of the total weight of the machine (not the surface area of the wheels or tracks in contact with the soil). There is usually less rutting with forwarders.

Soil disturbances resulting from the forwarding log removal method are similar to those disturbances described for the skidding method (see 6.2.2.1). However, because the equipment is much larger and heavier, there is potential for even greater soil disturbance. Another difference is that the logs are not dragged along the ground. The logs may contribute indirectly to compaction of the soil, but not scarification of the soil as in skidding. Also, the limbs and foliage will always be left in the forest stand, unlike skidding, where the limbs sometimes may be left on the logs until reaching the landing.

6.2.2.3 Elevated

Elevated harvesting systems (skyline cable), where the logs are suspended above the ground by strung cables, require that clear areas be cut through the forest. Clear areas are needed so that supporting structures can be put in place and the logs have a clear passage to the landing. Partial cuts can be harvested, but clear corridors must be constructed. An alternative is harvesting with helicopters (or balloons). This system is more costly, but does not require the cutting of corridors or clearcutting. In both cases, the logs have to be delimbed before being moved to the landing.

Soil disturbances related to the elevated log-removal method are significantly fewer than disturbances associated with ground skidding or forwarding, especially when using a helicopter or balloon. The majority of the disturbances occur when the corridors are being constructed (unless using a helicopter or balloon) and when the trees are being cut, delimbed, and attached to the elevated cable. Ground skidding and forwarding also require cutting trees, perhaps delimbing, and attaching to, or hoisting by, some equipment. So, corridor construction is the only component of elevated harvesting that might create an additional soil disturbance. These corridors are constructed using a tractor. Soil disturbances created by the tractor are similar to those discussed for ground skidding (see 6.2.2.1), but they do not have the same intensity of disturbance. The impact of the tractor creating the corridor is much less frequent than ground skidding or forwarding, plus there are no logs carried by the tractor. Thus, the disturbances are substantially less intense. Also, any limbs or small trees that are removed to clear the corridor are left to decompose near the site of removal. Therefore, the nutrients are retained locally.

6.2.3   Harvest systems

Harvesting can be done to promote regeneration or to thin the stand and promote growth of the uncut trees, or a combination of the two. In the past, when timber was considered the sole or overriding objective of forestry, silvicultural systems were named by the harvest system used to achieve regeneration. Harvesting systems have moved away from this practice and now think in terms of silvicultural regimes that include all the operations carried out in the forest stand during the rotation or cutting cycle. Silviculture is described in terms of the goals of management rather than how the trees are harvested. However, the harvest systems are one way of categorizing the effects of logging on the future of the forest stand because they are an indication of how much of the stand is removed.

Even-age management can be thought of as a gradient of cutting intensity. If all the trees are removed at one time and regeneration begins either from natural regeneration or planting after the harvest, we have a clearcut. If most of the trees are cut to release growing space for the establishment of regeneration, but a few trees are retained for seed source, we have a seedtree cut. The remaining trees will be cut soon after in a second cut or left to grow as residuals above the new stand. If only part of the trees are cut and the remaining trees are left, not only as a seed source but also to mollify the environment, the harvest is referred to as a shelterwood cut. In a shelterwood, the remaining trees may be removed in one or several additional cuts. If the cut is light enough that the growing space made available is taken up by existing vegetation and regeneration is not established, then the cut is a thinning. In all cuts, except for clearcuts, the remaining trees usually respond to the release of growing space by increased growth. All harvests that release enough growing space to establish regeneration may be augmented by planting to change the species composition or increase the number of individuals that regenerate.

6.2.3.1   *Clearcut*

Of the six major methods for managing for regeneration (Smith et al. 1997), the one that has created the most controversy is clearcutting because of its misuse during past forest harvesting operations which resulted in site degradation. Much of the clearcutting problem was generated when harvesting occurred immediately adjacent to streams and resulted in high erosion, nutrient losses at levels higher than what was acceptable by law, and the loss of species utilizing the stream for their habitat (e.g., salmon). The seed-tree method has not received the same negative attention as clearcutting. In fact, the seed-tree method is similar to the "green tree retention" method suggested for use to balance the need to harvest trees and at the same time retain the environment that is suitable for species survival (e.g., ecosystem management). Both approaches are similar in that live trees are left in an area that has been clearcut. In the former case, retained trees are selected for their characteristics and ability to produce viable seed for future regeneration, and in the latter case, trees are retained for habitat and reproduction sites for animals requiring coarse woody debris.

The most intensive harvesting method is clearcutting. Since clearcutting does not leave any residual trees that inhibit machinery, any method of log transport can be used. Site constraints, such as steep slopes may make some log transport methods impossible, but usually cost and equipment availability are the only factors affecting equipment selection. The method of log transport, however, will determine how the soil is influenced (see 6.2.2). Aspects such as rutting or compaction of the soil depend on the method of log transport used and the type of equipment.

Clearcutting can result in several changes in the forest landscape. Whether or not these effects degrade the environment is the question. It is not always clear how much the lack of acceptance of clearcutting is driven by the non-scientific values that people have for what they find to be acceptable in the forest landscape (see 5.3). Early studies implemented to examine clearcutting effects were not designed to test management practices, so the negative literature that appeared on clearcutting was not based on management practices in use at that time (Bormann and Likens 1979). This does not mean that clearcutting does not result in changes in the ecosystem. In fact, clearcutting *does* change ecosystem structure and function. But it is important to determine whether a threshold of change has resulted in the ecosystem and whether this change will cause the ecosystem to degrade. Some reported landscape effects associated with clearcutting are a change in the amount of edge environment (see 5.2.1.2.1) and an increase in the degree of forest fragmentation (see 5.2.1.2.2). These changes in the stand and its interactions with the landscape will result in the development of new legacies for that site. Society often does not accept these new legacies, and is therefore unwilling to accept the use of clearcutting, even though it is a useful silvicultural tool in some forests (See 5.3.1).

Some of the changes in the ecosystem and landscape that are a result of clearcutting are covered thoroughly in Chapter 5 and will not be further discussed in this chapter. These effects can be summarized as changes in:

- The abiotic and biotic environment of the soil (see 5.2.1.1)
- The landscape patterns (see 5.2.1.2) that can result in changes in species diversity (see 5.2.1.3)
- The incidence of insect pest and pathogens (see 5.2.1.4)
- The number of potentially invasive species (see 5.2.1.5)

In terms of disturbance to the forest ecosystem, clearcutting affects a very large area in the forest stand. One disturbance that occurs immediately is a dramatic change in the microclimate near the ground. The soil temperature will certainly increase without the canopy cover or the forest floor cover (Smethurst and Nambiar 1990b). Soil moisture will also increase initially due to a lack of transpiration by the trees (Smethurst and Nambiar 1990a). However, the soil moisture will slowly begin to decrease and maybe even become a deficit as plants grow back into the clearcut (Smethurst and Nambiar 1990a, Adams et al. 1991, Miegroet et al. 1992).

The immediate impact on the soil hinges greatly on whether the forest floor is removed or mixed into the soil. For example, when the forest floor is mixed into the soil and soil temperature and moisture increase, the rates of decomposition and nutrient mineralization increase (Covington 1981, Kimmins 1987, Edmonds and McColl 1989, Frazer et al. 1990, Smethurst and Nambiar 1990b). However, others have shown that decomposition rates and/or nitrogen mineralization may not increase (Yin et al. 1989, Binkley 1984, Prescott 1997). For example, Seastedt and Crossley (1981) showed that in a cable logging and clearcutting of a forested watershed at the Coweeta Hydrologic Laboratory in the southern Appalachians of North Carolina, temperatures at the litter-soil interface increased without a protective canopy. This lethal, or near lethal, increase was so great in the logged areas (daily maximum summer temperatures averaged over 40°C) that the microarthropod densities (regulators of forest litter decomposition and nutrient cycling processes) decreased in the top 5 cm of litter and soil.

If trees have been removed and there is no uptake of the nutrients by plants, the nutrients mineralized from the decomposing forest floor may leach from the forest ecosystem (see 6.2.1), especially if the ecosystem is fertile (Vitousek et al. 1979, Vogt and Edmonds 1982). But if the vegetation returns quickly, plant assimilation of nitrogen from deeper soil depths may be possible in some cases. As found by Matson et al. (1987) in a clear-felled and burned tropical forest in Turrialba, Costa Rica, there was a nitrogen loss from the surface soil (0–25cm), but not necessarily from the deeper volcanic soil profile where the anionic exchange properties retained the nitrate from leaching. In a positive light, the addition of organic matter to the soil will contribute to better soil structure and enhance soil aeration and water drainage. But if the forest floor has been dozed aside, then the cleared areas will lose the nutrients contained in the forest floor and the areas where the forest floor was moved or dozed to will gain additional nutrients (see 6.2.2.1). Also, soil structure may be lost in those dozed areas, rendering these areas even more susceptible to erosion without plant or forest floor cover, especially if on a slope and the soil is fine textured (see 6.2.2.1). How much carbon and nutrients are lost from the site or redistributed depends on the method of log processing used for transport (see 6.2.1). Some studies have shown the potential for nutrient losses after clearcutting (e.g., boreal mixed wood forests — Schmidt et al. [1996]; warm temperate conifer plantation — Vitousek and Matson [1985]; tropical lowland rain forest — Robertson and Tiedje [1988]). The severity of these losses is generally related to the time in which it takes plants to revegetate the cleared area (Bormann and Likens 1979; Mou et al. 1993; see Table 6.1). However, other areas measured no loss of nutrients after clearcutting (temperate hardwoods — Johnson et al. [1991a,b]; temperate conifers — Martin and Harr [1989]; see Vogt et al. [1997a]). The watersheds that had no measurable increases in nutrient leaching are found in areas where soil nutrient levels were lower than before the manipulation began (see Table 6.1). It is important to realize that, in all cases, results are dependent on the logging system used. For example, a clearcut conducted on the Coweeta experimental watershed using skidders resulted in nutrient loss (Mann et al. 1988), while a similar harvest using cable yarding had no loss (Knoepp and Swank 1997).

6.2.3.2   Shelterwood

Shelterwood involves entering the stand for harvesting at least twice, if not three times, during a five to fifteen-year period. Multiple entries increase the exposure time to potential damage, although the potential damage at each entry is probably less than that for a clearcut. Damage to the residual vegetation, however, becomes an important consideration since remaining trees and understory are subject to repeated passes due to the multiple entries over time. Sometimes the trees are harvested in strip cuts, which is often a hybrid between small clearcuts and shelterwood. Strips are clearcuts through the forest that are narrow enough and placed close enough together so that the overall effect is like a shelterwood cut. Sometimes there is also partial cutting of the trees in the "leave tree" strips. Any of the described log processing types and methods of log transport can be used with this method of harvest (see 6.2.1 and 6.2.2, respectively).

In many cases, the width of the strip and the amount of forest floor that is destroyed are not part of the silvicultural plan for the stand. Rather, it is a result of the particular type of logging equipment used during an operation. Usually the strips are not reused for skidding during the second or third entry because the regeneration that becomes established would be destroyed. Not reusing the initial skid trails greatly increases the amount of the stand that is subjected to soil compaction although the intensity of compaction on each skid trail might therefore be less (Fabos 1997). However, the compaction differences between infrequent and multiple uses of the skid trails may not be as noticeable at deeper depths (Fabos 1997). Depending upon which log processing method is used will dictate whether any nutrients are lost via limb removal to the landing sites (see 6.2.1). There will probably not be any significant leaching losses of nutrients from the decomposition of slash material because of root uptake by the plants from the neighboring "leave tree" strip and regenerating strips (see 6.2.1). Roots of mature trees can easily extend 20–30m from the leave-tree strip into the current shelterwood cut area (Vogt et al. 1997a).

6.2.3.3   Patch cuts

Patch cutting can also be used to regenerate a stand. This method is similar to strip cutting except openings are made as holes in the stand rather than as strips. However, skid trails or cableways must still connect the patches to the landing. Patches are subsequently cut into the remaining stand until the entire original stand is removed and regeneration has been established.

Processing and transport of logs are similar to those associated with shelterwood. (For a discussion of disturbances related to this type of harvesting, see Section 6.2.3.2.). The amount of disturbance associated with patch cuts is therefore related to the size and number of openings at each entry.

6.2.3.4   Selection cuts

Uneven-age management can result in cuttings of varying intensity. *Group selection* is similar to patch cutting except that the openings are much smaller. The intent of the system is to enter into the stand on a regular basis rather than eliminating the entire overstory over a specific period of time. Therefore, the openings are not done in a systematic, total coverage manner like patch cutting. The regular cutting cycle of uneven-age management almost necessitates that some skid trails are maintained on an permanent basis.

Single trees rather than groups can be cut. This is often called *selection cutting*, but should not be confused with thinning, which will be discussed later. The purpose of selection cutting is to establish regeneration, while the purpose of thinning is to spur the growth of the remaining surrounding trees. Of course, some of each effect occurs in both cases. It is difficult to use many of the different logging systems in selection cutting without causing great damage to the residual vegetation. This means that this type of harvest is restricted to fairly flat, accessible sites. It is often impossible to regenerate shade-intolerant species with the high degree of shading from the residual trees, so selection cutting can reduce the plant species diversity of the stand. In some areas, it has been misused to justify removing the largest and most valuable trees from an even-age, mixed species stand.

6.2.4   Post-harvest site management

6.2.4.1   Slash management

Except in cases where the crowns of the harvested trees are chipped and the material utilized (for example, as industrial fuel), the harvest operation must consider how the limbs are dealt with after they are removed from the trees. In the simplest case, they are left attached together as an intact crown. If none of the branches are severed some of the small branches can be left more than four meters off the ground. This leads to slow decomposition, but often keeps large mammals from eating the small vegetation growing under the crown. The branches can be partially lopped, which keeps the crown somewhat intact, but brings the small branches much closer to the ground. Generally, this practice is used in harvested hardwood stands with large crowns. If aesthetics are considered important, the branches are lopped closer to the ground.

If the branches are removed from trees where a strong central stem is utilized, such as in many conifer stands, the branches may be left scattered on the ground where they fell. In many cases, these branches are then broadcast burned. Burning is prescribed to reduce fire hazard from wildfire, aid in the control of disease infestation, and/or to facilitate reforestation by removing the slash and/or organic matter (Oswald and Brown 1993). The effects of burning at one site are frequently different from those at other sites. This is

TABLE 6.1

Effects of forest management activities on nutrient dynamics in different hardwood, conifer and mixed forest stands (grouped by region).

Location/Species	Management Activity	Time Period	Observed General Effects* (consult reference for specific results)	Reference
Newfoundland *Betula papyrifera*	Whole tree harvest (WTH), Clearcut (CC)	3 yrs.	↓ N, P, K, Ca, & Mg (via biomass removal)	Titus et al. 1998
Quebec *Picea mariana* *Pinus banksania*	Whole tree harvest		↓ N & Ca; (P, K & Mg	Weetman and Alger 1983
Quebec *Abies balsamea* *Betula papyrifera* *Picea alba*	Whole tree harvest	5-12 yrs.	↓ Ca, Mg in forest floor; ↔ Ca, Mg, & pH in mineral soil (dry sites); ↓ Ca, Mg, pH, & CEC in mineral soil (moist sites)	Brais et al. 1995
Quebec *Abies balsamea* *Betula papyrifera* *Picea alba*	Windrowing (after WTH)	5-12 yrs.	↓ Ca & Mg	Brais et al. 1995
Ontario *Pinus resinosa* *P. strobus* *Populus tremuloides* *P. grandidentata*	Whole tree harvest	1 yr.	↑ N, K & base cations; ↔ SOM	Hendrickson et al. 1989
Ontario *Pinus resinosa* *P. strobus* *Populus tremuloides* *P. grandidentata*	Clearcut, bole only	1 yr.	↑ Ca & Mg ; ↓ base cations; ↔ SOM	Hendrickson et al. 1989
Maine *Picea rubens* *Abies balsamea*	Whole tree harvest		↓ N, P, K, & Ca	Smith et al. 1986
Maine *Picea rubens* *Abies balsamea*	Whole tree harvest	3 yrs.	↓ Ca & K; ↑ SO$_4$ (via runoff); ↓ Ca & NO$_3$ (via leaching)	Mann et al. 1988
New Hampshire Mixed Hardwoods	Whole tree harvest		↓ NO$_3$, Ca & K (via leaching and streamflow)	Hornbeck and Kroplin 1982

Location	Treatment	Duration	Effect	Reference
New Hampshire Mixed Hardwoods	Whole tree harvest	2 yrs.	↓ Ca, K, & NO_3 (via runoff and leaching)	Mann et al. 1988
New Hampshire (Hubbard Brook) Mixed Hardwoods	Whole tree harvest	2 yrs.	↑ SO_4	Mitchell et al. 1989
New Hampshire (Hubbard Brook) Mixed Hardwoods	Whole tree harvest	3 yrs.	↑↓ OM	Johnson et al. 1991a
New Hampshire (Hubbard Brook) Mixed Hardwoods	Whole tree harvest	3 yrs.	↓ CEC in upper horizons; ↑ CEC in lower horizons; ↔ CEC in total soil profile; ↓ pH, base saturation, & base to acid cation ratio	Johnson et al. 1991b
New Hampshire (Hubbard Brook) Mixed Hardwoods	Whole tree harvest	2 yrs.	↓ K & NO_3 (via leaching and streamwater)	Romanowiz et al. 1996
New Hampshire (Hubbard Brook) Mixed Hardwoods	Whole tree harvest	3 yrs.	↑ exchangeable K	Romanowiz et al. 1996
New Hampshire (Hubbard Brook) Mixed Hardwoods	Whole tree harvest	8 yrs.	↑↓ K	Romanowiz et al. 1996
New Hampshire (Hubbard Brook) Mixed Hardwoods	Clearcut, bole-only	3 yrs.	↑ CEC	Johnson et al. 1997
New Hampshire (Hubbard Brook) Mixed Hardwoods	Clearcut, bole-only	8 yrs.	↓ Ca, Mg & K from Oa and E horizons; ↑ Ca, Mg & K in Bh and Bs1 horizons; ↑ Ca, Mg & K in total soil profile; ↓ CEC:OM ratio	Johnson et al. 1997
New Hampshire (Hubbard Brook) Mixed Hardwoods	Whole tree harvest	NA	↓ P (via biomass removal)	Yanai 1998
New Hampshire (Hubbard Brook) Mixed Hardwoods	Whole tree harvest	2 yrs.	↓ P	Yanai 1998
Connecticut Mixed Hardwoods	Whole tree harvest		↓ Ca & OM	Tritton et al. 1987
Connecticut Mixed Hardwoods	Thinning		↓ OM & base cations	Tritton et al. 1987
Connecticut Mixed Hardwoods	Whole tree harvest		↓ Ca, K & NO_3 (via runoff); ↓ NO_3 (via leaching)	Mann et al. 1988

TABLE 6.1 (continued)

Effects of forest management activities on nutrient dynamics in different hardwood, conifer and mixed forest stands (grouped by region).

Location/Species	Management Activity	Time Period	Observed General Effects* (consult reference for specific results)	Reference
Michigan (UP) *Acer* spp.	Whole tree harvest (on 3 stands w/different site indices)		↓ N & K; ↓ Ca/Mg; ↔ P	Mroz et al. 1985
Tennessee *Quercus* spp. *Carya* spp.	Whole tree harvest Clearcut, bole only		↓ N, P, K, & Ca	Johnson et al. 1982
Tennessee *Quercus* spp. *Carya* spp. *Acer* spp.	Whole tree harvest		↓ NO_3 (via leaching)	Mann et al. 1988
North Carolina *Pinus taeda* (plantation)	Whole tree harvest, clearcut, each with chop/burn and pile/disk site preparation		↓ N, P, K, Ca, & Mg	Tew et al. 1986
North Carolina (Coweeta) Mixed Hardwoods	Whole tree harvest Clearcut, bole only	3 yrs.	↓ N, PO_4, K & Ca (via leaching)	Mann et al. 1988
North Carolina (Coweeta) Mixed Hardwoods	Clearcut, cable yarding	17 yrs.	↑ Ca, Mg, & K (soil 0-10cm); ↑ Mg, K (soil 10-30cm)	Knoepp and Swank 1997
South Carolina *Pinus taeda*	Whole tree harvest, Clearcut, bole only		↑ NH_4; ↓ NO_3	Mann et al. 1988
South Carolina *Pinus taeda*, *P. palustris*	Prescribed burns at 1, 2, 3, and 4-year intervals	30 yrs.	↑ C:N with burn frequency	Binkley et al. 1992
Florida *Pinus elliotii*, *Eucalyptus viminalis*	Whole tree harvest		↑ N (via runoff); ↓ Ca & K (via runoff)	Mann et al. 1988
Alberta *Picea glauca*, *Populus tremuloides*	Clearcut, bole only	20 mos.	↑ C:N; ↓ pH, N (total, mineralizable), P, & exchangeable base cations	Schmidt et al. 1996
Alberta *Picea glauca*, *Populus tremuloides*	Clearcut, with various mechanical site preparation	15 mos.	↓ C:N, N, & P; ↑ pH, base saturation, & exchangeable base cations	Schmidt et al. 1996

Location / Species	Management practice	Time	Effects	Reference
Montana *Pinus contorta*	Clearcut w/residue left, removed and burned (exp. comparison)	2 yrs.	↓ available macronutrients; ↔ micronutrients	Entry et al. 1987
Idaho *Pinus ponderosa, Psuedotsuga menziezii*	Clearcut, helicopter yarding, slash burn		↓ N (dissolved); ↓ N & K (biomass)	Clayton and Kennedy 1985
Washington *Pseudotsuga menzeisii*	Whole tree harvest, Clearcut, bole only	1 yr.	↓ Ca; ↑ K & NH_4 (via leaching)	Mann et al. 1988
Washington *Alnus rubrum*	Whole tree harvest, Clearcut, bole only	2 yrs.	↑ Ca, K, & NO_3 (via leaching)	Mann et al. 1988
Costa Rica Secondary Rainforest	Felling, mulching, burning	11 wks.	↓ S, C, P, K, & Ca (result of felling); ↓ K & P (during mulching period); ↓ C, N, & S (result of burning); ↓ N, K, Ca, & Mg (via leaching, after burning)	Ewel et al. 1981
Costa Rica Secondary Rainforest	Felling, burning, and planting with successive monocultures, and 3 diverse successional communities	5 yrs.	↓ base cations, available P, & CEC; ↑ acid saturation; ↔ OM, N, & S	Ewel et al. 1991
Peru Secondary Rainforest	Felling, burning, planting to crops, fertilizing/liming	1 mo.	↑ pH, K, Ca, Mg, & available N; ↓ exchangeable Al	Sanchez et al. 1983
Peru Secondary Rainforest	Felling, burning, planting to crops, fertilizing/liming	6 mos.	↓ N, K; ↓ available S, Cu, & B	Sanchez et al. 1983
Peru Secondary Rainforest	Felling, burning, planting to crops, fertilizing/liming	1 yr.	↓ N (total), C (soil org.); ↓ available P, Mg, Ca, Zn, & Mn	Sanchez et al. 1983
Peru Secondary Rainforest	Felling, burning, planting to crops, fertilizing/liming	8 yrs.	↑ pH, Ca, CEC, & P (fertilizer effects); ↓ Al	Sanchez et al. 1983
NE India *Picea smithiana, Abies pindrow*	Clearcut, shelterwood, selection (experimental comparison)		↑ N (with increased cut intensity); ↓ available P (with increased cut intensity)	Gupta et al. 1990a, b
Tasmania *Eucalyptus* spp.	Clearcut, slash burn		↑ pH, NO_3, exchangeable cations, & base saturation; ↓ exchangeable acid, & NH_4	Ellis et al. 1982

* ↑ = increased levels, ↓ = decreased levels, ↑↓ = initial increase followed by decrease, ↔ = unchanged levels.

mostly due to the very uneven effects of burning attributed to the uneven distribution of the slash (also see 6.2.1). For very hot fires, much of the nitrogen is volatilized but the remaining ash contains other available nutrients (especially base nutrients such as calcium and magnesium). However, these available nutrients are now potentially leachable, erodible, or even wind-transportable from the ecosystem. In some cases, when the fire is very hot, the volatile organics may infiltrate the topsoil and cause the soil surface to even become hydrophobic for a short time. A good discussion of the ecological effects of forest fires at the forest community and ecosystem level is given in Agee (1993) and will not be further discussed here.

The severed limbs can also be piled. Sometimes they are placed by hand into small piles, but they are often piled by bulldozers into long rows called windrows. Little vegetation grows under the piles and they concentrate the nutrients as the branches decompose (see 6.2.2.1). Roots from other trees grow into these areas and proliferate. If machines are used for the piling, topsoil is often moved into the piles, as well. These piles, large or small, are sometimes burned like the broadcast burns just discussed. On nutrient-poor soils (e.g., sandy soils) these piles may contain a significant proportion of the site's nitrogen capital. Studies by Morris et al. (1983) and Gholz et al. (1985) showed these windrows contained from 10% to 17% of the N capital to a 1-m soil depth. In a *P. radiata* plantation on a sandy podzol soil, Smethurst and Nambiar (1990a) found that the slash contained 12% of the total N capital to a 30-cm soil depth. Removal of slash could therefore contribute to a significant reduction in nutrients in the ecosystem and has led forest managers to consider leaving slash when economically feasible (Squire 1983).

6.2.4.2 *Planting of tree seedlings*

In some parts of the world, regeneration can be achieved by using natural regeneration, but planting is often used. In systems using natural regeneration, seeds, sprouts, or advanced regeneration may be relied on. Planting is used when natural regeneration would take too long (in some dry areas the regeneration period following wildfire may take 100 years or more), would not be at a uniform spacing or density, or be of an undesirable species. Planting can also be used to introduce genetically improved individuals. Planting is used in many industrial settings in order to reduce the period of low productivity that occurs after timber harvest and to either increase yields or reduce rotation length.

Because of the high cost of planting and the fact that the return on the investment will not be returned for many years — until the end of the rotation — the trees must be grown as fast as possible. For this reason, planting is usually accompanied by other silvicultural activities such as competition control or site amelioration. The disturbances that accompany the planting activity itself are minor. However, the activities that prepare the site for planting (see 6.2.4.1 and 6.2.4.3) or follow planting to improve survival and/or seedling growth (see 6.2.4.4) may be significant in terms of disturbances to the ecosystem.

6.2.4.3 *Competition control for planted seedlings*

Much of the controversy about forest management is based on practices used to control vegetation that competes with the target tree species. This competition may be from any plant life-form, herbaceous, shrub, or tree species. Although the target species may grow and survive to an older age, they are often unable to compete well with other species at early ages. In fact, the growth attributes, which might be most desirable at maturity, can be detrimental at early ages. If a tree is out-competed for growth resources when it is young and dies, it will obviously never display these older age attributes (such as grand tree height).

The major controversy for this topic exists over the methods used to kill the competing vegetation. Herbicide chemicals are usually the cheapest way to kill the competing vegetation, especially if broadcast application can be used. Many certification systems do not permit chemicals, but do allow mechanical killing of the unwanted vegetation.

Many different types of chemicals are used to control growth of competing vegetation. They vary greatly in their specificity and the rate at which they break down in the ecosystem. As well as problems that may arise from the use of specific chemicals, the method and care of chemical application have a major impact on their effects in the environment. In many cases, application rates are used that are higher than needed to accomplish the competitive shift. Extremely high dosages may even leave a legacy of stunted plants and dead stems for years to follow. Application is often from the air, either by helicopters or airplanes. If the chemicals could be applied only to the targets, this application method would not be so controversial.

Studies have shown significant growth responses by planted trees when competing vegetation is controlled. In fact, Munson et al. (1993) even suggested that the effects of vegetation control will become more prominent with time as competition intensifies for nutrients and other resources. Their study showed that vegetation control might even contribute more to tree growth than fertilization in the long term.

6.2.4.4   Site amelioration for seedling growth

6.2.4.4.1   Fertilizer

Fertilizer is sometimes used at the time of planting. Although various forms of nitrogen are the most common types of fertilizer (e.g., urea compounds), phosphates and potassium are also used. Application can either be done by helicopter or small packets of chemicals can be added to the planting hole.

The application of fertilizer can be a disturbance, especially when an imbalance in nutrient availability occurs on the site (see 5.2.1.1). In the past, nitrogen fertilizer was the most common application to a forest. More recently, the push has been to use a balanced application of fertilizer so that other nutrient deficiencies do not appear (Vogt et al. 1997a).

The timing of the fertilizer application is an important consideration as is the type of fertilizer nutrients applied. Nutrients from the fertilizer may leach from the ecosystem just as nutrients may leach following the mineralization of disturbed slash and litter layers (see 6.2.1, 6.2.2.1, 6.2.3.1). This may especially be the case if the ecosystem has not had a chance to revegetate and, thus, assimilate the available nutrients before potentially leaching. Smethurst and Nambiar (1990a,b) even found that the nutrients released from the decomposing slash and litter on a sandy podzol soil in southeastern Australia were sufficient for *Pinus radiata* growth for the first two years after planting and that the fertilizer application was not effective until the third year. A review of fertilizer effects on plant carbon allocation and production exists in Vogt et al. (1997a) and will not be further discussed here.

6.2.4.4.2   Bedding

In much of the coastal plain of the southeastern United States, there is poor horizontal drainage, especially in the spring. After clearcutting, the problem of high water table and surface water is exacerbated by the lack of evapotranspiration. The seedlings would die during the wet spring if the planting sites were not bedded. In this case, a bulldozer is used to pull a plow that creates raised ridges where the seedlings are planted. This activity will affect the ecosystem in different ways. The plow itself may create a plow-pan, which may become impermeable to root penetration and may even inhibit water infiltration.

Any disturbance to the soil, such as scarifying, plowing or bulldozing it, will destroy the soil structure (see 6.2.2.1). This will certainly increase the decomposition of the soil organic matter, which then decreases the soil structure and decreases soil aeration and water infiltration.

Mounding of the soil is also used at higher latitudes and elevations to enhance forest regeneration. Planted seedlings must become established during a relatively cool, short growing season. By mounding the soil in these climates, soil temperatures were increased which then allowed for increased mineralization of organic matter (releasing nutrients) (see 6.2.1, 6.2.2.1, and 6.2.3.1) and improved root growth and uptake (Teskey et al. 1984). However, Bassman (1989) showed that spruce seedlings in the Cariboo Forest region in British Columbia, Canada were under water-deficient stress during the first two growing seasons due to the greater soil surface area exposed to evaporation and radiation. But by the third year, the seedling roots had grown beyond the drying influence of the mound and responded with substantial increases in biomass.

6.2.4.4.3   Drainage

In areas with a very high seasonal water table, sites are drained even when trees have reestablished their function of recycling soil water as evapotranspiration. This draining is also a very controversial practice because the digging of ditches and channels significantly changes the water table and can impact the species diversity of herbs and forbs. Drainage is a common practice in agriculture and, because of its high cost, is used in forestry situations that bear close resemblance to agriculture. Drainage systems also provide conduits for rapid removal of chemicals, particularly fertilizers, to other sites.

6.3   Forest management activities not associated with regeneration

6.3.1   Thinning of trees

Not all timber harvests are intended to lead to regeneration of the site. Thinning is frequently used in forest management. In every case, the stand structure of the residual stand is different from the stand before the harvest operation. In some cases, the primary goal of the cutting is to receive an immediate monetary benefit. In these cases, the residual stand is not always of the same vigor and health of the preceding stand. In other cases, the primary goal is to create a faster growing stand where trees of particularly high quality are given more growing space and unwanted trees are removed because of their vigor, quality, or species.

All the harvesting practices and concerns previously described apply to thinning, as well (see 6.2.1, 6.2.2, 6.2.3, and 6.2.4). Thinning always leaves many residual trees in the stand, so damage to these trees is an important consideration.

6.3.2   Fertilization of trees

Few forest soils provide an optimum supply of the essential nutrients needed for the fast growth rates desired by the forest managers. Sometimes marked deficiencies may exist because of improper land management in the past or merely because of inherently low natural fertility of the site (Smith et al. 1997). Fertilization of these forest stands then becomes an economic benefit.

A major problem might occur if the fertilizer used is not balanced for the specific nutrient status of the site (see 6.2.4.4.1). Another potential problem is leaching of the fertilizer. This may occur if too much fertilizer is applied or heavy rainstorms follow its application. Besides wasting the fertilizer two other problems may occur. If the nutrients leach into the ground water, eutrophication in streams or lakes may occur or increased concentrations of some nutrients may become toxic (e.g., nitrate) (Edwards and Ross-Todd 1979). Also when anions leach from the soil (e.g., NO_3^-, PO_4^{3-}, SO_4^{2-}), they are paired with cations (e.g., Ca^{2+}, Mg^{2+}). So if the applied fertilizer has an anion (or a cation that converts to an anion — e.g., ammonium to nitrate) that may leach, the ecosystem may slowly lose its bases (Likens et al. 1969). This explains how acid rain depletes bases from the soils. It does occur naturally, but human activities can speed up the process. This problem can be rectified by adding lime.

6.3.3   Competition control for tree growth (chemical)

Competition control is based on shifting the competitive balance from the species that are not wanted to the target species. Often, this is most easily done by killing these trees. Although this mortality frees up growing space allowing unwanted trees to regenerate, they will be small and pose less of a threat to the target trees. Sometimes the competing trees will not kill the target trees, but will retard their growth by using valuable resources (such as light, water, or nutrients). In this case, killing the trees will prevent growth from slowing down in the target trees allowing them to become much stronger competitors for the growing space.

Mechanical killing is done either by cutting the trees down or by girdling them. A compromise solution is sometimes used where the trees are either injected with chemicals or chemicals are added to the girdling cut (see 6.2.4.3). The use of chemicals requires care. For example, studies have shown negative influences on soil microbial communities, especially mycorrhizal associations (Vogt et al. 1997a).

6.4   Infrastructure related to forest management activities

Often, the infrastructure needed for forest management activities has a bigger impact on the ecosystem than the activities themselves. Harvest equipment must be moved to the logging site and harvested materials must be moved from the site. Roads are one type of infrastructure required for these types of operations. Other necessary infrastructures for harvest operations are landings (log-concentration areas) and skid trails for skidding and forwarding (see 6.2.2). These infrastructures not only have an impact on forest management activities and the environment by occupying a significant portion of the landbase, but they may also influence forest productivity and the environment by exposing mineral soil which frequently leads to erosion and sedimentation of water courses.

6.4.1   Roads

Haul roads are necessary for hauling logs from the forest to a collector or arterial road, and then on to a wood-processing plant. As necessary as they are for forest management activities, haul roads can cause some of the most serious impacts on the environment. For example, to reduce cost, roads are often constructed in a straight line rather than following the best route. This means that much earth is moved from cuts to fills and more stream crossings are probably required than would have been with a better-designed road network. When a road is constructed, land is taken out of the land base for vegetative growth. Roads can also significantly impact wildlife populations by imposing barriers to animal movement.

Roads often degrade water quality in streams either by poorly constructed stream crossings or by erosion that occurs from the road itself. Stream crossings that are not perfectly perpendicular to the stream can lead to significant undercuts in the bank. If bridges and culverts are improperly sized and impede water flow, they may be blown out during storm events. When this occurs, a large amount of road-building material may be dumped into the stream or river. Also, if the road climbs steeply after the stream crossing, sediment in the road drainage may flow into the stream even during minor rainfall events. As with skid trails, proper location, as well as water diversion structures, such as water bars, can prevent the sedimentation of the watercourses.

Factors that are considered when evaluating soils for road building potential are soil wetness, texture, and soil slope. When the duration of the water table above 30.5cm (12in) is more than one to three months, the limitation of the soil for road-building suitability is moderate to severe (Soil Survey Staff 1993). If the slope is more than 15% to 30%, the road-building suitability is again moderate to severe. If the soil texture has excess fines to a depth of 0.6 to 3m (2–10ft) depending upon the 15% to 30% slope, the suitability is moderate to severe. Moderate to severe ratings may require extreme measures or conservation practices designed to overcome the limitations. If all of these factors are present together, the impact of roads will be even more severe.

6.4.2 Landings

Landings are the areas where logs are stacked and stored before being trucked from the site. The size of the landing depends on the number of grades and material being sorted, as well as how quickly material is moved from the site. In some operations, all the material from a harvest area may be stored at the landing before trucking begins. In these cases, landings may be up to a hectare in size, but they are usually much smaller.

An obvious impact is that landings are completely denuded of vegetation. In many cases, to protect truck tires, even the stumps are removed. The constant dragging of material and driving of equipment removes the entire forest floor. If the ground is wet, deep rutting can occur.

Another major impact is soil compaction. The constant driving of equipment compacts the soil and prevents trees from growing well on the landing site for many years. Herbaceous plants will either invade naturally or are often planted for stabilization. Before the herbaceous plants become established, runoff and erosion are major problems because of the compaction.

The problems of landings can be significantly mitigated by their careful location. Landings near watercourses or wetlands can lead to major sedimentation.

6.4.3 Skid trails

When ground-based logging systems are used, skid trails are used to bring logs from the forest to the landing. The use of skid trails concentrates the impact of the equipment to a smaller portion of the land base. However, tractor trails may still account for 25% to 35% of the area of a tractor-logged site (Froehlich 1979).

The majority of soil compaction and disruption occurs during the first few passes of the equipment. Additional trips may not greatly increase the impact unless the logging operation is conducted on easily compacted soils, such as soils derived from volcanic ash (Fabos 1997).

The direct impacts of skid trails have previously been discussed (see 6.2.2.1). Locating skid trails on firm, dry ground or frozen soil reduces the damage to roots from rutting. Keeping turns and bends to a minimum will minimize damage to standing trees. This is less of an issue in clearcuts. Both the number and location of trails determine the total impact of skid trails.

Disturbances of skid trails on an ecosystem are not only expressed directly (e.g., soil compaction and scarification), but also indirectly, through the effects on the survival percentage of planted seedlings and their subsequent growth and even effects on the growth of the residual trees. The survival of planted seedlings may be reduced by two-thirds and the growth of the established seedlings may be reduced by as much as 40% or more (Foil and Ralston 1967, Pomeroy 1949, Youngberg 1959). Moehring and Rawls (1970) even found that soil compaction due to skid trails negatively affected the growth of the nearby residual forty-year-old loblolly pines (*Pinus taeda* L.). In fact, the growth of the heavily impacted trees was reduced by as much as 40%.

6.5 Summary

The ecosystem effects of specific forest management activities cannot be generalized without respect to the regime of actual operations and the site where they occur. Certification rules seldom take into account that there are *combinations* of management activities that produce negative results. For example, clearcutting alone cannot explain most of the deleterious effects attributed to it. In fact, it is usually the combination of operations often used in the process of clearcutting such as skidding, wholetree removal, site preparation, and road building that leads to soil compaction, nutrient leaching, habitat and biodiversity loss, and alien species invasion.

How then do we distinguish "good" forestry practices from those that are unsustainable? The answer is based on the time and the place. Presently, certification systems either endorse or condemn forest management systems based on broad descriptions that disregard the large number of combinations of activities that could be employed under these management regimes. Consequently, silvicultural practices such as clearcutting cannot be used as indicators of bad forestry. Indicators must be chosen to identify specific activities that must be avoided under particular circumstances. The evaluation of good forest management should focus on whether a system can be sustained economically, ecologically, socio-politically, and silviculturally.

7

Synthesis Discussion of Issues Relevant to Certification

7.1 Necessity of assessing the landscape matrix within which a management unit is embedded

Brooke A. Parry, Kristiina A. Vogt

Forest edges, boundaries, and other discontinuities in the forest landscape can present a direct challenge to forest structure and function (Saunders et al. 1991). In spite of this threat, certification protocols give little attention to the matrix landscape of a forest and the threats to its functions that may result from the management practices of adjoining properties. Land managers must look beyond using sustainability indicators within a particular forest and include the matrix landscape in their assessments.

Presently, the size of a forest has been approached as being simply a size issue with a minimum land area needed to economically pursue certification (see 5.2.1.2). The certification process has not yet attempted to appraise the impact of what occurs at a forest's property line or at a forest edge. However, individuals involved in the development of certification tools are aware of the problems that neighboring lands can introduce. For example, Mankin (1998) argued that if "something next door is clearly challenging sustainability" and that event is prohibited in the guidelines, the forest would not be certified regardless of whether it is the landowner's fault or his neighbor's. However, Mankin (1998) and Ervin (1998) stated that, in general, landowners should not be responsible for what happens on their neighbors' lands. Currently, this argument is not an issue, since consideration of the matrix landscape is largely missing from certification protocols. Certification tools should be elastic — collapsing or expanding to match property size and its boundaries.

In the United States, most non-industrial landowners by 2010 will have 6.9 ha of land. Among the many repercussions of that fact is that these land parcels will have significant edge effects and impacts from neighboring lands. These sizes of forestlands should not be simply approached as an issue of having a large-enough land base to be able to afford the pursuit of certification. Instead, the issue should focus on the functioning of the management unit within the matrix landscape. In addition to considering chain-of-custody and economic issues that are scale dependent, certification protocols will need to incorporate the impact of changing scale of property size on ecological functions.

The same set of criteria will not work the same way for landowners owning 10,000 ha of land as for those owning 100 ha of land because both economic and ecological indicators of system function are scale dependent (see 5.2.1.2). Economically, the benefits of economies of scale often place a lower limit on the size of the managed property. Ecologically, the impact of a smaller parcel of land is the increasing significance of the edge environment because as property size decreases, the relative extent of the edge increases (Hunter 1990). There are two solutions to the scalar problems faced by small landowners pursuing certification:

1. If possible, adjoining landowners with similar management objectives can consolidate their forests for certification purposes.

2. If consolidation is not possible because of different management goals or because contiguous properties are not forested, indices of sustainability relevant to certification will need to change with changing scale of the property.

If the property is small, the influence of the edge as the driving variable in the certification process should be greater, particularly if the property is embedded in a degraded landscape. Support for this argument is derived from the literature on edges, edge effects, and forest fragmentation. This body of literature suggests that the impact of edges and fragmentation on forest function and resilience is significant (see 5.2.1.2, 5.2.1.3).

There is no question that edge zones are heterogeneous regions. Dynamic abiotic and biotic gradients are present that often produce a diversity of environments and species that are differentially influenced by an edge. However, the presence of diverse species at an edge is not necessarily an indication of community biodiversity and, in fact, may endanger it (Meffe and Carol 1994). Thus, certifiers must be aware when the character of the edge environment changes sufficiently enough that an edge destabilizes an ecosystem or does not allow for the maintenance of what we value in a particular forest. Since FSC-type certification protocols place such great emphasis on the conservation of biological diversity, certifiers will have to be sensitive to conditions that will affect this diversity. For example, the sixth principle of the FSC Acadian Regional Guidelines states that "Forest management shall conserve biological diversity and its associated values…" and lists over 60 subprinciples outlining how biodiversity must be maintained (see 4.2.1). Similarly, Woodmark's Section 5.1 outlines nine principles in "Planning for Conservation of Biodiversity" (see 4.2.1).

Because our understanding of edges is limited enough to prevent generalization about edge impacts, certifiers cannot rule out the possibility of positive edge influences. A good certification protocol will allow for a case-by-case assessment of whether the edge is a zone that threatens ecosystem function or fosters it (see 5.2.1.2). Similarly, a good protocol will identify which indicators will demonstrate when an edge is a zone of susceptibility to disturbance or serves as a zone of resistance to disturbance. Despite our incomplete knowledge of the impact of forest fragmentation and edges, certification protocols should incorporate the impact of the matrix landscape on the property being certified. But we should mention here that the SFF protocol already includes a landscape approach to certification. As suggested by this latter protocol, GIS maps, aerial photos, and field observations of property boundaries should help certifiers assess the possible threats to sustainability due to degrading influences external to a property.

References to smaller-scale properties and the importance of the matrix landscape are not entirely absent from these certification protocols. However, specific protocols either have a limited view of the impact of the matrix landscape on property sustainability or do not consider it at all. The FSC-type protocol states that small landowners must cooperate with adjoining property owners. For example, "If the management unit is less than 500 hectares, the owner/manager must participate, in a voluntary and significant way, in the development of landscape-level forest planning in the local community" (FSC 1997). ISO also refers to the influence of the matrix landscape in their protocol, by suggesting that certifiers reference external impacts on the property and cite appropriate articles in the certification report concerning the particular external input (ISO 1997).

However, the majority of references to property scale in certification protocols mainly recognize that properties of different sizes will be certified. For example, "Appropriate to the scale and diversity of the operation, monitoring of plantations shall include regular assessment of the potential on-site ecological and social impacts..." (FSC 1997). The bias of certification protocols for large-scale properties or corporations can be seen in which lands have been certified (see 2.5).

The current strong concern that forest certification protocols seem to have for larger-scale properties will not serve the forest industry nor the consumer well, as trends toward management of smaller forest parcels become the norm. Certifiers must recognize that the indicators chosen to be incorporated into protocols will have to better reflect the fragmented matrix of small properties and edge effects. Evidence from the literature on forest fragmentation and edges suggests that external forces, as well as interactions among ecosystems, have a significant impact on the processes occurring within ecosystems. Thus, the certifiers must be aware that the influence of the matrix landscape may have more of an impact on forest sustainability than the influence of the internal management and stewardship of the property, however good that management may seem to be. Understanding that influences external to forest boundaries are significant should not, however, discourage small landowners. Studies of edges have shown that external impact (and ecosystem interactions) can modify ecosystems in ways that we value, as well as in ways that we do not. Further, in some cases, the influence of the matrix landscape is insignificant and is therefore irrelevant to the certification process.

It is essential that heed be taken in the evaluation of the edge influence on properties. The current debate on the nature and significance of edges stems from the judgment that edges are often spatially and structurally heterogeneous and that edges can appear to be diverse zones. However, an edge is often on a fast trajectory of change, one that certifiers should attempt to assess and predict. Certifiers should try to answer the following questions: Will the edges of the property be zones of stability in the future? Does the presence of certain species at an edge pose any threat to the diversity or sustainability of the rest of the ecosystem? What is the influence of the matrix landscape in which the property is embedded? What is the impact of the property size on its sustainability? How should the indicators vary when a management unit abuts a park or a national forest, versus suburban sprawl or an industrial park? Certification protocols need to evolve to include guidelines for making these kinds of assessments for properties. Whether a property undergoing certification is large or small, certifiers need to evaluate the variables that are driving the processes of that ecosystem. It is critical to remember that the forces affecting ecosystem change, or the challenge to ecosystem sustainability that the certification process is designed to identify, may come from outside the boundary line of their traditional region of focus.

Further, the assessment of a management unit should change to reflect how the property's resilience or resistance to disturbances changes depending on neighboring lands. Invasive and species, insect or pest outbreaks will not be confined to one landowner's area. Management of such a problem will require the use of a landscape-level approach. Such issues often cross property boundaries and, therefore, require solutions that incorporate coordinated efforts (NRC 1998). Fragmentation and inadequate management by landowners has been implicated in increased losses of forestlands due to insects and pests (NRC 1998). It is thought that invasive species are pervasive enough in the forestlandscape that they cannot be managed by focusing on one's individual management unit.

Certain certifiers (e.g., FSC) do not feel that landowners should be held accountable for activities occurring on adjacent lands (Ervin 1998, Mankin 1998). Similarly, land managers themselves feel that they should manage their land to best provide for habitat and timber, but not to restrain their management because of neighboring activities. Thus, certifiers are faced with a challenge. How can one assess a given management unit in which the function and sustainability are strongly controlled by activities occurring outside of their land? Landowners cannot be held responsible for their neighbors' land management. However, if management practices of adjacent lands affect potential ecological constraints of a given piece of land, analysis of the adjacent landscape should be considered prior to making a decision concerning a given property (Vogt et al. 1999ab). The issue of the matrix landscape should be of greater consideration for the small landowner who has a relatively greater border or edge with their neighbor.

7.2 Social- and natural-science links

Kristiina A. Vogt, Joyce K. Berry, Toral Patel-Weynand

7.2.1 Necessity for linking the social and natural sciences

Even though by definition sustainable development, ecosystem management, and certification require integration of the social and natural sciences (see 3.2.2), it is worthwhile discussing why this need exists. One reason this needs to be discussed is that in practice this integration does not occur, although this philosophy is integral to all of them (see 4.2.3). Collection of some social science information as part of an assessment protocol does not address what type of social information is needed to really determine social sustainability, nor does it identify when the social science factors are the driving variables controlling natural-system functions. Part of this difficulty has occurred because of the impediments to conducting interdisciplinary research and a classification of the social sciences as "soft" by natural sciences because the tools were considered to be qualitative and not quantitative in nature (Vogt et al. 1997a, Brewer 1998). It has not helped that there is a:

- General lack of good examples demonstrating the importance of social variables in driving natural resource issues
- Low priority has been given to the social sciences (except economics) when managing natural resources
- Lack of projects demonstrating the successful application of interdisciplinary research to address natural-resource problems

The only successful integration of the social sciences with natural resource management has occurred in following fields:

- Economics: where quantitative tools are used to value and to determine the trade-off of natural resources.
- Geography: where population and demographic information have been linked spatially to the natural resources (Miller 1994, Constanza 1996).

When defining sustainability (see 3.1.2), it is accepted that this means both the social- and the natural-science systems. At present, information needs for both the social and natural sciences are not linked, but are collected as independent and separate pieces of information which are summarized to conduct the whole analysis. Because of the weighting of the information, the social-science side is not weighted as much, since it ends up being a smaller part of the data needs for existing certification protocols (see 4.3.5.4).

This lack of interlinking of the social and natural sciences also occurs because of the problem of identifying indicators and tools specific to the social sciences (the exception being economics). Superimposed on top of this is the lack of good examples documenting effective mechanistic linking of the social system to the natural system (see 3.2.2.1). Because of these factors, there is a tendency to suggest that if we take care of the natural system, then the social system will track it and both will be sustainable. For example, a small woodlot owner managing forests as sustainable in Nova Scotia stated that a balance between the social- and natural-science needs in certification is unneeded and is "bogus" (Drescher 1998). Drescher stated that if a healthy system is being managed as a priority objective then the benefits from producing an ecologically healthy system follow to the social system (1998). Unfortunately, this logic does not deal with the fact that different human driving forces (frequently values) may force a particular management style that may not be ecologically sustainable but satisfies public values. The management of sugar maple is an excellent example of this since its management has been predominantly driven by human values for sugar maple products (Vogt et al. 1999ab).

One of the problems with social science research is the fact that it has not effectively been integrated into natural resource management issues. It also has not always been taken as a serious constraint to natural resource management. Social scientists have lagged behind natural scientists in developing a substantial, well-recognized body of knowledge that can be used in the process of assessing the social science part of sustainable development, ecosystem management, or certification. Most efforts have focused on economic analyses of forests that are based on quantitative information related to specific products that are available or harvested from forests (e.g., timber and non-timber forest products, wildlife, recreation, and aquatic resources) (Constanza 1991).

Economics has had difficulty evaluating and placing a monetary value on the "human driving forces" (Ewert 1996) which determine how the public values natural resources, and how they understand and react to natural resource management issues. Stern et al. (1992) defined "human driving forces" as:

- Economic growth
- Population and other demographic changes
- Technological change
- Political-economic institutions
- Attitudes and beliefs held by the citizenry

Despite the importance and relevance of understanding human driving forces, the economic perspective continues to predominate. It is the main tool used to integrate the social system into natural-resource management issues. There are several reasons why an economic approach will be the primary tool in certification assessments. For example, there are very few examples linking the application of human driving forces to natural resources. And available data are not at the level where they can feed into decision-making apparati (Ewert 1996).

Research studying "human driving forces" (e.g., human dimensions research) has to be recognized as an important research field that strongly constrains natural resource management. Similar to the natural science, Ewert (1996) suggested that human dimensions research needs to be analyzed using a framework that recognizes the "cumulative and systematic effects of human actions." To be effective, this research will have to focus on understanding the links between the human driving forces and the natural system at different scales (e.g., individual, family, community, region, country, etc.) very similar to the issues existing in the natural sciences (Ewert 1996, Vogt et al. 1997a). Human dimensions research focuses on understanding how human driving forces at the local level can have global-level environmental impacts (e.g., acidification, ozone depletion, greenhouse warming, and loss of natural habitats) (Ewert 1996). Further developments in understanding human driving forces will eventually make it easier to link the social and natural sciences when the mechanistic connections are recognized.

Despite the need for more research in the human dimensions area, our understanding of how the social sciences link to natural resource management is becoming increasingly tangible and is at a point where useful tools are developing which are implementable. For example, there have been significant advances made on understanding how people value resources and ecosystems (Rolston 1986, Ewert 1996, Kellert 1996, 1997). We are developing a better understanding of the relationships among institutions, policy and decision-making, and natural resource management (Loomis 1993, Brewer 1998). In addition, we are beginning to document the relationships between social impacts and the natural resources themselves (Burch and Deluca 1984, Burch 1995, Machlis et al. 1997, Grove and Burch 1997).

7.2.2   How certification integrates social and natural sciences at the values level

Part of the social- and natural-science links occurs at the level of human values driving what landowners and other organizations attempting to regulate management activities on land they do not own (see 3.2.2.1).

> Landowners participate for several reasons in certification: access to markets, to improve their management techniques, or to be recognized for good forest management (Drescher 1998).

For many landowners, it is an independent verification by a third-party certifier that their forests are being well managed. For many landowners, the ability to certify forests that are well managed is the primary reason for their interest in forest certification (Puller 1998, Drescher 1998). The public is interested in how forests are being managed because of concerns and distrust of how industry is managing land. They are also concerned about whether management activities result in a spillover of negative impacts on the landscape outside of the management unit (e.g., soil erosion into streams) or the loss of species (e.g., marbled murrelet).

Certifying organizations are bringing many values into the process. Some of these values are:

- Healthy forests
- No loss of biodiversity
- Wildlife habitat
- Riparian-zone maintenance
- Relationships of landowners to the community (e.g., job creation, safety of operations, right to unionize, etc.)
- Economic viability of the operations and how these feed back to affect the adjacent communities

It is quite apparent that individuals and organizations interested in certification are bringing different values to the discussions.

Small landowners also bring to this process other values that influence how they will interface with a certifying organization or to public demands that they maintain their lands for social values (e.g., preservation or conservation of species). Some of these values can be summarized as follows:

- Aversion to government regulation and involvement on their lands
- Strong support of private property rights

These values are also affected by the small landowners having less interest in investing capital in management, since their primary ownership objectives are for aesthetics or recreation instead of timber management. They are also affected by being isolated from the knowledge needed to manage lands sustainably because of the limited success of existing outreach efforts to reach them.

As part of the Canadian Standard Association's Sustainable Forest Management system, the social sciences have been integrated with the natural sciences because of the high requirement for public participation (CSA 1996ab). Public participation is a significant element of this system and a formalized, operational system is required to obtain and inform the public. This approach also focuses on human values at the local level in determining what should be managed, and on the identification of which indicators should be used. An interesting aspect of this protocol is a greater emphasis at the local level, instead of exclusively fusing national- or international-level values as part of the indicator-identification approach (approach seemingly used by most of the other protocols). Social-science variables are also considered at the level of implementing the ISO-based management system elements. This management system includes a review of all regulations impacting that management system and a forecasting of the long-term quantifiable results of using this management system on a specific land unit. Understanding the regulations impinging on a given land unit is common to all of the certification protocols.

7.2.3 Linking social and natural sciences at spatial scales

Keely B. Maxwell, Kristiina A. Vogt, Daniel J. Vogt, Bruce C. Larson

Traditional approaches to scale used by both the social and natural sciences results in an inability to integrate both fields (Vogt et al. 1999a). Once it is recognized that different scales of analysis are being used by different disciplines, the ability to integrate information into the discipline will be greatly facilitated (Vogt et al. 1999a). Miller (1994) stated that previous linking of the social and natural sciences has occurred in geography and psychology where similar scales of analysis were being used. These examples, however, do not mean that for integration to be successful similar scales have to be used (Vogt et al. 1999a).

Scale describes the size of the area which is being studied or managed. It encompasses how much area is covered, as well as what the resolution of the smallest distinguishable unit is. Scale is being increasingly recognized as important to research and management. The scale of examination helps configure the information available to us. It affects conclusions about what the driving forces and relations are in that system. For example, at the individual leaf surface — stomata regulate transpiration — it becomes apparent that climate is the primary control at larger scales (Wiens 1989).

The potential effects of scale on research conclusions and management outcomes are not often mentioned. However, scale issues can place constraints on our ability to obtain and utilize information about the potential sustainability of forest management practices. Most significantly, there is a mismatch between common natural- and social-science scales. This mismatch impedes our capacity to devise and best utilize timber certification protocols. Without understanding the existence and potential impacts of such scale problems, the risk exists of continuing to make inappropriate management decisions without understanding why the decisions are not correct.

So, what precisely are the issues of scale mismatch with respect to certification? Let us look at how natural and social science each deal with scale, and what the implications are when they are both required to make judgments in timber certification.

Natural science research often takes place on very small scales relative to the scales at which management decisions are made (Vogt et al. 1999a). Many field experiments are conducted at scales of one or ten meters (Kareiva and Anderson 1989). These choices of scale are often dictated by manageability issues and historical standards. The question then arises of how well research at these scales can answer questions about a larger area, given that organisms perceive and respond to the universe at different scales (Milne 1992). In Amazonia, for example, it has been shown that even vegetation measurement in one-hectare plots is not a representative sample of the vegetation present in a larger area (Campbell 1994). Timber certification practices place an added burden on the extrapolability of information. Timber certification is often carried out on lands that are larger than 1,000 ha in size (see 2.5). The scientific information gathered for certification is often at the scale of less than 0.01 ha (Vogt et al. 1999a). Are the management strategies to be certified ultimately sustainable at larger scales?

Social hierarchies include household, family, kin group, tribe, community, village, state, region, nation, and globe. These units often have no implicit scale measurable in land-area units. Social scientists have not historically incorporated scale as an explicit element in their study (Fox 1992). This incorporation has not even taken place with respect to human interactions with the environment, where understanding the full range of causes and effects is crucial. The scale of social science research is, again, not necessarily influenced by the relevance of the social patterns examined to fully answering a given question. Often, it is the result of historical standards or political factors (Miller 1994). Some studies are beginning to look at the study unit, not as if it is in a black box, but how larger-scale influences affect it. Social science scales in timber certification often involve scales that range from the family unit to a region.

In both social- and natural-science issues, larger-scale events (government policies, climate) affect smaller-scale patterns and processes. If these larger-scale events are not visible on the scale examined, the full array of contributory factors will be missed (Clark 1985). This issue is actually less problematic than aggregating up from small to large scales, given the potentially non-linear transitions that occur between scales (Wiens 1989). Common forest management divisions are by ownership or other administrative units that do not necessarily correspond to ecological units, such as a stand or watershed, lending added difficulties to the necessities of extrapolation.

To obtain social data relevant to the natural-resource management issues, natural scientists may often demand of social scientists information that would take too long to collect or be too difficult to analyze due to its large scale (Molnar 1989). Social scientists tend to research site-specific problems that are perceived by natural scientists as answering questions which are too in-depth, and not broad enough in scope (Miller 1994).

The ways to reduce this mismatch do not necessarily stem from organizing all studies on the same ecologically or socially determined units, as has been suggested by some (e.g., Montgomery 1995). There is no single, universally appropriate scale (Vogt et al. 1999a). The driving forces of human behavior and nutrient cycles are not likely to originate from the same scale. Some efforts are being made to identify indicators at different scales (Hellawell 1991). Given our current understanding, the first step may be to pay attention to how patterns and processes may be dependent on scale, and to be aware when making conclusions during timber certification processes.

It is important that forest certification not assume that the same scale of analysis (i.e., the management unit) should be used in all cases. Different sites and circumstances and past land-use legacies (Vogt et al. 1999ab) will make some scales of analysis more appropriate than others. The scale of analysis for the social and the natural systems will also probably be different and should not be conducted at the same scale (see Vogt et al. 1999a). It will be important to identify the appropriate scale within the social and natural sciences. Examples of natural-resource management problems and the most appropriate scale of analysis are given in Vogt et al. (1999a). Others have also attempted to specifically determine what the most appropriate scale of analysis is for their research (Fox 1992, Freudenberger 1997). More research is needed to specifically inform the natural resource manager on which scales might be appropriate for particular natural resource problems which are ecosystem based.

7.3 Public participation in certification

Joyce K. Berry, Kristiina A. Vogt

There has been a strong interest in identifying and determining who the stakeholders are and who the stakeholders should be who contribute to natural resource management (Colfer 1995, Colfer et al. 1995, Colfer et al. 1996, Decker et al. 1996, Merino 1996, Prabhu et al. 1996). Much of the discussion generated on this topic has focused on public lands (Ewert 1996). These are now being transferred to private lands, with insufficient understanding about whether the same approaches relevant for public lands are also appropriate for private lands.

In the past, the stakeholders were easy to identify on public lands and were usually the clients or the special-interest groups who utilized these resources (e.g., hunters, trappers, anglers, hikers, campers, timber companies, etc.) (Colfer 1995, Decker et al. 1996). Today, the stakeholders are more difficult to characterize and are a diverse group of people with many different values which a land manager is expected to satisfy. Since most certification protocols require considerable public participation in many stages of the analysis process (see 4.1.3.1), it becomes important to identify who the stakeholders contributing to these dialogues should be. Since the "public," or stakeholders, also have values, it becomes important to ensure that the values of a more "vocal" public do not control the process and drive management to satisfy specific values at the expense of a sustainable ecosystem. For example, human values for harvesting specific tree species or for managing for deer hunting on public lands motivated land managers to produce a landscape that now is recognized as more susceptible to fire and insect attacks. The resulting forested — landscape produced by management guided by values — is now classified as being unhealthy (Rapport 1997, Vogt et al. 1997a, 1999ab).

Most owners of forestlands pursuing certification are private, with some government lands being certified (see 2.5). Since much of the debate on public participation has been driven by the conflicts occurring on public lands, there is a need to identify who the appropriate stakeholders should be when evaluating private lands. Much of this debate, and the conflicts over the values that are to be obtained from publicly owned lands, have carried over to private lands, where the public is requiring that more values be produced from a piece of land — in addition to those values held by the private landowner. These debates, of course, raise many questions related to other issues, such as private-property rights and how much the public should infringe on private rights in the name of the common good.

A question that has not been sufficiently addressed is: What is the exact role of the public and what is its influence over an evaluation of management plans and management activities on private lands? If the pubic is insufficiently informed, narrowly focused, or driven more by achieving their personal values from a given piece of land, an actively participating public can have a detrimental impact on sustainable forest management.

A strong role of public participation in controlling or influencing the certification process and how management is conducted is very relevant when the lands are publicly owned. For example, the Canadian certification standards have a strong and extensive public involvement in all phases of managing forests (CSA 1996ab). This is logical considering that 94% of the forestlands in Canada are publicly owned. These lands are also subject to all of the federal and provincial-level environmental laws and regulations that exist in Canada. Identifying the varying roles of the public in certification, especially on private lands, and creating reliable and effective processes for involving the public are critical.

7.4   Importance and participation of non-industrial private forests

Kristiina A. Vogt, Jennifer Heintz, Christie Potts, Allyson Brownlee, Heidi Kretser, Luisa Camara

Non-industrial private forest (NIPF) landowners cannot be ignored in any discussions of future timber supplies, especially in the United States. In many places, NIPF contribute significantly to the total wood supply and they play a vital role in the economic welfare of rural communities. Since nearly 70% of the rural communities are highly dependent on earnings that result from selling natural resources (Janik et al. 1998), ensuring the competitiveness of these communities is essential. If certification does confer an economic advantage in the marketplace, this label would greatly facilitate NIPF landowners obtaining an economic return off their land, and to work towards maintaining sustainable forest ecosystems. Development of a certification approach that effectively assesses small tracts of land typical of NIPF owners should, in general, improve the economic viability of rural communities. However, current certification approaches under the FSC label do not appear to assess NIPF lands, because of their structure which requires large amounts of data on social and natural systems, and the chain-of-custody requirement (see 4.2.1, 7.7). Other programs (e.g., the National Forestry Association's Green Tag Forests, see 2.5) are able to certify smaller acreages. However, even this program will not be able to provide certification for about a fifth of the forestland owners (Table 7.4) whose acreage is smaller than the 16 ha limit set by the Green Tag Forests program. Therefore, it becomes important to understand what limits the ability of small (NIPF) landowners to become certified or labeled as practicing good forest management (see 7.5).

In the United States, the importance of small NIPF landowners in contributing to wood supplies cannot be overemphasized. This fact is quite apparent when one examines who owns forestlands and who are the major suppliers of wood. For example, non-federal forestland are extensive and comprise over two-thirds of the forested land in this country

TABLE 7.1

The size of acreage owned by private forestland owners and the number of owners in the United States in 1994 (NRC 1998).

Size of Land	Total Acreage (% of total)	Number of Owners (% of total)	Average Acreage in ha/Individual (Average Acreage in ac/Individual)
0.4–19.8 ha (1–49 ac)	31,174,089 ha, 77,000,000 ac (20%)	8,557,000 (86.5%)	4 (9)
20–40.5 ha (50–100 ac)	19,028,340 ha, 47,000,000 ac (12%)	717,000 (7%)	27 (66)
>404.9 ha (>1,000 ac)	61,943,319 ha, 153,000,000 ac (39%)	27,000 (0.3%)	230 (567)

TABLE 7.2

Forestland ownership objectives of NIPF owners (Birch 1995).

Primary Reason for Owning Forestland	Percentage of Owners (%)	Percentage of Acres (%)
Part of farm or residence	40	17
Recreation and aesthetic enjoyment	23	16
Increase in land values	21	16
Timber production	3	29

(NRC 1998). Today, about 69% of non-federal forestland areas are owned by 9.9-million non-industrial private landowners or the NIPF owners. However, most of this land is individually owned in small tracts which are not contiguous to one another and that certification protocols do not assess efficiently. For example, data summarized for the United States showed that most NIPF owners have less than 20 ha (50 acres) of forestland (Table 7.1). Based on 1994 figures, less than 0.3% of the private forestland owners owned more than 405 ha (1000 ac) of forested land (NRC 1998). This size of forest ownership is still less than half of what is typically certified by the FSC-type protocol (see 2.5). Current statistics show that the proportion of individuals owning more than 405 ha (1000 ac) has not varied significantly from 1978 to 1994, since the same percentage was recorded during both years (NRC 1998). However, future trends of the size of forestland ownership suggest that ownership sizes will even be smaller. All of this data suggests the need to adapt certification assessments to deal with small tracts of forestland.

The importance of NIPF contribution to the United States' wood supply is quite apparent when using the total acreage being managed economically, and not just the number of landowners. For example, despite only a quarter of the individual NIPFs using their lands for economic gain, 45% of the forestland areas are, in fact, being managed for economic purposes when summing the total NIPF acreage (Table 7.2). Therefore, almost half the NIPF lands are being managed for timber or for some economic gain. In terms of the potential for forestland acreage to supply economic products, a sizable amount of timber can be supplied from lands owned by NIPF owners. In fact, based on 1992 data, NIPF owners involved in timber production contributed significantly to the forest-products industry. During 1992, they supplied 51% of all timber harvested in the United States, including 32% of the softwood timber and 72% of the hardwood timber (Powell et al. 1993). The role and importance of NIPF owners in timber production may continue with the pattern of reduced timber harvesting of federal lands in the United States.

To understand the problems being faced by small NIPF owners pursuing some kind of certification or green labeling for their lands, it is useful to characterize these small landowners. Understanding how they differ in character from the large, private landowners helps to explain some of the reasons they cannot be treated in the same way as the large industrial or non-industrial landowners. To begin with, most NIPFs do not own their land primarily for harvesting forest products. Most of NIPFs own land for recreation, for limited timber production, as an investment for descendants, or for many other non-industrial uses. Many of NIPF owners do not own or operate a sawmill have or other secondary processing capabilities. Because harvesting forest products is secondary to these landowners, most are not actively managing their lands to increase their potential for harvesting forest products or for making improvements to their forest structure and composition. This also means that good forest management is probably not occurring on these lands since sustainable forest management is not a goal of most of the landowners.

In addition, these small NIPF landowners are not dependent on their forestlands for their primary income. For example, a study conducted by the USDA Forest Service Forest Inventory and Analysis unit (Birch 1995) showed that most landowners use their lands for passive enjoyment (63%) while a fourth may use it for economic gain (Table 7.2). When economic gain is not the primary reason to own the land, decisions being made on these lands will not be based on economic factors. This also means that forestland owners will not be able to afford to invest much capital in improving the condition of their forests or to deal with insect or pathogen outbreaks that might occur on their lands.

Despite the fact that most NIPF owners do not manage their forestlands as their primary source of income, they have become an important timber source in the United States. Many NIPF owners sell timber from their forestlands at specific times for income generation (e.g., pay for children's college education, etc.).

The incomes that can be generated from forested lands do vary by the size of the ownership. This means that small NIPF owners have more constraints on earning financial returns from their lands. For example, small landowners using their lands as secondary or supplemental sources of income do not have the advantages of efficiencies in processing and marketing that exists for large companies. Currently, negative externalities on poorly managed forestlands are not accounted for in the price of extracted forest products.

Consequently, the marketplace creates disincentives for lands to be managed sustainably.

With these marketplace-induced disincentives for sustainable management of lands, companies and landowners with large land holdings, processing facilities on site, and management plans geared toward production, will be able to monopolize markets with lower costs. For example, Kane Hardwood is certified, but is unable to sell in the European markets because of competition from cheaper Asian softwood imports (Puller 1998).

The issues of small landowners and their ability to be certified using the present assessment protocols is not just an issue for developed countries. It is also a topic of importance for tropical forestlands in developing countries, because many private landholdings are small. For example, Brazil has 1,726.5 million hectares of privately owned forestlands that are 20 ha or less in size (Table 7.3). An almost equivalent acreage exists in tree plantations (1,935.8 million ha) in this land-size category in Brazil (Table 7.3). As a percentage of the total, 27.2% of the tree plantations were in this smallest land-size category, while native forests had less than 2% of the total in this category. Although half the total native forestlands and tree plantations were in the greater-than 1,000 ha category, from 42% to 29% of the forests were smaller than 1,000 ha in size (Table 7.3). Therefore, a large proportion of the forestlands and tree plantations in Brazil would probably not be certifiable under the current FSC-type system.

TABLE 7.3

Total acreage by land use in native forests, tree plantations, and other (crops, pastures, marginal lands) in Brazil in 1996 (IBGE 1997).

Size of Land, hectares	Native Forest	Tree Plantations	Other
	hectares × 1,000,000 (% of total by land use)		
<1–20	1,726.5 (1.9%)	1,935.8 (27.2%)	15,738.0 (6.1%)
>20–50	4,350.8 (4.9%)	290.0 (4.1%)	20,784.1 (8.0%)
>50–100	6,015.2 (6.8%)	223.6 (3.1%)	21,205.7 (8.2%)
>100–1,000	25,211.7 (28.4%)	1,063.6 (14.9%)	97,252.7 (37.5%)
>1,000	51,587.3 (58.0%)	3,608.9 (50.7%)	104,298.1 (40.2%)
Total, by land use	88,891.5 (100%)	7,121.9 (100%)	259,278.6 (100%)

7.5 Certification: constraints and opportunities for non-industrial private forestland owners

7.5.1 Size of management unit being certified and reasons size limits ability to be certified

Kristiina A. Vogt, Bruce C. Larson, Daniel J. Vogt, Jennifer Heintz

Any discussion of the constraints and advantages of land size and certification must remember that small NIPF owners are an important source of timber, and own more than half the forestland in the United States (see 7.4). So, it becomes important to identify the minimum land size that can be certified and to determine what constrains the ability of small landowners to pursue certification. Typically, FSC-type certification protocols have given preference to large land areas when assessing forests for their sustainability (see 2.5). Data presented earlier (see 2.5) showed that most certified forestlands are over 1,000 ha in size and in the five-digit-number range.

The ownership patterns of forestland in the United States suggest the need to understand why there are so many obstacles to the certification of small, non-industrial landowners. Most of these landowners are currently excluded from certification (see 2.5, 7.5). This is important, because in 1990, 57% of the United States timberland was owned by non-industrial private landowners (Haynes 1990), making them a significant portion of the forestland owners in the United States. This was in contrast to the 15% owned by the forest industry and the 28% owned by federal, state, and local governments in 1990 (Haynes 1990). In 1992, the non-industrial private landowners owned 59%, while industrial forest owners owned 14.6% of the total non-federal forestland. Comparing the 1990 data published by Haynes to the NRC (1998) report showed that the amount of non-federal forestland owned by private non-industrial landowners has increased during this period.

The NRC (1998) report showed the importance of industrial and non-industrial forestland owners in satisfying the wood needs in the United States. In this report, it was stated that one-third of timber supplies came from industrial lands and half came from non-industrial private landowners. Characterization of the private, non-industrial forestland owners becomes relevant because this group owns close to 48% of the total forestlands in the United States (Figure 7.1). However, this group is the least likely to become certified using current protocols (see 2.5, 7.4, 7.5).

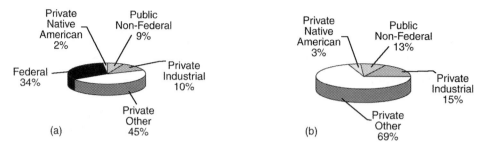

FIGURE 7.1
Percent ownership by area (a) for total forestlands (federal, non-federal) and (b) of non-federal-owned lands in the United States in 1992 (NRC 1998).

This potential for non-industrial forestland owners to contribute to increasing the amount of carbon sequestered exists because they own almost 48% of forestland area and because they have not been managed to optimize biomass accumulation. Most of these forests have low stocking density and have low volumes of wood (NRC 1998). These private forestlands also are important habitats for many of the threatened and endangered species and public lands are insufficient to provide for these species. For example, of the 712 species currently listed as threatened or endangered, 609 are found on industrial and non-industrial, private forestlands (NRC 1998).

In the United States, this lack of participation by private industrial and non-industrial forestland owners in certification has occurred because of conflicts in balancing societal values and the management needs of forestland owners. It is crucial for these conflicts to be resolved so that the major players contributing to wood supply can participate. If certification can be a tool that allows this to occur, it will serve an extremely important function. This highlights the importance of continuing to develop certification approaches that are capable of effectively and accurately assessing and analyzing the trade-offs inherent to forested ecosystems

Some attempts have been made to identify what size of a forestland area is too small for it to be certified. LaPointe (1998) stated that, in general, you could not deal with a 20 ha (50 ac) woodlot as sustainable, because it is too small. The National Forestry Association's Green Tag Forestry program has stated that it is not financially worthwhile to certify parcels that are smaller than 16 ha (40 ac). Donovan (1998b) stated that one needs at least 1,000 ha to be certified because of the costs of certification. The problem with accepting these land-area values as the lower limit for certification is the significant number of forestland owners who would be excluded from pursuing certification. For example, the average size of forestland owned by a non-industrial private forestland owner in Vermont is 7.7 ha (19 ac), yet they own 77% of the forestlands in that state (Brynn 1998).

Several issues are highly relevant in determining the minimum land area that can be certified and whether certification can occur on small land areas:

- Chain of custody (see 7.7)
- Price premiums necessary to pay for forest certification (see 7.6)
- The influence of adjacent land areas on affecting the decisions being made on your management unit (see 5.2.1.2, 5.2.1.3)
- Satisfying both the natural- and social-science part of sustainability, and obtaining all the human values from a given piece of land (3.2, 5.2)
- Number of criteria and indicators needed to address social- and natural-science links (see 5.1, 5.2)

- Knowledge base and a lack of training of the landowner on the different elements of certification, their management options, and how to achieve sustainable management (7.2, 7.5, 7.8)
- Costs of certification for forestlands not previously managed or poorly managed, so that lower financial returns accrue from natural resources extraction (7.5, 7.6)

It is interesting to determine the relevance of the seven previous issues for determining the minimum area that can be sustainably managed for forest products. Since some of these issues have been well covered earlier or will be presented later in this chapter, the cost of certification and its large requirement for information will be discussed in this section.

Several questions need to be asked when discussing what the minimum lower limits of forestland are that can be certified without creating hardship for a small landowner. Even though some of the seven issues are present in different certification protocols, they have not all been given the consideration that they deserve, and one or two have been used as the rationale for what land-area size is realistic to certify. It is important to determine whether or not most of the rationales are socially based. Do natural science factors contribute to deciding what the minimum certifiable area should be? How much do the costs associated with certification drive and determine the minimum land area needed for certification? What is the role of the "chain of custody" in this assessment? How do the different values desired from a managed piece of forestland determine the land size needed to achieve sustainable management? An examination of the literature suggests the following hypothesis:

> Many of the lower limits of what land area is sufficiently large to be certified have been established by social science factors (mainly economic and human values for different resources), without considering natural science factors or constraints.

The cost of certification is frequently discussed as one of the primary factors affecting whether small landowners pursue certification (see 7.5, 7.6). It is clear that the cost of certification will initially determine whether a landowner decides to pursue certification. Therefore, cost will be a primary determinant driving landowner options with respect to certification. Other factors (natural science variables), however, need to be added to the formula being used to determine whether a landowner can satisfy the requirements of certification and sustainability. These other factors will constrain whether sustainable management can be achieved in the management unit.

Two costs that a landholder will have to cover are (Heaton 1994):

- The costs associated with the certification assessment itself, which is paid to the certifying company
- The costs of implementing sustainable forestry on the lands in order to become certified

The cost of certification and implementing management practices are not obstacles for mid-size to large forestland owners, because they can recover incurred expenses in a couple of years. However, the costs do become obstacles for small landowners because they cannot obtain high enough returns on forest-resources extraction to cover their costs (see 7.6). These lower financial returns are, in part, due to the difficulty of small forestland owners to reach markets and/or have enough variety of products to satisfy demands. If implementing sustainable forest management really creates mechanisms for these small forestland owners to reach niche markets, then forest certification will become attractive. However, for these landowners, certification schemes under the FSC-type protocol still can be too expensive due to the insurmountable requirements (see 7.6).

It is difficult to obtain good numbers on how much certification organizations will charge a landowner to conduct an assessment. The costs of certification will vary with the organization conducting the assessments. For example, the National Forestry Association's Green Tag Forestry program is cheaper for a forestland owner to pursue than the FSC-type certification. Costs associated with the certification itself are summarized below for some certifying organizations.

The National Forestry Association's Green Tag Forestry program has the following costs (Argow 1998):

- $15 to set up a permanent record base
- $1 to $3 per-acre (average) expense for field inspections

The costs for the Green Tag Forestry program are reduced by the size of the tract and the availability of complete inventory and management data prior to the assessment. Landowners wanting to participate in this program must also be members of the National Woodland Owners Association which has $25 annual dues.

In 1994, Smartwood (FSC-accredited) published the following costs in U.S. dollars (Heaton 1994):

- $1.30/ha when 5,000 ha is being managed
- $0.24/ha when 100,000 ha is being managed
- $0.08/ha when 600,000 ha is being managed
- $0.01/ha when multi-million hectares are being managed

The costs diminish dramatically as the size of the land area being managed increases. However, no data was given for small landholdings. Simula (1996) summarized the following information for the Smartwood certification costs:

- $5,000–$75,000 per assessment, based on the consultancy time and travel costs
- $0–$3,000 per assessment when subsidized by Rainforest Alliance (Smartwood)
- $3,500–$45,000 for a full cost assessment reflecting actual costs incurred by Rainforest Alliance
- $500 for annual source and company audits; costs can range from $100–$2,000 (Heaton 1994)
- $500 annual fee for all applicants to support program management (this can be reduced or waived) (Heaton 1994)

SGS charges $31,000 annually for large operations (in this case, for a 500,000 ha tract of land) (Simula 1996).

Soil Association (Woodmark scheme) charges 1% of the sales value of large forest tracts as an annual fee (Simula 1996). Smaller operations have an annual fee of $225 that increases as the value of the forestlands increases (Simula 1996).

Based on the figures given above, the minimum costs for a landowner to certify 20 ha of forestland can vary from a low of $190 (under the National Forestry Association's Green Tag Forestry program) to several thousand dollars. The exact costs for certification cannot be given (but see 7.6) because that information does not exist. However, it appears that the cost an NIPF owner would have to pay upfront could make them decide not to pursue certification. These costs also do not consider what costs would be incurred by a forestland owner — who has never managed or who has poorly managed their forestlands — to begin to implement good forest management and to develop a management plan.

When social science factors (especially values) predominate in assessments, it can be hypothesized that the land area will need to be larger to manage for forest products and sustainability. It would be an interesting analysis to determine how satisfying different values (see 3.2.2) simultaneously from the same piece of land would change the minimum size of the land area needed for certification. Satisfying some values (e.g., preserving species, clean drinking water, etc.) would probably require much larger minimum land areas for those values to be continuously maintained. Other values (e.g., obtaining financial returns from harvesting timber to pay for college education) can be achieved from a smaller land base because they occur infrequently.

Several natural-science factors will constrain the ability of forestlands to be managed for good forest management and sustainability. An ecological approach to determine the minimum land area needed for certification must address the following topics:

1. How does the management unit maintain its functions and structures because of the landscape matrix within which it is embedded?
2. How does the amount and type of information needed to assess a site change with spatial and temporal scales (e.g., scaling up from smaller to larger scales)? How do the changing information needs modify the indicators that need to be used to assess the management unit?

Utilization of only natural science variables to define the lower space limit for certification can result in two scenarios, depending on the values that are included as part of management. The size of the land area will vary from small to extremely large (e.g., landscape), depending on how values define the intensity of management intervention needed in the system. The minimum land area needed will also vary with the ecosystem and landscape characteristics of the management unit.

If a forestland is being managed for timber and a certain natural mixture of regenerating species is desired, it must to be ensured that sufficient land area will be allocated to maintaining the different developmental stages particular to that system. Moreover, if a forest stand can persist for 50+ years in the same developmental stage — and, in general, four stages are possible (Oliver and Larson 1996) — the restoration of all developmental stages within the landscape will be extremely difficult. If the existing landscape has all the developmental stages present, it will be much easier to manage using this ecological approach, but it will still require assessing the landscape instead of the management unit by itself. Therefore, the amount of land area that a certifier has to consider will vary depending on the area of existing forest present within the landscape and the amount of contiguous forestland that can be considered when evaluating the potential of the management unit to sustain itself within the landscape.

If there is not the need for all the natural developmental stages to exist within the landscape because intensive management produces the desired forest stage, the amount of land area required could probably be considerably smaller. In this case, the ability of maintaining a healthy and sustainable forest will still be related to the matrix landscape and how much atypical edge environment exists adjacent to the management unit (see 5.2.1.2). Therefore, although the management unit can be smaller, an assessment at the landscape level will be required to determine that the landscape matrix will not influence the sustainability of the management unit being certified. This scenario requires different types of information to be collected as part of an assessment, compared to the scenario which requires the restoration of all the stand developmental stages particular to a forest type.

The smaller the land area owned by an individual or family, the greater the edge environment that abuts their land (see 5.2.1.2). This makes it essential for small landowners to consider the implications of management occurring on neighbor lands, because they feedback to affect the sustainability of their system. The larger a given piece of land, the less land area is associated with neighbors and the easier it is to satisfactorily control many of the human values that are being managed (e.g., biodiversity, wildlife, etc.). Much of the land-size issue will revolve around how much edge environment exists for any land unit and when increasing the amount of edge environment begin to detrimentally affect the ability of small landowners to sustain their forests.

One approach to counteract the constraints of small landholding sizes has been to conduct group certifications, as opposed to certifying individual landowners. Presently, only 2.6% (Table 2.1) of the FSC-certified forestlands are communal or group certified. Despite this low percentage of group certifications, the idea of forestland owners becoming certified as a group is not new. In fact, group certification guidelines have been developed by FSC (FSC 1998) because they recognize the limitations of small landowners becoming certified as individuals. Guidelines for group certification are the same as the requirements for individuals. The only difference between individual and group certification will be that several landowners must agree to follow similar rules and practices.

The different philosophies presented by the FSC and SFI approaches (Figure 2.1) suggest that the SFI approach is more amenable to small landowners, while the FSC approach is more amenable to large land ownership or to group certification. The basic philosophies of the different certifying organizations help to explain how they approach the certification of small landholdings. For example, the FSC approach to forest certification uses a narrow selectivity approach that makes a few forestland owners improve a lot, in comparison to the SFI approach that uses wide selectivity so that many may improve a little.

The Canadian Sustainable Forest Management system recognized the problem associated with the size of the land area being certified (CSA 1996ab). They stated that the defined forest area being certified could vary in size from a few hectares to more than a million hectares. However, they immediately stated that it might be necessary to combine small forest areas into larger units because of the scale-related issues that would apply. They perceived a system where small woodland owners would be part of a larger, defined forest area to which the criteria and critical elements would be imposed when they could not be imposed at the small scale. They also felt that this would make the economics less cumbersome to the small landowner because the program would be more cost-effective. However, the Canadian system will be cumbersome and difficult for individual small landowners to pursue for certification because of its strong requirement for public participation. As part of its assessment, it requires the forestland owner to help create formal structures to get the local public involved in the management process and to help identify the local values for a given piece of land (CSA 1996ab).

Other issues relevant to small landowners are the amount of information needed to pursue certification and whether a small landowner can realistically obtain all the information needed to assess social- and natural-system sustainability (see 4.2, 5.2, 7.5). A small landowner will have difficulty in acquiring and paying for the collection of the required information on species diversity or the wildlife surveys (see 4.2, 4.3). In most cases, the landowner is not trained to collect the types of information that are required. The costs of obtaining the required social information may be unrealistic and in many cases, may not be relevant (i.e., the right to form unions) for their lands.

The cookbook approach to selecting indicators does not adapt to changing ownership land sizes. For example, the large-scale analysis emphasize landscape-level parameters, while the small scale focuses on the detailed plot information. The certification approaches do not deal with the changing information needs that are scale dependent for ecological systems. Today, it is quite apparent that the information requirements of an assessment will vary depending on the scale of analysis (Vogt et al. 1997a). For example, much less ecological information is needed to predict ecosystem structure and function at large scales than when analyses occur at the small-plot scale (Gosz 1993). This suggests that if landscape-level analyses were being conducted, a smaller database would be needed to predict what is happening at a site. What is interesting about the changing data needs is that the FSC-type certification assessments appear to be designed for larger landowners; however, the data needs have not been adjusted to deal with the changing spatial scales. Whether this ecological scale phenomenon is also relevant for social systems needs to be researched since it has implications for the types and amount of indicators required for an assessment. The scale-dependency of data has strong implications for which indicators are minimally needed for an analysis to be effective, reasonable, and to sensitively reflect the site (see 5.1).

7.5.2 Constraints to private landowners: balancing public and private rights and interests in the United States

Kristiina A. Vogt, Joyce K. Berry, Christie Potts, Allyson Brownlee, Jennifer Heintz, Heidi Kretser

In the United States, public attitudes and interests in how private forests are being managed have been increasing dramatically over time. Much of this interest and many of the constraints placed on private landowners have evolved along with the debates over the management of federal forestlands that occurred in the 1980s and 1990s (see 2.1, 2.2). Before this time, private landowners were not factored in as part of the solution to providing the goods and services traditionally obtained from public lands (e.g., saving species, clean water, hunting game species). Because of this interest, it is worthwhile to analyze the other constraints and limitations which control management activities on private lands.

7.5.2.1 Federal, state, and local regulations relevant for private landowners in the United States

Many constraints are being placed on private landowners through regulation, especially at the state and local levels. Since private landowners must satisfy these regulations, it is important to compare the impacts of these laws with how well forestlands are being managed, and what added value accrues to a private landowner by becoming certified. Certification is, of course, voluntary, while regulations are not. It becomes interesting to ask what forest condition the many laws produce if a landowner satisfies all of them.

An issue that arises with regulations and with certification is the expectation that private landowners should cover the cost of providing and maintaining the goods and services from their lands that are valued by the public. This expectation has created conflicts between private and public rights. Recent attempts to address these conflicts have generated new approaches to resolving these problems. For example, one approach provides compensation to private landowners when they provide the goods and services that the public values, instead of opting to pursue financial returns from extracting forest resources (i.e., The Natural Conservancy's Forest Bank program). The purpose of the Forest Bank is to provide liquidity to non-industrial, private forestland owners in exchange for permanent control over how their forest resources are managed and harvested (TNC 1998). As part of this program, the bank will accept deposits for the right to grow, manage, and harvest trees, and the bank will pay annual interest on the value of the deposited trees. The Nature Conservancy has a history, similar to The Conservation Fund, of paying for public goods. These approaches differ from the Audubon and Sierra Club-type approaches which do not use incentives, but instead use regulations to force landowners to meet social objectives.

There are many regulations which control how private landowners can manage their lands (see 5.3.2, 5.3.3). Many of these regulations are having major impact on private rights and interests, and are constraining the management options available to private forestland owners. Laws affecting forestland owners exist at the federal, state, and local levels, with some of the stiffest and highest numbers of regulations existing at the local level (Martus et al. 1995). Most of the FSC-type certification protocols are strongly structured to satisfy two of the federal regulations: the Endangered Species Act and the Federal Water Pollution Control Act, later amended to the Clear Water Act.

> In fact, many of the indicators selected for the FSC certification protocols directly, or indirectly (see 4.2.1), satisfy the intent of the Endangered Species Act — which is to save species and their habitats.

The federal laws do not really address management practices on private lands, so they are not structured to deal with which guidelines are good for forest management. However, state and local laws focus on forest management practices and regulate activities which occur directly as a result of management.

Interestingly, most of the values embedded in the FSC-type certification protocols attempt to satisfy the federal laws, and not the state or local-level laws. This is obvious from an examination of the data requirements of the assessment protocols (see 4.2.1), where indicators relevant to sustaining endangered or threatened species reappear frequently in different places in a protocol. In fact, the value for species appears to be the central focus determining which criteria and indicators are integral to a certification assessment (especially for those certification protocols under the umbrella of the Forest Stewardship Council) (see 4.2). These certification protocols do require that appropriate laws be satisfied, but not how they might satisfy some of the indicators being used in any assessment. Most certification protocols state, in one sentence the requirement that all appropriate laws need to be satisfied, but give no further explanation. The details are not give regarding what these laws require private landowners to do on their lands; nor is it stated how satisfying these laws would move a landowner towards a sustainable natural system. This fact is interesting, because the state and local laws have very specific requirements for how forests should be managed.

This raises the question: if landowners have satisfied these laws, how much would this contribute to satisfying many of the reasons certification is being pursued? A comparison between the types of sustainable systems produced by following the existing laws, and those resulting from certification could be quite informative. It may be that the public cannot comprehend all the laws that private landowners need to satisfy when managing forestlands, and certification is a way to summarize all that information for the public. If this is the case, then there is a great need to compare what kind of forested landscape is produced by laws, compared to certification itself.

> Do both approaches result in forests with similar sustainability? Or is certification more sustainable because the analysis is holistic compared to the many individual laws that regulate parts of the system?

In some cases, the private landowner must be knowledgeable about a significant number of laws that potentially affect land management (see 5.3.2, 5.3.3). For example, the state of New Jersey alone has 100 local forestry ordinances which a forestland owner must satisfy (Martus et al. 1995).

State forest regulations have been documented to successfully alleviate public concerns about forestry and its practices (Ellefson and Cheng 1994). In addition, there appears to be agreement between special-interest groups and the public that this regulation has improved natural-resource conditions (Ellefson and Cheng 1994). Since most landowners' attitudes about forests mirror the public's views, forestland owners recognize the link between regulation and improvements in the condition of the natural resources and the development of a new ethic of forest conservation (Gangloff 1998).

One of the important goals of forest certification is to show the public that forests are being managed in a sustainable manner. It appears that state and local regulations in some parts of the United States may be providing the public with a similar type of confidence that natural resources are being well managed. The survey by Ellefson and Cheng (1994) showed that most landowners are complying with regulations and do not feel that they are overburdened by them.

It is still not clear how much regulation affects the decisions being made by private landowners regarding whether or not to manage their lands for timber. One study in western Oregon and Washington suggested that less than a fourth of the private, non-industrial forestland owners considered regulation as an important variable in their decision to harvest timber (Johnson et al. 1997). Unquestionably, forest regulations, especially more recent ones, incur additional costs and administrative burdens for private landowners, even when they do not believe them to be burdensome. Most likely, regulations will not be considered a burden for many non-industrial private landowners who own their lands for purposes other than timber production. However, this does not consider the fact that any extraction of resources from private forestlands will require a landowner to satisfy several different laws, which will force the landowners to become quickly educated on these issues. These laws will place another layer of documentation that a landowner will need to maintain — similar to what is required with certification.

When a landowner is able to comply with all the regulations relevant to their lands, the real problem becomes:

> Should private forestland owners bear all the costs of protecting environments and providing the goods and services that have typically been the products of public lands?

This is an important question, since landowners typically sell their lands when they cannot afford to maintain them under the regulations which they must satisfy. Regulations and laws are impacting the capacity of forestland owners to maintain ownership of their lands (see 5.3.2, 5.3.3). This problem is apparent when one examines the length of land tenure in many parts of the United States. For example, the average land tenure is seven years in Vermont, which is a high-turnover rate. Many forests and farms are being converted to higher-value uses (e.g., development) and the sizes of the larger blocks of forestlands are shrinking. In many cases, owners have to subdivide and sell part of their land to pay property taxes in the commercially zoned areas regardless of their open-space value. "Federal and state death-tax burdens cause disruptions in management and demand may cause heirs to abandon timber-production programs" (Peters et al. 1997). Vaux (1986) concluded that the California Forest Practices Act increased landowner costs and resulted in industries selling entire holdings between 1978–1982.

It is imperative to determine how private forestland owners can be part of the solution to existing environmental problems without disrupting their ability to own land. It cannot be ignored that private landowners do provide a vital portion of the timber needs and habitat for endangered species in the United States. This will be a challenge, since non-industrial private forestland owners do not primarily manage their lands for timber, but own the greatest proportion of forestlands in the United States, albeit in small tracts (see 7.2). NIPF owners are important sources of timber, despite the fact that their primary interest is not in timber extraction, but in aesthetics, wildlife, and property values. Almost half the NIPF owners think that they will harvest some timber on their lands within the next ten years (Sampson 1998). How much non-industrial forestland owners should be regulated by laws and when their primary land management is for other reasons, pose important questions related to balancing the rights and interests of the public and private sectors.

7.5.2.2   *Inheritance tax and private landowners*

It has been suggested that the changes in the size of private forestland ownership will work against the use of a holistic approach to land management (Gangloff 1998). This problem is being driven by shrinking land ownership for most private non-industrial landowners. This pattern is also being driven by deaths in families, causing land to become subdivided between a larger number of heirs who are inheriting that land, or heirs who are having to subdivide their land to pay the inheritance taxes (Gangloff 1998). (See sections 4.6.6 and 4.6.7 for a discussion of the laws affecting private landowners) It has been estimated that by the year 2010, 38% of the private forestland owners will own about 6.9 ha (17 ac) of forestland (Gangloff 1998). The size of this land area, however, is considered too small to manage for timber (Gangloff 1998, see 7.1, 7.5).

7.5.3   Benefits of certification to small, non-industrial private forestland owners
Jennifer Heintz

Several benefits potentially accrue to a small, non-industrial landowner by becoming certified. These benefits can be summarized as follows:

- Visible confirmation (to self and community) that a landowner is using the highest standards of good forestry practices.
- To be ready, in advance, if price premiums for green wood become a reality.
- With Green-Tag certification, a nationwide database is maintained, which includes data on the species, volume, and location of certified timber ready for harvest. Participating landowners can list their timber on this register when they are ready to sell.
- Many NIPF owners are very concerned about increasing regulation of forest practices and reduction of property rights (Cubbage 1991, Ellefson 1994, Johnson et al. 1997). Participation in third-party certification may serve as a much-needed public relations campaign for NIPF owners, which could lead to a reduction in the need for additional regulations. "Environmentalists distrust the willingness of private landowners to act in the public interest" (Cubbage 1995). If this is true, NIPF certification may alleviate some of these tensions between public interests and private rights.

- Generally, receiving higher value for both timber and non-timber forest products, by virtue of having a comprehensive management plan and involvement in the certification process.
- The goals of ecosystem management could be addressed by NIPF owners if the process of certification stimulates the landowner:
  1. To partner with adjacent landowners or, at least, be aware of management effects across the landscape (beyond their property boundary).
  2. To utilize landscape-level scales in management.

7.5.4 How to make certification useful for small private landowners

Kristiina A. Vogt, Jennifer Heintz, Christie Potts, Allyson Brownlee, Heidi Krestser, Luisa Camara

7.5.4.1 Acquiring information needed for certification

If the information needs which are so integral to many of the current certification protocols are deemed necessary as part of assessments, the public will need to provide private landowners with access to the information required to practice good forest management. Most small, private landowners cannot afford the costs of conducting research on their lands.

> Forestland owners should not be expected to become "quasi" research stations before pursuing certification.

This suggests the need for many existing environmental and forestry agencies to be able to supply the different types of information required by these assessments. Even some of the certification organizations have perceived this problem of information acquisition and are compiling data to be made available through their organizations to private landowners pursuing certification.

Without the efforts of other organizations to help supply the regional types of information needed by landowners, many small, private landowners will be unable to access or collect the information required during certification. If more lands are to become certifiable, information must be made readily accessible to landowners in centralized locations. This information needs to include data on:

- Local and regional natural and social systems
- All regulations (e.g., federal, state, local) for that region
- General information on how to improve forest management and to manage for sustainable ecosystems within a landscape framework

Of course, access to regional information does not eliminate the need for landowners to obtain specific information on their management unit. However, this information will begin to direct the landowner to where they should invest their time and resources.

Currently in the United States, several programs exist to facilitate private forestland owners in acquiring information and to provide incentives to help defray the costs of achieving environmental stewardship on their lands. Cubbage et al. (1993) conducted a comprehensive review of the major incentive programs that exist for private forestry — two of these programs are directly involved in information transfer. Cubbage et al. (1993) divided these programs into the following categories:

1. Taxation: exemption, remission, or deferred payment of taxes (i.e., capital gains reduction, modified property taxes, reforestation tax incentives).
2. Financial Assistance: subsidization of production through cost sharing, provision of materials, etc. (i.e., stewardship incentive programs, conservation reserves, state cost-share programs, loans, reduced cost for seedlings, etc.).
3. Technical Assistance: on-the-ground advice, extension programs (i.e., provided by state [Natural Resource Conservation Service] and federal programs [Forest Service]).
4. Indirect: government research, training, marketing information (i.e., provided by state and U.S. Fish and Wildlife agencies and public cooperative forest-protection programs).

Cubbage et al. (1993) concluded that almost all research indicated that state technical-assistance programs are effective and contribute to both private and public goals in the United States.

Technical assistance is especially crucial in an age when we are asking landowners to think holistically of their land resources. The ability for private landowners to acquire the most current and relevant information and to educate themselves on how to holistically manage their lands is a significant challenge. It has become critical for private landowners to educate themselves about issues which are relatively new for most of them (e.g., ecosystem management, how satisfying ecosystem sustainability or species survival values may conflict with their own desire for resource extraction). This is further complicated by the fact that landowners are being asked to implement management that is based on an environmental ethic when the tools for assessing those values are still evolving. No clear consensus has been reached on the tools to achieve ecosystem management or system sustainability. These are issues into which many scientists are currently placing a considerable amount of time in developing, and are still daunting challenges (Vogt et al. 1997a).

One of the weaknesses of the existing certification protocols has resulted from this lack of clarity in how to best assess ecosystems and the human activities occurring within them — our science in this area has not yet crystallized. This has resulted in the current approach being used by most certification protocols, to requiring the collection of an enormous amount information on each site (see 4.2.1, 4.2.2). However, having this large amount of data does not mean that the sustainability of that site can be assessed, since the important driving variables for the site are frequently diluted among all the irrelevant data. This happens because the approach for indicator selection is based on collecting information on all components of a generic ecosystem.

Because of these challenges, the information that is acquired by private landowners must be current and reflect the best science that exists at that moment. That information must also be synthesized in such a way that landowners can manage their lands and be buffered from the uncertainty in our information base. State and federal agencies are ideally suited to providing information acquisition for private landowners.

Some organizations are ideally suited to helping private landowners understand their forestlands as an ecosystem. For example, the U.S. Forest Service has already had to wrestle with problems related to balancing the different values that can exist with public lands. In 1992, they also officially adopted the ecosystem management paradigm as their management approach. They have actively provided extension services to other forest owners on issues related to water and watershed conditions, wildlife habitat, and insects/diseases across the landscape. The U.S. Forest Service has recently initiated a new action strategy to forge collaborations and improve services-delivery through new partnerships with state forestry organizations, tribes, nongovernmental groups, and private citizens (Janik et al. 1998). One of the priorities for this new strategy is entitled "Forest Information for Landowners and Managers" which is an agency-wide effort to improve and manage available information on forests for landowners, managers, and communities (Janik et al. 1998). The focus of this strategy is to manage agency-wide information systems to make information more available and to build on this information base by testing criteria and indicators being used to monitor the sustainability of forests in the United States.

A forestland owner can locate assistance on specific problems encountered while managing their forestlands from a variety of sources ranging from county foresters, county extension agents, forestry extension specialists, individual consultants, and industry consultants (Olmstead et al. 1989). Some of these sources of information are free, while others require the owner to pay for the information (e.g., individual forest consultant). All of these sources will provide information on forest management, assistance programs pertinent to forests, and taxes. However, only the county foresters, individual consultants and industry consultants will provide services to help a landowner develop a management plan. Information related to silviculture, forest management for stand improvement, timber harvesting and marketing, and stand appraisals in general, are available only from the individual or industry consultant. The county foresters and the individual and industry consultants will provide a landowner with information on multiple-use management, prescribed burning, and insect and disease control. From an examination of who provides information, it is obvious that most private landowners will have to hire an individual consultant in order to manage their lands for silvicultural sustainability. Access to information is not centralized and readily available to a private forestland owner, and would necessitate that landowners spend considerable time tracking down information. Much of the information that private landowners do access is quite specific and does not provide useful information to them on how to manage for sustainable social or natural systems at a holistic level (the requirements of certification protocols; see 4.2).

Group or communal certification will not eliminate other social constraints which may still make the certification process difficult to implement (see 7.8 for more discussion of certification in the tropics). For example, the land-area sizes and land-tenure issues are not the factors limiting the certification potential of forestland in many tropical countries in the Americas. In most cases, newly formed community forestry programs do not have the experienced, trained individuals who are qualified to manage and understand all aspects of sustainable development of forestlands. For example, the community forestry project in Quintana Roo, Mexico has approximately 35,000 ha of seasonal tropical forest that they are managing. These forests are comprised of 22 communal ejidos where each community has tenure rights on their land (Bray et al. 1993, Snook 1998). In this community forestry project, overcoming the social obstacles to successful certification of forestlands will require a considerable investment of resources from outside the community for education and training. The communities do not have the resources to acquire the training that is needed. In addition, the forest resources are not sufficiently valuable. This means that they cannot be used as collateral for developing the infrastructure and knowledge needed to manage the forests in a sustainable way, and at the same time to "alleviate poverty" (expected as part of social sustainability; Merino 1996).

This community forestry program in Mexico is not ready to be certified, despite having been organized for more than ten years and having a forest management plan that is being implemented. The communities are unable to self-administer and manage their forests, they lack knowledge of what certification is all about, and they do not understand why they need to practice sustainable forest management nor why they need to have long-term planning.

Most of the community members require training in all aspects of administration, accounting, and good forestry practices, such as protection of riparian zones, road location, silvicultural practices, and education on how to design long-term forest management plans. This suggests that for these groups to even consider certification of their forest resources, several issues have to be dealt with. For example, there is a need to invest time and finances to educate indigenous groups to sustainable development (Sandoval 1996). The immediate short-term issues faced by indigenous groups need to be satisfied, such as obtaining resources to maintain their livelihood (Camara 1998). These communities also need to be empowered economically. This can be facilitated by providing the technology (e.g., dryers, better sawmills) that adds value to the forest products being harvested so that they are able to obtain better prices for them (Bray et al. 1993, Carreon 1994).

7.5.4.2   *Creative collaborations*

For small, private landowners to become certified, greater emphasis needs to be placed on financial, technical, and educational incentives. In addition, information needs to be made available on how to form collaborative associations between landowners, managers, scientists, and the concerned public (Sample 1994, Seventh American Forest Congress 1996, Larson et al. 1997). These collaborations are needed because of the difficulties encountered by small landowners pursuing certification on their lands (see 7.5), and the fact that ecological processes do not stop at the borders of a property (see 5.2.1.2).

Numerous examples of successful collaborations have occurred because of gridlocks on public lands. These creative collaborations were formed among small and large private landowners, federal and state organizations, environmental organizations and university scientists. They were responding to the need to remove perceived impediments which had stopped the extraction of natural resources from public forestlands, and also to the need to obtain from a larger land space the goods, services, and society's values. These collaborations were formed to create institutional structures that share in the decision-making of managed environments. Some of these collaborative efforts developed in response to problems resulting from timber extraction from public lands: the Applegate partnership in Oregon, the Quincy Library Group in California, the Ponderosa Pine Project in Colorado, and the Minnesota Forest Resource Council. Other collaborations resulted from the fear that timberlands being sold by companies would be bought for development in the northern United States, including Maine, New Hampshire, Vermont, and northern New York (Levesque 1995).

Currently, much thought is being given to creating ways for private landowners to overcome the difficulties in becoming individually certified. Small landowners have to deal with the constraints of ownership rights of their neighbors (see 7.5); the difficulties of the "chain of custody" needed for FSC certification (see 7.7); and the financial difficulties of becoming certified as an individual owner (see 7.6). One solution has been for small landowners to become more knowledgeable about the issues related to certification. There is a need for the formation of some kind of organization, such as a Landowner Stewardship Association that could promote improved stewardship activities with or without the added impetus of a certification procedure (Potts et al. 1998). Such an association should be "locally initiated and regionally specific. The aim of members would be to create groups of landowners who would work on joint alternatives to increase the values of the land itself and of products obtained from their land. Such an organization should utilize the knowledge and skills of resource professionals, and one another to address overarching goals of good management on a landscape level" (Potts et al. 1998). Others (Norberg-Hodge 1996) have suggested that long-term solutions to social and environmental problems will require

TABLE 7.4

NIPF statistics for NIPF landowners in Vermont (Brynn 1998).

Average land tenure	7 years
Forestland owned in Vermont by NIPF owners	77% of total
Contribution of wood to the economy of Vermont	$2 billion annually

a range of small, local initiatives similar to the Landowner Stewardship Association proposed by Potts et al. (1998).

Programs coordinating the certification of many landowners located within proximity to one another are developing as a way to counteract the extreme limitations of individual landowners pursuing certification. These organizations are helping small landowners retain a land ethic, achieve good forest management on their land, and obtain an economic return from their land. A good example of this is the Vermont Family Forests organization which has developed a forum for private landowners with no forestry background to learn about good management practices and about resources available to them to manage their lands. The Vermont Family Forests is an organization that was developed "to cultivate local family forests for economic and social benefits, while protecting the ecological integrity of the forest community as a whole" (Brynn 1998). This organization was formed to reach the non-industrial family foresters who periodically harvest land to pay taxes or to secure income. By informing and educating non-industrial foresters, sustainable forestry practices can become the management style that protects society's values for biodiversity, water quality, and site quality, and at the same time respects the economic values associated with forest ownership. The Vermont Family Forests have 31 participants in their certification program that covers 2,024 ha (5,000 acres) of forestlands (Brynn 1998).

An approach similar to the Vermont Family Farms is critical in those parts of the country where small NIPF landowners are a significant part of the forest business. Frequently, NIPF owners are common in areas where they own a large part of the forestlands and are important suppliers of wood, but where their land tenure is not long enough (Table 7.4) for them to pursue certification.

Most of the certification protocols presented in Chapter 4 expect or assume that long-term tenure of a given land base is needed to have sustainable management. The reality for most NIPF landowners is that their land areas are too small, and their records for continuous ownership are too short to be able to verify that the land is being managed in a sustainable manner. This suggests the need for a creative way to rapidly inform new landowners, as they become part of the potentially important source of wood (Brynn 1998). The family-owned forests that have been successful in becoming certified (Kane Hardwood, Seven Islands Land Company) are those which have had long-term ownership of their lands where their land ethic could be sustained (see 7.5).

7.6 Estimating price premiums necessary to pay for forest certification

Joe Taggart

As a voluntary action, forest certification proposes several benefits to the landowner. Some of these benefits can be summarized as follows:

- Insurance to the landowner of responsible land management
- A promise to the consumer of a sustainably-manufactured product
- Access to new markets
- A price premium paid to the landowner of the certified product

Given the additional cost of certification to the landowner, the price premium paid for the certified product offers one method by which to fund forest certification. There is considerable question, however, as to what exactly that premium should be. This section will examine the price premiums necessary to pay for forest certification, using three hypothetical forests. These forests each have different total acreage and estimated annual revenue generated from harvesting activities on them. Certification costs were estimated for each forest and added to the annual operating budget. A price premium was generated for each forest using the net present value approach (all other costs were assumed to be equal among the three forests). This price premium is the percentage increase in stumpage prices that must be achieved for a forestland owner to break even on the added cost of forest certification. The hypothetical forests will be described in three different locations in the United States: the Yankee Woodlot in the Northeast, the Family Fir Block in the Northwest, and Southern Pine, Inc. in the Southeast. If the reader is interested in seeing the data that was used in these analyses, please e-mail us and we will send the spreadsheets.

The Yankee Woodlot

The first hypothetical forest is a privately owned parcel in New England which consists of 40 ha (100 ac) of high-quality hardwoods that have been under family ownership for several generations. The landowner ranked recreation as the main reason for owning forestland, including aesthetics, wildlife habitat, and access to the woodlot. The landowner is also interested in generating some income from the forest in order to help pay property taxes. With the help of a consulting forester, the landowner has developed a management plan that meets all the objectives of land ownership.

Although the owner owns 40 ha of property, only 32 ha (80 ac) are under management for timber production. Of the remaining acreage, four hectares (10 ac) surround the house and are kept in their present state. Another four ha are unsuitable for timber production (e.g., wetlands, riparian management zones, and special areas). The forest is managed on an 80-year rotation that requires three entries into the stand during that period. The first entry occurs at age 57 and generates for the owner 1500 board feet of mostly fuel wood and small sawtimber. The initial entry removes smaller, suppressed trees, and opens up growing space for the dominant trees in the stand. The second entry occurs at age 68, and produces about 2000 board feet of medium-grade sawtimber. This second entry is a harvest in preparation for the final overstory removal that takes place when the stand is 80 years old. This final harvest removes the overstory trees, releasing advanced regeneration in the understory. This harvest generates about 3000 board feet of high-value sawtimber. For each entry into the forest, the landowner harvests 1.2 ha (three ac), one of each treatment type.

The net present value (NPV) of the Yankee Woodlot, under the current management regime, can be calculated without the additional cost of forest certification. Net present value (NPV) is presented net of the 25% tax rate that the landowner pays, and is evaluated over a 15-year period at a discount rate of 6%. The NPV is also evaluated under two scenarios. The first scenario generates the NPV according to stumpage values, which vary depending on the size and grade of timber removed from the property. Three values are presented, one for each silvicultural treatment employed. The NPV of the Yankee Woodlot, given three different stumpage values, is $12,373.

The second scenario employed to value the Yankee Woodlot uses only one stumpage price. This value is the average of the three values used in the first scenario. The NPV of the Yankee Woodlot with one average stumpage price is $11,041.

The reason for the two scenarios will be made clear later when price premiums are determined on forest products (stumpage prices). The first scenario, with three stumpage prices, assumes that the wood taken from the Timber Stand Improvement (TSI) and prep-cut treatments is of low value and will be used for utility markets (e.g., pallets, barn boards, pulp). Given the low visibility of these low-grade products, one would not expect to receive a price premium for green certification. Conversely, the wood taken off in the final overstory removal is of a premium grade that is suitable for veneer and finish stock. This is the forest product for which consumers will presumably be prepared to pay a premium. Therefore, the first scenario assumes a price premium to be paid only on the products generated from the overstory removal. The second scenario, which assigns an average stumpage price to all forest products, assumes that every wood product removed from the certified forest can be sold at a price premium.

The Family Fir Block

The Family Fir Block is a 405 ha (1000 ac) privately held forest in the Pacific Northwest region of the United States. The property is a family treasure and is therefore managed for multiple purposes. The property is bisected by a small stream that is stocked with trout. As an avid fisherman, the landowner has particular interest in keeping the water quality suitable for fish. In addition to the stream, there are two bass ponds on the property. Both the ponds and the stream are protected with deep buffer strips, resulting in a 40.5 ha (100 ac) set-aside area. Furthermore, there is a 40.5 ha (100 ac) patch of old-growth forest on top of the highest point on the property that is not included in the harvesting activities. This old-growth forest is a unique site that the landowner uses for day trips and overnight camps.

The remaining 324 ha (800 ac) are harvested on a 50-year rotation. The majority of the income goes either into a trust fund for the education of the landowner's children or the landowner's personal retirement account. The land is of site value 125, which according to USFS technical report 201, yields 35,520 board feet of timber at age 50. The landowner does not believe in clearcutting due to his concern for water quality. Subsequently, only 75% of the standing volume is harvested from the forest, leaving the largest overstory trees for erosion control and to provide some structural diversity. Each year, 6.5 ha (16 ac) are treated in this manner.

The NPV of the Family Fir Block, evaluated at 6% over 15 years, under the landowner's current management regime, can be calculated as follows. This analysis includes refores-tation costs of $300/acre or 0.4 ha for planting under the residual canopy. The landowner also incurs a $4/acre/year cost for annual administration and management due to the size of the forest. The stumpage value of $400/MBF is the 1994 average of West Side stumpage cut. The NPV of $1,248,493 is given net of the 25% income tax that the landowner pays.

Southern Pine, Inc.

Southern Pine, Inc. is the owner of 40,486 ha (100,000 ac) of plantation pine in southern Georgia. The land is approximately 10% of the total 404,858 ha (1,000,000 ac) owned by Southern Pine, Inc. in the southeastern U.S. As a publicly traded company, Southern Pine,

Inc.'s primary objective is to maximize shareholder value. The land is open to hunting and hiking, but no value supercedes the main goal of the company.

Of the 40,486 ha (100,000 ac), only 36,437 ha (90,000 ac) are suitable for timber production. The remaining 4,049 ha (10,000 ac) unsuitable for timber production are in wetlands or roads. On the land managed for timber production, the company has adopted the harvesting regime developed by the Plum Creek Timber Company. This plan calls for 32-year rotations with three stand entries to deliver three different wood products: pulpwood, chip and saw (CNS), and sawlogs. Pulpwood is roundwood sold by the ton to a pulp mill to be converted to chips for the production of pulp and paper. Chip and saw is a term used for small-diameter logs that will be partially chipped for pulp and paper, and partially sawn for the production of framing material. Sawlogs are large, high-grade logs that are used for larger-dimensional lumber, and face stock for flooring, furniture, and cabinets.

The first entry into the stand is a low thinning at age 17. This first entry generates 24.4 tons of pulp per acre or 0.4 ha and 2.25 cunits (100 cubic feet) of CNS material. The second entry into the stand is another low thinning at age 22–25. This harvest yields 15 tons of pulp, 2.9 cunits of CNS, and 1 cunit of sawlogs. The final treatment is a clearcut of the overstory trees at age 32. This final removal yields 16.5 tons of pulp, 10.7 cunits of CNS, and 15.2 cunits of sawlogs. Each year, Southern Pine Inc., treats 911 ha (2,250 ac) of each stand type.

The NPV of Southern Pines, Inc. forestland at 6% over 15 years can be calculated. The values calculated are net of the 35% corporate income tax that Southern Pine, Inc. pays annually. Southern Pine, Inc. also incurs planting and site-preparation costs for the 911 ha (2,250 ac) that are clearcut annually. In addition to planting and site preparation, Southern Pine, Inc. has expenses of $7/ac/yr or 0.4 ha/yr for administration and management.

Similar to the Yankee Woodlot, two scenarios are given. The first scenario, with an NPV of $75,516,030 has three different prices for pulp, CNS, and sawlogs. This scenario will assume that Southern Pine, Inc. will only receive a price premium on its high-value products (sawlogs) for green certification. The second scenario, with an NPV of $56,971,798 assumes one average price/ton for all materials harvested from the land. This will yield a price premium that assumes that all materials harvested from the forest will be sold as green-certified.

7.6.1   Estimated cost of certification for the three hypothetical forests

In general, the direct cost of forest certification consisted of the initial expense of being certified and the annual audits. Both the initial certification and the annual audit entail an on-site visit by forestry professionals, and an office write-up of the results. The number of certifiers on site, and the number of days that they spend in the field, as well as in the office, depends on the size of the land holding and the intensity of the operations. Once the number of field and office-person days is determined, an estimated cost of certification can be obtained.

Travel costs and daily consulting costs were held constant when estimating the cost of certification for each of the three forests. Each forest was assumed to require a 500-mile round-trip drive for the certifier, thus eliminating airfare. The eight to nine hours of driving time constituted one full working day of driving, which was billed at $300. Each certifier was paid $50 per diem. Thus, the fixed cost of delivering a certifier into the field (at $0.30 per mile) was $500. While in the field, the certifiers were paid $350 per day. An additional $110 was charged per day for expenses and lodging. In the office, certifiers billed $250 per day for the write-up. Clerical, photocopying charges, and mailing expenses varied, based on the size and scope of the certification.

TABLE 7.5

The initial and annual audit costs of certification
for the Yankee Woodlot.

	Initial Certification Cost (in U.S. dollars)	Annual Audit (in U.S. dollars)
Travel	1,000	0
Field work	920	460
Office work	700	250
Clerical/General	100	75
TOTAL COSTS	$2,720	$785

TABLE 7.6

The initial and annual audit costs of certification
for the Family Fir Block.

	Initial Certification Cost (in U.S. dollars)	Annual Audit (in U.S. dollars)
Travel	2,000	500
Field work	7,360	920
Office work	2,000	250
Clerical/General	300	150
TOTAL COSTS	$11,660	$1,820

In order to determine the number of person-days required, both on the ground and in the office, John Landis of Smartwood was contacted. After being given a description of each forest, Landis was able to estimate the number of person-days required for each job, both on the ground and in the office.

7.6.1.1 The Yankee Woodlot

Given the relatively small area that the Yankee Woodlot covered, it was estimated that the fieldwork would require one day for two people to conduct on the ground. Once the initial fieldwork was complete, the write-up would take an additional day in the office by both certifiers. Once the initial certification was complete, the annual audit would require only one person for one day in the field and one day in the office (Landis 1998). The total cost of certification for the Yankee Woodlot is given in Table 7.5.

Notice that in order to reduce costs, it is assumed that the annual audit can be contracted out to someone near the forest, requiring no travel expenses.

7.6.1.2 The Family Fir Block

The Family Fir Block is a forest ten times the size of the Yankee Woodlot, and therefore requires significantly more effort to carry out the certification. It is estimated that it would require four certifiers in the field for four days in order to carry out the on-site process of certification. After the initial fieldwork, it would require the same four people to spend two additional days in the office to prepare the appropriate write-up of their findings. The annual audit could be carried out by one person, spending two days in the field and one day in the office (Landis 1998). The total cost of certification for the Family Fir Block is given in Table 7.6.

TABLE 7.7

The initial and annual audit costs for certification for the Southern Pine Inc.

	Initial Certification Cost (in U.S. dollars)	Annual Audit (in U.S. dollars)
Travel	2,000	1,000
Field work	13,800	1,840
Office work	5,000	1,000
Clerical/General	1,250	300
Total Costs	$22,050	$4,140

7.6.1.3　Southern Pine, Inc.

The area of Southern Pine, Inc. forest is one-hundred times the size of the Family Fir block, and one-thousand times that of the Yankee Woodlot. Subsequently, the certifiers must spend considerably more time, both on the ground and in the office. It is estimated that the certification of 100,000 acres would require four people to spend seven and a half days on the ground and an additional five days in the office. The annual audit would be quite labor intensive, as well requiring two people working for two days each in the field, as well as an additional two days in the office (Landis 1998). The total certification cost for Southern Pine, Inc. forestland is given in Table 7.7.

The values of $22,050 for Southern Pine, Inc.'s initial certification and $4,140 seem to agree with empirical data. Some estimates have placed certification costs at about $0.16/ac or 0.4 ha for initial certification, and $0.06/ac or 0.4 ha for annual audits (Upton and Bass 1996). When Collins Pine had two individual 40,486 ha (100,000 ac) forests certified, they estimated their costs to vary between $0.40 to $0.50 per acre or 0.4 ha (Hansen 1997). If this is the case, then the estimate for Southern Pine, Inc. appears to be conservative.

It is of particular interest that although Southern Pine Inc. was one-thousand times larger than the Yankee Woodlot, its cost for independent certification was only eight times higher. This is largely due to the fixed cost of transportation. Although evidence points to certification as costing $0.16 to $0.50 per acre or per 0.4 ha, these figures are applicable only to larger acreages. It would be foolish to assume that the Yankee Woodlot could be certified for $50 (barring outside subsidy), given that it would cost $5 acre to simply get a certifier on-site at the Yankee Woodlot.

7.6.2　Results from the price-premiums analyses

For all three forests, the benefit of forest certification was treated as a capital improvement on the property. Therefore, the cost of certification is capitalized and depreciated over a five-year period. At the end of five years, the landowner has the option of renewing his or her certification (recertification). In all three cases, it is assumed that the landowner chooses to be recertified. The cost of recertification is the same as the initial certification cost in all three cases. This is not necessarily true, since recertification can often be less labor-intensive than the initial process. This is a reasonable assumption. However, certification costs are likely to rise with or above inflation over time. Annual audit costs are held constant as well, and are written off each year as an annual operating expense.

The initial analyses conducted here do not include the other costs that can accrue to a forestland owner. "The costs of certification include both the direct costs of obtaining certification and the indirect costs of implementing any measures that are required to achieve sustainable forest management" (Mathias 1996). Eric Hansen stated that "There

TABLE 7.8

Price premiums necessary to cover certification costs.

Scenario	Current volume	25% Reduction	50% Reduction
Hardwood 3	119%	210%	391%
Hardwood 1	87%	150%	275%
Douglas Fir	2.3%	36.4%	104%
SYP 3	0.123%	65.5%	196%
SYP 1	0.115%	33.4%	100%

have been other added costs, i.e., updating forest inventories and adding geographic information-system technology" (Hansen 1997). Given these additional costs, each forest was also evaluated under three specific harvesting regimes: current management, a 25% reduction in harvest, and a 50% reduction in harvest. Although each regime is stated in reference to current harvest volumes, it could also reflect the increased cost of management, which would erode harvest revenues.

The net present value of the Yankee Woodlot with forest certification is given next. The "green premium" was determined by holding the NPV at pre-certification levels, and solving for the stumpage value which holds the NPV constant. (Note: To cover the cost of certification, the landowner would have to be paid $767.57/MBF for the high-quality hardwoods or an average of $412.72/MBF for all grades of wood from the various harvests) These figures represent increases of 119% and 87% respectively.

The NPV of the Yankee Woodlot with 25% and 50% reductions in harvest levels can also be calculated. The NPV of the Family Fir Block with certification under current management can also be calculated at a 25% reduction and a 50% reduction in harvest. Finally, the NPV of SPI with forest certification under current management can be calculated at a 25% reduction, and a 50% reduction in harvest. The results of the price premiums necessary for each scenario are summarized in Table 7.8.

The five forest scenarios under three separate harvesting regimes are given in Table 7.8. The scenarios labeled "Hardwood 3 and Southern Yellow Pine 3" assumed that the Yankee Woodlot and Southern Pine, Inc. were only paid a price premium on the high-value forest products generated from overstory removal. "Hardwood 1 and Southern Yellow Pine 1" assumed that a green premium was paid on all forest products harvested, including utility wood products. Since only one stumpage price is assumed for Douglas fir, only one scenario is necessary.

A contingent valuation study attempted to determine the consumer's willingness to pay for certified forest products (Ozanne and Vlosky 1997). A total of 768 respondents stated that they would be willing to pay an average 4.4% price premium on a $100,000 home made of certified products (Ozanne and Vlosky 1997). The 4.4% would more than cover the 1/10th of 1% premium necessary for Southern Pine, Inc. to pay for green certification under its current harvest regime. This would assume that this home was made of southern yellow pine framing, flooring, paneling, and siding.

The same 768 individuals also stated that they would be willing to pay an 18.7% premium on a 2x4-framing stud (Ozanne and Vlosky 1997). This more than covers the 2.3% premium necessary for the Family Fir Block to cover the cost of certification. However, the outlook is not so bright for the Yankee Woodlot. The article by Ozanne and Vlosky determined that consumers would be willing to pay a 14.2% price premium on a certified hardwood dining room set. This doesn't even come close to meeting the 87% to 119% price increase necessary to cover the cost of forest certification.

The prospects become even bleaker when harvest volumes are reduced. At a 50% reduction in harvest, the Yankee Woodlot would have to see an increase of 391% on their high-value hardwood sawlogs. Even the high-volume Southern Pine, Inc. would have to receive a 100% price increase on their pulp and paper products if their harvest volumes were halved.

7.6.3 Concluding comments on price premiums needed to cover the costs of certification for the three hypothetical forests

The purpose of this section was to determine where green certification could be implemented with the greatest economic efficiency. The results shown here, though discouraging in some cases, show a positive future for green certification in others. Several points are raised by this analysis:

- The bigger the land base the greater a landowner's ability to cover their costs
- Travel costs are a stumbling block
- Small landowners do better with group certification
- A better price premium would result if timber could be exported from the United States

These points are further discussed below:

The analyses conducted here suggested that the model developed is tremendously sensitive to land size. If the landowner can amortize the cost of certification over a larger landscape, the challenge to recover those costs becomes smaller. Similarly, volumes of available timber are critical. Although the Family Fir Block is a relatively small forest, its species composition produces tremendous yields of high-value products. If volumes are reduced, however, the goal of covering certification with a price premium may quickly slip out of reach. Similarly, the increased costs of information that the landowner must gather can also prove prohibitive (see 7.6).

The Yankee Woodlot, for example, may be certifiable under its current management practices. The landowner, however, may be forced to pay for the collection of costly growth, yield, and inventory data, which in effect reduces the harvest income. Conversely, Southern Pine, Inc. probably already has detailed growth, yield, and inventory records. This is common practice for large, industrial forests. Southern Pine, Inc.'s cutting regime, however, is another story. It is likely that certification would require Southern Pine, Inc. to scale back on its clearcutting and use of genetically engineered stock. This would effectively reduce the company's harvest volume.

In the case of the Yankee Woodlot, travel costs of the certification team made up 37% of the total certification cost. This contrasted with the travel costs for Southern Pine, Inc., which had only 9% of the total certification cost invested in travel. Even to send one person to a small forest for one day makes the costs prohibitive.

John Landis of Smartwood (1998) was quick to point out that smaller landowners should group together with surrounding landowners for a "team" certification. This helps to reduce travel costs by allocating them among several acres and multiple landowners. If these forest "co-ops" can pool large enough acreages, they can greatly reduce the direct costs associated with certification.

Due to the lack of resource base in many wealthy nations, it is common to receive a tremendous price premium on wood exported from the United States. It is not uncommon to hear of price premiums of 100% to 200% being paid on Douglas fir, red oak, and other high-value species headed to the Asian Pacific-Rim region. Unfortunately, many certification systems discourage the exporting of forest products. The export of some forest products could help to mitigate some of the costs of forest certification.

In general, this analysis has helped determine under what conditions forest certification is appropriate economically. Forest certification offers a package of benefits to the landowner and the consumer. Economic return to the landowner is not the only benefit which can be achieved by the landowner, nor in many cases is it the most important. It is a crucial element, however, which must be studied when looking at the viability of forest certification. By evaluating all of the benefits which forest certification offers, it allows forest managers and landowners to decide the form which certification will take on our landscape in the next century.

7.7 Chain of custody as an impediment to certification

J. Scott Estey

Chain of custody is the term applied to the process of tracking a unit of wood as it is harvested from a specific certified forest. This tracking follows the piece of wood through the various distribution mechanisms and manufacturing processes to which it is subjected while being transformed from a tree to a finished product (whether wood products — such as lumber or panels — or paper products). For a producer to make claims of forest certification in association with its products in the form of a product label, the chain of custody must be established and certified, as well.

There is much on-going debate as to whether the "chain of custody" certification should be required. There are many logistical and economic difficulties associated with the chain-of-custody certification and it is important to understand whether this can be an impediment to advancing the certification of forests in the United States. The issue may ultimately determine whether small landowners pursue certification using the FSC approach. The cost of tracking the small number of trees being harvested from their lands would add costs for the small landowner already paying to be certified and in implementing good forest management on poorly managed lands (see 7.5, 7.6). Since the chain of custody is particular to the FSC approach to certification, the ensuing discussion is only relevant to FSC-type certification.

Chain of custody is one of several factors that will influence the decision made by a small or large forestland owner to pursue certification using the FSC approach. The chain-of-custody certification process does contain elements that are within a landowner's ability to make their own decisions and to control. However, it also embodies elements that are largely outside of the landowner's control. This becomes a critical issue of concern for large wood companies since chain-of-custody certification is a requirement in order for claims regarding forest certification to be made in association with a company's products. Any large company would need to have a relatively high degree of certainty that chain of custody could be achieved prior to committing time and resources to pursuing certification.

Some of the important impediments to chain-of-custody certification will be discussed next and can be summarized as follows:

1. Fragmented fiber-supply system
2. Woodyard operations
3. Production economics
4. Finished-product handling
5. Product quality

TABLE 7.9

Statistics on total annual harvest in cubic meters (m³)
by different landowners in the United States
(USDA Forest Service 1994, Jaakko Pöyry Consulting 1996).

Total annual timber harvest	490 million cubic meters
Harvest for industrial purposes	400 million m³
Harvest for fuelwood purposes	90 million m³
Timber harvest from NIPF lands	245 million m³ (~50% of total)
Timber harvest from industrial lands	162 million m³ (~33% of total)
Timber harvest from public lands	83 million m³ (~17% of total)

The most significant drawbacks to chain-of-custody certification for the majority of primary forest product producers will be the fragmented nature of wood supply systems and the potential for costly woodyard modifications. The other impediments that exist for chain-of-custody certification are also important, but are not as critical as the first two topics mentioned above.

7.7.1 Fragmented fiber supply system

The fragmented nature of wood supply systems in the United States creates an enormous barrier for larger operations consuming large volumes of raw wood material and is probably the most significant impediment to chain-of-custody certification. This fragmented nature occurs because of the different ownerships that exist for forestlands in the United States which cause timber companies to have no control of the majority of their wood-supply needs. This occurs because only 28 million hectares of the total 300 million hectares of forestland in the United States are owned by the private forest industry (USDA Forest Service 1994). In addition, 67% of the wood supply for industrial needs comes from noncontrolled sources, such as the NIPF owners and public lands (Table 7.9). The number of landowners a timber harvest for industrial purposes would have to satisfy would be in the 4 to 5-digit range, and would be cumbersome to monitor or track using a chain-of-custody approach (see Boxes 7.1, 7.2).

Box 7.1
The number of individual landowners that a kraft pulping operation would need to interact with in acquiring timber supplies.

A rough estimate of the high number of individual landowners that a company might have to interact with is demonstrated by using an example from the kraft-pulping operation (the predominant pulping process in the United States). In the calculations, it is assumed that the wood supply is harvested using sustained-yield basis and that the growth rates approximate the national average of 3 m³/ha/yr as given in USDA Forest Service (1994). The national averages for timberland ownership and harvest by ownership were used to calculate how much of the volumes came from different owners (Table 7.9). The number of NIPF owners was given as being over 9 million in the United States (Jack 1998), and estimated each owner to have an average land area of 13 ha.

Statistics on a typical bleached-kraft mill-pulping operation in the United States

- Production of pulp by mill –
  300,000 metric tons/annually
- Consumption of pulp by mill –
  4.5 m³/metric ton of pulp (1.35 million m³ per year)
- Mill-required forestland base to sustain production –
  450,000 ha
- Ownership of forestlandbase needed to sustain production –
  266,000 ha from NIPF;
  121,000 ha from public lands;
  63,000 ha from company lands
- Number of NIPF wood sources for the mill –
  20,462

Since it is safe to assume that many of the smaller NIPFs are not managing their lands for timber production, the number of landowners required to supply this mill is likely to be quite a bit less than the calculation would suggest. It is very common for facilities of this nature to obtain wood from hundreds of sources each year with many of the sources changing from year to year. It is also important to note that, in practice, pulp mills do not source their entire wood requirement in roundwood form directly from the landowners. A large portion of their wood is sourced in the form of sawmill residues, which further complicates the task of monitoring wood movement from the forest to the mill.

Box 7.2
The number of individual landowners that a commodity wood products company would need to interact with in acquiring timber supplies.

A company with commodity wood products (e.g., softwood dimension lumber, softwood plywood, oriented strandboard, medium-density fiberboard, and particleboard) would have to interact with a high number of individual landowners. Several different statistics were used in these calculations. For example, the national averages for timberland ownership and harvest by ownership were used to calculate how much of the volumes came from different owners (Table 7.9). Number of NIPF owners was given as being over 9 million in the United States (Jack 1998) and estimated each owner to have an average land area of 13 ha. Lumber yields were assumed to be 50%.

Statistics on a typical softwood sawmill operation in the United States:

- Production of lumber by mill –
  200,000 m³/annually
- Requirement of wood by mill –
  400,000 m³/annually
- Mill-required forestland base to sustain production –
  133,000 ha (330,000 acres)

- Ownership of forestland base needed to sustain production –
  66,500 ha from NIPF;
  43,890 ha from public lands;
  22,610 ha from company lands
- Number of NIPF wood sources for the mill –
  5,115

Since it is safe to assume that many of the smaller NIPFs are not managing their lands for timber production, the number of landowners required to supply this mill is likely to be quite a bit less than the calculation would suggest. It is very common for facilities of this nature to obtain wood from hundreds of sources each year, with many of the sources changing from year to year.

The estimates produced in Boxes 7.1 and 7.2 also do not consider the fact that the quality of the landbase varies significantly for different forestlands. This means that less wood quantity and quality can probably be harvested from many of the NIPF lands. For example, of the total forestland that could potentially supply wood, only 66% of it has been classified as being appropriate for maintaining productive, commercial timberland (~ 200 million ha) (USDA Forest Service 1994). Many of the NIPF lands are unable to achieve the forest production rates that can be found on well-managed forestlands. This factor would increase the number of NIPF lands that might need to be harvested to maintain the operations of many of the mills given in Boxes 7.1 and 7.2.

The impediments that exist for chain-of-custody certification are strongly highlighted in the paper-production industry where the industry does not control their primary sources of wood suppliers nor do they frequently produce the primary material (pulp) used in paper making. Typically, many mills purchase some percentage of their fiber requirements in the form of market pulp. This means that they neither control the timberlands from which the wood is harvested, nor do they make their own pulp. Faced with this type of supply structure, it is highly unlikely that the forest products industry will be able to source its wood totally, or even substantially, from certified sources. In order for a product to contain an FSC certification label, FSC requires that wood products contain 100% certified wood and that pulp-and-paper products contain at least 70% certified wood (FSC 1999b). In order to meet these levels, most producers would be required to make modifications to their production systems in order to accommodate the production of small, pure (or 70% pure for pulp and paper) volumes of certified product, and to swing the process between certified and non-certified production runs.

7.7.2 Woodyard operations

Other impediments exist after satisfactorily dealing with the issue of fragmented wood supply system and the ability to secure sufficient volumes of certified wood to produce a portion of the output in the form of a certified product. The next challenge to the chain-of-custody certification will occur in the facilities woodyard. It is here that the inventory of raw wood material occurs prior to the production process. The requirement to track certified wood all the way to finished product means that wood from certified sources will need to be inventoried separately from wood from non-certified sources. Depending upon the volume of certified wood that a given facility will receive and the number of different quality and species sorts that it keeps, this may or may not require a doubling of the inventory sorts, but additional sorts will be necessary. For many facilities, additional yard space is limited, or even non-existent, and lack of space alone may hinder a mill's ability to become chain-of-custody certified.

In addition to the space needs required for more sorting of wood materials, the need for additional sorts will require modifying the wood receiving-and-handling systems in the woodyard. Modifications of this type are often very costly due to the need for additional wood-handling equipment, conveying and reclaim systems, and expenditures for items such as concrete pads under chip piles for quality purposes.

Sawmill and plywood operations will need to be concerned not only with the logistical problems associated with managing separate inventories of wood raw material going into a facility, but also separate inventories for wood residues, such as chips and sawdust coming out of the facility. Wood residues are a major component of a sawmill's revenue stream, and serve as an important raw material for other products, such as oriented strandboard, particleboard, medium-density fiberboard, and wood pulp.

While the capital requirement for woodyard modifications will vary from facility to facility, it is likely that expenditures could easily amount to multiple millions of dollars for many mills.

7.7.3 Other issues relevant for chain of custody

7.7.3.1 Production economics

Many commodity-oriented forest products are produced on a continuous basis where operating rates and uptime are essential factors driving profitability and the cost-competitive positioning of a facility. In this regard, it will be a challenge for facilities to manage the production process between certified and non-certified product with as little-as-possible negative impact on the cost performance and revenue line. For example, in the 300,000-ton-per-year kraft mill example used in Box 7.1, assuming trend prices for the products it produces, this mill generates revenue at a rate of $23,000 per hour (Jaakko Pöyry Consulting 1998). Any additional downtime or slowback that arises from grade changes between certified and non-certified products can quickly become very costly.

7.7.3.2 Finished-product handling

Similar to the logistical problems associated with handling raw wood material, once a facility is producing both certified and non-certified products, there will be a requirement to modify the handling and inventory systems to accommodate additional finished product sorts prior to distribution. The significance of these modifications will be determined on a facility-by-facility basis, but will likely cause much less difficulty and expense than the wood-handling operations.

7.7.3.3 Product quality

Product quality is an important component of a company's product offering. While chain-of-custody certification is not likely to result in significant quality problems, several issues will need to be addressed to ensure that negative impacts are minimized. Quality considerations will be more significant for pulp-and-paper operations, particularly those producing products where variation in fiber furnish can result in changes in the end product's performance characteristics.

Many mills carefully manage species mixes to achieve certain characteristics, and it is conceivable that in some cases, the species mix of certified wood available to a facility will differ from the species mix of non-certified wood. In these situations, the requirements to make pure (or even 70% pure) certified products may result in an inability to make products with performance characteristics that conform to the required quality specifications.

7.7.4 Concluding comments on chain-of-custody certification

If certification is viewed as a tool to widely promote, initiate, nurture, and measure improvements in forest management practices towards some generally agreed-upon vision of "sustainable forestry," a clear objective would be to maximize the harvest available from each piece of forestland being managed. It is possible that incorporating and requiring chain-of-custody certification will serve as an impediment to meeting this objective by limiting the level of acceptance among the major wood-consuming entities in the United States.

An alternative approach is the concept of product labels (see 2.3) containing proportional claims of certification. In order for a company to accomplish this it would still be required to monitor the wood flow up to the mill gate. However, the more costly and problematic issues associated with separate inventory systems and dedicated production runs could be eliminated. Essentially, a company would be able to communicate to the consuming public what percentage of the wood raw material in its products is from certified forests. The simplest approach would be to allow companies to make claims based on the volume of certified wood as a percentage of total receipts over a predetermined period of time — for example, one year. This could be done at either the total company level or at the mill level. While this would not allow the consumer to know the percentage of certified wood in a specific product, it would allow for more informed buying decisions based on the relative levels of certified wood utilized by one company compared to another.

If the ability to make proportional claims stimulates a higher degree of participation on the parts of large, forest-products companies, the consuming public would begin to recognize these labels as more products began to have labels attached to them. This increased exposure would help to stimulate consumer awareness of forest certification and would give an incentive for forest products companies to continually increase the certified-wood content in their products if consumers are attracted to products with higher proportions of certified wood. If this were to happen, product labeling with proportional claims could potentially be more effective in creating consumer pull for certified wood products than strict chain-of-custody certification. This would also create an environment more conducive to continual improvements in forest management practices — in other words, it would make adaptive management integral to forest management.

If chain-of-custody certification is required — or if the standards regarding the proportion of certified wood required for product claims to be made are not significantly relaxed — FSC certification may end up being relatively limited to niche and higher value-added wood products. This is what is currently happening. This means that certification would never successfully penetrate the commodity wood products or pulp-and-paper industries. FSC states that its goals are to promote environmentally responsible, socially beneficial, and economically viable management of the world's forests (FSC 1996). However, this goal is not achievable if the major players in the forest products industry are not active participants in the process.

TABLE 7.10

Forest types and areas of each type in Mexico
(Forestry Department SEMARNAP Report/95).

Forest Type	Land Area, millions of ha
Pine-Oak Forest	30.4
Humid Tropical Forest	26.4
Arid Vegetation Zones	58.5
Hydrophilic and Halophytic Vegetation, Degraded Areas	26.4

7.8 Challenges and opportunities for tropical timber certification: Mexico's experience

Heidi Asbjornsen, Enrique Alatorre

Timber certification efforts in tropical developing countries have unique challenges requiring special consideration and analysis within the global context of certification and its impacts on long-term forest management practices. Mexico was one of the first developing countries to initiate the certification process, with the first Mexican forestry operation, Sociedad de Productores Forestales del Estado de Quintana Roo (SPFEQR), being certified first in 1992, and again in 1995. The location of FSC's central office in the Mexican state of Oaxaca has played an instrumental role in sparking certification initiatives within the country. The challenges and opportunities confronted by Mexico during the certification process provides a useful model — both in terms of lessons learned and progress made — for other tropical countries interested in certification.

Forests are a dominant feature in Mexico's landscape, representing 72% of its land area or 141.7 million ha (Forestry Department, SEMARNAP, Report/95). Of this total, 30.4 million ha are classified as temperate forests, 26.4 million ha as lowland forests, 58.5 million ha as arid-zone vegetation, and 26.4 million ha are dominated by hydrophilic and halophytic vegetation and degraded areas (see Table 7.10). The total potentially manageable forest area in Mexico is approximately 50 million ha (including temperate and tropical forests). Of this total, almost 15 million are considered to be commercially valuable and only 7 million ha (14.5% of the total forested area) are currently under authorized management.

Forestry in Mexico is distinguished from many other tropical countries in that its communal land tenure system has created a situation in which *ejidos* and communities collectively manage their forest resources. The strong presence of the social sector is reflected in ownership patterns. For example, ejidos and communities together possess and inhabit 80% of the forested area, while only 15% is privately held (companies and individuals), and 5% is nationally owned (including national parks) (Figure 7.2). Approximately 7,000 ejidos and indigenous communities, and 100,000 private landowners own forest timber resources (Forestry Department 1995). The majority of the privately owned properties are small land holdings which are often small farms; in contrast, the indigenous ejidos and communities have land holdings on the scale of several thousand hectares.

The communal land tenure system in Mexico, combined with the large proportion of forested lands under communal control, has precipitated an approach to forest management that is unique from many other tropical countries where land ownership is typically

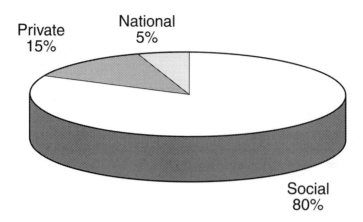

FIGURE 7.2
Land tenure and ownership in Mexico (SARH 1994).

more centralized. Sometimes referred to as "social silviculture," the collective forest management approach often sustains a social development system where the profits from the forestry industry are either distributed among community members or reinvested into the community.

The following section describes the advances and difficulties encountered during Mexico's four-year experience with certification. This analysis will draw upon three forest operations which have had their forest management practices certified within the last five years: the Union of Zapotec-Chinantec Forestry Communities in the Sierra Norte (UZACHI); the Civil Society of Ejido Forestry Producers of Quintana Roo (SPFEQR); and the ejido of Echeverria de la Sierra in Durango.

In all of these cases, some form of social silviculture is practiced. For example, communities belonging to UZACHI have reinvested profits to establish and maintain health centers, municipal buildings, public parks, schools, churches, and roads. Decisions relating to the community forestry industry are made by the local assembly comprised of *comuneros* (heads of the family and persons older than eighteen). The assembly also nominates forestry coordinators who are responsible for the day-to-day operations. The forestry sector is an important source of employment for people within the community, where salaries are often greater than in other sectors. Forestry training of individuals from the community — through taking courses or pursuing other educational opportunities — is generally supported by the community forestry industry. Due to the high level of social consciousness that permeates the forestry sector, evaluations of social criteria have generally received highly positive scores by certification teams.

However, following the modifications of constitutional Article 27 promoted in 1996 by the ex-president, Carlos Salinas de Gortari, a series of important changes has occurred in land tenure. These changes have had significant impact on the forestry sector in both the interior and the exterior of the ejidos. According to these modifications, ejidos can now privatize their lands if the decision to do so is voted on by the Ejidal Assembly. Forestry ejidos have responded by firmly establishing that their ejidal forestlands (i.e., the common good) cannot be divided up into many ejidos. Consequently, a more common scenario has been to split the annual timber extraction volume among a series of groups that have formed within the interior of the ejidos. In some cases, this has permitted a better distribution of the annual benefits from the sale of timber. However, in many other cases, it has halted the reinvestment into public works and forest management at the community level. Apart from the benefits or limitations that the formation of groups generates, a crucial aspect is the existence of consensus agreements within the ejidos which permit harmonious fellowship and production, and which assure the sustainability of the forest resource. Therefore, in the certification-evaluation process, much attention has been placed on the existence or absence of these mechanisms.

The issue of "who pays" for timber certification is often of greater significance in tropical developing countries than in western countries. This is especially true in Mexico, where a major portion of the timber is produced by small, communally owned land units. To overcome this potential obstacle, many of the initial certification initiatives in Mexico were partially or totally funded by external funding sources (e.g., NGOs, international organizations, and private companies). A central motive for investments by donor organizations has been to contribute to improving the management and conservation of tropical resources. This contrasts with private companies, whose interests lay in enhancing their market options by creating possibilities for selling certified products. For example, Oaxacan communities belonging to UZACHI completed their certification process with financial support from private foundations. In contrast, certification of forest management in the ejido Echeverria de la Sierra in Durango was obtained through sponsorship by the primary charcoal purchasing company, NORAM, who was interested in providing certified charcoal to its clients. Ejidos in Quintana Roo forming the organization SPFEQR also received funds from private foundations to pay for their costs of certification. In the latter case, the ejidos manage their tropical forests primarily for mahogany and tropical cedar.

Under the current situation, ejidal production units are required to obtain some form of subsidy to pay the costs resulting from the certification process. The support provided by international agencies has been instrumental in initiating the certification program. However, for the long-term self-sustainability and security of the program, it is important that the certification process be linked to specific market niches and articulated as business agreements between the producers and the buyers. In some cases, this may entail investments on the parts of the transforming businesses or commercial interests to certify their sources of primary materials.

The ability of the ejido associations to develop a superior organizational structure has also enhanced their options for obtaining certification. For example, in the case of the National Union of Communal Forestry Organizations (UNOFOC), the union was able to solicit funds from international agencies to cover the costs of certification in the ejidos that were better organized. Once the more-organized ejidos become certified, it is important that these efforts translate into improved market conditions for them. It is also important that improved market conditions can be shown to be a result of certification in order for the less-organized ejidos to become motivated to improve their forestry practices and thereby be able to pursue certification.

These diverse cases illustrate the variety of options available for spurring certification initiatives in the tropics. Yet they also give warnings about the financial implications and potential constraints for certifying less-profitable, small-scale operations, despite their high degree of ecological and social importance. There are still many problems hindering the wide-scale acceptance of the certification process which are related to the production and commercialization of forest products. A major factor determining whether certification will be accepted as an approach worth pursuing is its potential for enhancing profits for those who have become certified.

One factor influencing the consumer demand and market value of certified products is their quality. When forestry operations are currently evaluated for chain-of-custody certification (see 7.7), it does not include an assessment of the quality of the products produced. Consequently, the certified producers must often, in order to be competitive in international markets, dedicate substantial effort to enhancing the quality of their products, whether of raw timber or finished products, and promoting their availability. Restrictions to market access have been growing because of the globalization of product quality standards and the commercialization of services. In general, Mexican forestry producers are not aware of the international market conditions. Historically, they have been dependent on large intermediary businesses which retained most of the profits. In response to increasing national and international demand for certified products, producers are confronting a tremendous challenge to reaffirming their forestry management practices and to restructuring themselves as commercial enterprises. Market access also requires that consumers be willing to pay the price of producing quality products. Consumer willingness to pay for quality products will greatly influence the long-term success of the certification process — as a stimulus for enhancing the sustainability of the forests and their community owners.

A growing requirement within the global certification process is proof of chain of custody. However, this requirement poses special problems for tropical timber operations due to its creation of added infrastructural and logistical demands. When the ejidos in Quintana Roo were certified, an evaluation of the chain of custody was not carried out in their sawmills since they only requested certification on their raw timber. Currently, these evaluations are being carried out in the sawmills and timber collection centers. A floor-manufacturing company of tropical woods is already obtaining its wood from those ejidos in Quintana Roo which have chain-of-custody certification. Other regional companies and sawmills are also requesting that chain-of-custody certification be conducted on their operations. In addition, companies manufacturing furniture and doors in several other states and in the capital, Mexico City, have also asked for their operations to be evaluated for a chain-of-custody certification.

Despite these concrete advances, few Mexican companies are familiar with, or interested in, chain-of-custody certification. The internal markets have not yet generated a demand for these products in Mexico. Only companies that are linked to export markets have shown more interest in certification. This is because the countries to which they export their products are beginning to request proof of certification, thereby transforming certification into a tool with increasing power to affect business transactions. Thus, the chain-of-custody certification will require additional effort and development to increase its utilization. No broad-scale campaign has been launched in Mexico to inform the public that forest certification is a guarantee of good management of temperate and tropical forests. Depending on the extent to which consumers acknowledge the label of the Forest Stewardship Council and what it represents will determine whether manufacturing companies and commercial interests in Mexico will pursue certification to enhance their competitiveness.

Another particularly relevant issue for certification initiatives in tropical countries is the overwhelming and disproportionate emphasis on a single — or no more than a few — species as the dominant valued resource from the forests. This emphasis on extracting a limited number of tree species occurs despite the typically high tree-species diversity that is particular to tropical forests. Limited market development and/or lack of initiative to explore new market opportunities are generally cited as the primary obstacles to diversifying forestry operations. The situation is further exacerbated by the high levels of investment and capital resources required for enhancing the processing capacity in communities for handling a variety of tree species. When local processing capabilities are absent, intermediaries often take most of the profits of forestry operations, leaving little behind in the communities for profits or reinvestment.

For example, in the high-elevation, moist, pine-oak forests in the Sierra Norte region of Oaxaca where UZACHI operates, pine species have been strongly favored in forestry operations to the exclusion of the co-dominant oak species. Only recently have communities begun to acknowledge the potential ecological importance of non-pine species, and are taking steps to promote forestry practices that conserve biodiversity. In predominantly pine-producing regions of northern Mexico where oak is also an important co-dominant species, significant progress has been made in developing markets for charcoal production from native oak species. This highlights the importance of developing stable and profitable markets for other ecologically important species, such as oak, as a mechanism for ensuring the maintenance of species-diverse forests in this region.

In the tropical forests of southern Mexico, forestry operations have historically concentrated on the extraction of mahogany and tropical cedar. This practice has caused an imbalance in the natural regeneration capacity of these species in many areas and, in some cases, has led to their disappearance. To date, the general rule is that the market is primarily controlled by mahogany and tropical cedar, without considering the broad spectrum of timber species that exist in the tropical forests. As a consequence, high-grading of the tropical forests has occurred, combined with a waste of large volumes of species that are poorly known, but which, nevertheless, possess attractive characteristics for the wood industry. A prevalent problem with marketing other tropical species is the reduced volume available of the various potentially marketable species. To confront this problem, the southeastern region of Mexico has carried out several initiatives to form collection centers to accumulate the volumes of at least ten lesser-known species from each associate ejido member. In this way, it is possible to offer the market a greater volume of each species than would have been possible by each ejido operating individually.

In the certified ejidos in the Mexican tropics, advances towards the diversification of marketable species and their management are occurring, although mahogany and tropical cedar continue to dominate. The diversification process is slow and is closely tied to fluctuations in market demand. Of the total annual possible harvest, 100% of the mahogany and tropical cedar are actually harvested, 70% of softwoods are harvested, but only 30% of other hardwood species are harvested.

The long-term ecological implications of single/few species management are immense. They can include loss of species diversity; reduced ecosystem resiliency and resistance to disturbance; and loss of nutrient cycling and production capacities. Significant improvements may be achieved by enhancing the utilization and sustainable management of a greater diversity of species which are present in tropical forests. However, to effectively process these species, this requires generating demand for the lesser-known hardwood and softwood species, as well as the appropriate infrastructure in the sawmills. A positive trend is the growing accumulation of knowledge by producer communities about the best methods for reforesting and enriching their tropical forests. Many communities have experienced the disadvantages of inappropriate practices, such as planting mahogany or tropical cedar plantations in close rows with insufficient light for adequate plant establishment. Consequently, communities are experimenting with effective means of facilitating forest regeneration. For example, reforestation of mahogany and tropical cedar is occurring in areas that are more open so that guide species can become established.

Juxtaposed to the single-timber-species dilemma are the myriad non-timber products that are often extracted from tropical forests and either utilized by local people for personal consumption or sold in local markets. While the stamp of "certified good forest management" applies to all operations conducted on the forested lands, only very recently has there been an increased effort to develop specific guidelines for evaluating the management and impact of non-timber forest product (NTFP) extraction. The general public, and even many researchers, believe that any NTFP is inherently sustainable. However, these activities can often pose significant ecological risks; for example, in chicle extraction, where bad practices may lead to tree death; or in the over-extraction of several species of palms, orchids, and bromeliads without any understanding of their regeneration capacity. An important challenge for the future of certification will be to develop standards for evaluating the social, ecological, and economic implications of non-timber product management, for which the issues surrounding sustainability are likely to be more complex. In addition to the development of general standards for certifying the silviculture and management of NTFP, certifiers should give special attention to developing specific standards for each product incorporating potential regional variability in accordance with particular ecological and socio-economic characteristics. For example, the extraction of chicle requires the permanent establishment of chicozapote trees as part of management (e.g., single-tree management within the stand). On the other hand, the collection of orchids forces management to ensure that the integrity of the entire ecosystem is maintained if they wish to sustain orchid production (e.g., ecosystem management). Obviously, these two activities will have very different socio-economic contexts.

Linking forest management practices to natural disturbance regimes is an important, yet often overlooked factor in tropical regions where intensive management for timber production is a recent activity. Information about natural-disturbance cycles can provide a useful basis for designing management interventions that create the necessary conditions for the regeneration of desirable tree species. In the pine-oak forests of the Sierra Norte, fire is a reccurring phenomenon and concern during the dry season, and local communities allocate large amounts of resources to combat them. Fire, however, also plays an important ecological role in these forests by creating favorable conditions for the natural regeneration of pine: scarified soil with a sparse litter layer and low competition from herbaceous growth. UZACHI's silvicultural system does not adequately prepare harvested stands for pine regeneration, as the thick litter layer and large amounts of slash generally remain on the site and interfere with natural pine regeneration. Therefore, costly planting of seedlings produced in nurseries, combined with post-treatment interventions to reduce competition, are often required. Greater use of prescribed burns could both encourage natural regeneration and reduce the risk of severe fires during the dry season. Examples of the incorporation of prescribed burns into management can be found in Durango, where controlled fires are applied in order to remove debris from forest operations and enhance pine regeneration.

Another critical aspect in terms of ecological impacts of forest management in the tropics is the tremendous need for — yet often lacking — ongoing monitoring and evaluation as a basis for the continued improvement and adaptation of silvicultural practices. Certifying teams consistently cited this requirement as a "condition for future improvement" in their evaluations of community forestry operations in Mexico. Although this problem is often directly related to both the small land-holding size and limited financial resources of communally owned forestry operations, lack of skills and experience pose equal, if not more important, constraints. However, since the unknowns surrounding ecosystem processes and harvesting impacts in tropical forests are numerous, determining appropriate extraction levels, rotation times, regeneration methods, and other silvicultural techniques urgently necessitates an integrative approach to management and research. As more knowledge is acquired, management practices can be adjusted, allowing an adaptive management approach to evolve.

Currently, most information on ecological responses to management is collected by the technical staff with assistance from the timber producing communities. For example, in the certified ejido of Noh Bec in Quintana Roo, the technical advisors have established a system of permanent-monitoring plots. These plots are being used to determine the levels of regeneration and growth of the most important commercial species with respect to variables such as light, soil, stand structure, and disturbance intensity. This system has already generated substantial information which is currently being analyzed in order to develop guidelines for forest management. The establishment of this monitoring system assuring the continual and systematic collection of data is a significant achievement for the communal forestry sector.

Other initiatives in research and monitoring are taking place in the form of collaborative links between ejidos or communities with external universities and institutions. In the state of Durango, the certification process has catalyzed agreements between the technical groups responsible for the management of several of the certified ejidos and the Ecological Forestry Area of the Institute of Silviculture and Wood Industries, Juárez University in Durango. These agreements contemplate the participation of investigators and their students to conduct inventories in the certified forest ecosystems and evaluate forest responses to different silvicultural treatments. Further, several government institutions in the forestry sector (e.g., the National Institute for Forestry and Agricultural Research [INIFAP]) are developing research projects that can provide valuable information to the technical-support groups. Coordination between state institutions and interested technical-service groups is, therefore, an essential mechanism for establishing links that strengthen the scientific basis of forest management.

Monitoring is also an important issue confronting CCMSS, the primary certifying agency in Mexico, since many of the certified forestry operations are accompanied by "conditions" for improving forestry operations within specific time limits, and therefore require follow-up evaluations. This creates the risk of a potentially disastrous backlash on the entire certification process if adequate follow-up and vigilance is not applied, and a certified forestry operation is discovered to be actually carrying out inappropriate or damaging practices. During the process of certifying an ejido, a dialogue is initiated with the technical service staff and the ejido members by the CCMSS evaluation team to ensure the technical and organizational feasibility of achieving certification goals. The purpose of this dialogue is to determine whether the conditions established in the evaluation reports are realistic. If the technical service staff or the ejido members are not in agreement with any of the conditions, they may be modified or adjusted until there is consensus that permits their accomplishment. As part of this, it is important to consider temporal scales since the improvement of forest management practices cannot be automatically achieved upon receiving certification. However, certification can provide a mechanism for establishing an acceptable period for meeting the specific conditions that are presented for each certified production unit. These conditions are generally established for the first, second, third, fourth, and fifth year during which the certification contract is valid before it needs to be renewed. CCMSS is responsible for verifying what accomplishments have been achieved by each certified production unit during each annual audit. So far, the level of compliance with the established conditions has been high.

An important question remaining is: what has the response been to certification by the various players in Mexico — the government authorities; the producer organizations and ejidos; as well as the technical services and companies? Historically, the forestry sector in Mexico has received little attention in the state and private development plans. Therefore, there is a high degree of receptivity today within the forestry sector for the new alternatives and initiatives being presented. Within this context, certification has encountered fertile territory for its development in Mexico. However, as discussed previously, the direct benefits that certification can offer ejidos still have not materialized. Thus, although significant advances have been made in improving forestry practices and in community organization of production systems, it has not translated into more access to markets and better prices for certified products. This shortcoming represents an urgent challenge that must be overcome in order to consolidate the certification process in Mexico and in other countries.

Despite the problems related to the market value of certified products, there are several important benefits that have accrued to ejidos and communities which are certified. In Durango — the principal timber-producing state in the country — certification has developed in an exponential fashion. In the case of Echeverría de la Sierra, certification has served to augment the prestige of the technical service group responsible for elaborating and implementing the management plan. It has also spurred the successful commercialization of secondary products of oak — a species which was often left in the forests and provided fuel for disastrous fires. Recently, these secondary products have entered the international market in the form of certified charcoal. In the case of UZACHI in Oaxaca, certification has served to elevate the regional prestige of this organization and its technical-service group, and has enabled the organization to obtain improved conditions for collaborating with government agencies. In addition to increased prestige, SPFEQR in Quintana Roo has gained greater access to government programs and has gained empowerment in its capacity to negotiate politically. In this region, certification has also attracted important international buyers interested in tropical forest products. Recently, the ejidos obtained chain-of-custody certification, which now permits them to use the certification label on their sawtimber and on craft products produced in community workshops. The experience up until now demonstrates that many forestry producers — especially those who are organized — are interested in certification despite the uncertainty revolving around increased prices for their products. Their willingness to improve management practices is based on their conviction that both the sustainability of the forest resource and the source of their livelihood — and their children's — are dependent upon this.

A strong limitation (or better yet, a great challenge) of the certification process in Mexico is that it has focused primarily on the more successful production units. However, these groups are also a minority among forestry ejidos in the country. Therefore, certification may be favoring those groups which apparently need less help, leaving behind the less-well-off producers who often remain at the mercy of corrupt companies and authorities. Considering this, the CCMSS has promoted certification not as an end in itself, but as a tool for facilitating the improvement of forestry practices and strengthening the organizational development of the communities active in forest management. Certification is perceived to be a global movement that will contribute to creating more ecologically and socially sensitive markets, thereby increasing both consumer preferences and willingness to pay higher prices for certified products. To the extent that certification operates in this global environment, the conditions and limitations of this process will reverberate within each of the countries currently promoting certified forestry operations.

The model being proposed by the CCMSS is to encourage less-well-managed production units to improve their practices by seeing the successes being achieved by the more-organized communities. The extent to which the more-advanced communities are receptive to improving their forestry practices and the sale of their products will determine how motivated the less-organized production units will be to pursue certification. In this way, the CCMSS would be complying with its mission to increase the number of communities which are carrying out good silviculture that protects the forest's ecological functions and assures their sustainability. Certification facilitates the construction of formal agreements based on current knowledge of sound silvicultural management practices appropriate for each region and forest type. In addition, it will create and secure possibilities for generating economic benefits for the rural communities that possess the forest resources.

8

Challenges and Benefits of Certification

8.1 Summary discussion of the advantages and challenges of certification

Kristiina A. Vogt, Bruce C. Larson, Daniel J. Vogt, John C. Gordon, Anna Fanzeres, Jennifer L. O'Hara, Peter A. Palmiotto

The movement toward certification has developed as a response of society to the perceived lack of sustainable management of the world's forest resources. Society's perception of what acceptable forest management is has been strongly influenced by the rapid increase in tropical deforestation and cutting of old-growth forests on public lands (i.e., spotted owl controversy in the Pacific Northwest United States; Vogt et al. 1997a). As a result, society's values have changed and have brought about regulations which are meant to control forest management practices that are more in line with society's desire to see forests managed in a strictly sustainable manner. The challenge for certification is to meet and satisfy the changing societal values of what they perceive to be acceptable management practices on forestlands. Adding to this challenge is the fact that the regulations borne from society's desire to see better management of public lands are now being imposed on private landowners.

One of the challenges to certification is that we exist in a time where there is a changing society but a stable biology in the temperate zones. This is contrasted by the situation in the tropics where there is a changing society but an unstable biology. The challenge is to meet and satisfy the changing societal acceptance of what they perceive should be allowed to occur on forested lands. Much of this has been driven by the changes which have occurred with public lands regarding what society found acceptable, and the resulting regulations to control the expression of societal values on these lands (see 3.2.2, 4.3.5.3). These same values are now being imposed on private landowners. During this same period — when public values have been evolving — our knowledge base in biology and silviculture has not changed dramatically. What has changed in forest management has been the introduction of technology so different tools have been developed (see Chapter 6).

Problems arise because the public expects more values to be obtained from a given piece of private forestland than the private landowner who owns the property does. These different values and the expectation of what private landowners need to provide to society are pivotal points in the conflicts that exist over land management. For forestlands, this is especially important, because many small, private forestland owners are an important part of the wood supply, especially in the United States (see 7.4). More than half the forestland in the United States is owned by small landowners, most of whom do not own their land primarily for the extraction of timber or non-timber forest products (see 7.4, 7.5).

A well-managed forested landscape has a range of ecological services it can provide simultaneously, such as water quality; carbon sequestration and its role in mitigating climate change; soil quality; and wildlife habitat for non-game species and aesthetics. In the United States, some of these services are regulated by the states' Best Management Practices; however, only soil and water quality are addressed as part of these practices. Because small private forestland owners do not primarily own the land for resource extraction, the lands are not managed — or are poorly managed — to obtain these different values. This is compounded by the fact that many of the environmental services provided by a given unit of land do not have economic values that can be realized in the marketplace. Therefore, the incentives for a small, private landowner to provide other values desired by society are difficult to balance with the potentially decreased ability to obtain the benefits of satisfying their own values from the land.

The difficulty in balancing the values desired by society with their own values should not automatically exclude this significant ownership of forestland in the United States from the opportunity to become certified. As discussed in Chapters 2, 4, and 7, the small landowners are effectively excluded from certification because of the many values embedded in the protocols that require large land ownership sizes to even consider pursuing certification. In order not to exclude small forestland owners from the process, there is a need to:

- Determine when a piece of land cannot provide society's values because the matrix landscape within which they are embedded will not allow a value to persist on the small piece of land (e.g., forest surrounded by agricultural lands) or because maintaining the value would require a landscape approach (i.e., valued animal species have wide habitat ranges).

- Change the structure of the existing protocols to determine when good forest management is occurring and realize that a range of societal values cannot be obtained from every tract of forestland, but good forest management can be assessed on small land holdings.

- Explicitly identify when providing most societal values from a forestland can cause degradation of that land or when a forestland owner will cause degradation of their land because of their values for their ownership.

- Balance and weight the choice of indicators in assessments to reflect the social and natural science constraints controlling the functioning and resilience of a forested ecosystem from those values specific to agencies developing protocols (i.e., acceptance that values do not necessarily reflect the best indicators of system health).

- Determine how small forestland owners can be compensated when they provide the values from their lands that society would like to have maintained.

It is critical that the values which may become embedded in a certification assessment are transparent and understood by all parties involved with certification. It is important that forestland owners pursuing certification are aware of what they are "buying into" and that the certifying organizations present the assumptions and values relevant to their approaches. These values can then be defined on the label. For example, the certifying organizations under the FSC label do abide by principles that are strongly environmentally based and these values are an integral part of their assessment structure. Those private landowners desiring to show that they are capable of obtaining these values from their land may want to pursue the FSC approach to certification.

This is not, however, just a simple case of identifying a certifying organization that matches your values, or to conduct group or communal certification to obtain a large enough land base to be certified. *It is important to understand that the values embedded in all of the different certification assessments will determine how effectively an assessment can determine social and natural system sustainability.* In addition, the high subjectivity that exists in many of the protocols creates a situation where different evaluators will assess the same site with totally different results (see 4.3.4). Using an approach where the results of an assessment can be questioned in the future (and even having to withdraw the certification label, which has happened) will only discredit the certification process. This possibility is real and it is not unheard of to hear discussions about whether certain groups should or should not have been certified (Kiekens 1997). It is important that these types of discussions do not detract from the benefits of certification.

A possibility exists for a protocol to not assess a system holistically so that unpredicted factors may reduce the resilience of that forest which eventually negatively feeds back to affect the social and/or natural system (see 5.1). The possibility also exists that forestland owners pursing certification can satisfactorily produce all the information needed as part of assessment and still manage their lands so that they degrade.

> Just having a large amount of data on a site does not mean that a system is being well managed or managed in a sustainable manner. The high data demands of assessing several different values for a site may result in emphasis being placed on the wrong variables during an assessment (i.e., the sensitive indicators being overwhelmed by all the other indicators) (see 5.1). There is a need to be able to weight the importance of different indicators being used in an assessment for each site, and is currently not part of most assessment protocols (see 5.1).

The lack of a framework to guide an evaluator as to what the most appropriate indicators are is quite apparent when examining the criteria and indicators in current protocols (see 4.2.1). For example, assessment protocols do not use the same indicators as researchers studying the specific effects of different management activities in forests (see Chapter 4.2.1, Chapter 6). As currently structured, the certification assessments do not define the parts of the ecosystem affected by different management activities nor do they explicitly state which variables and thresholds (and why) are appropriate for specific management activities. This suggests the need to amalgamate both approaches into current protocols that are either performance- or system-based (see 2.4).

Some management practices will probably never be acceptable and will create controversy for those certifying organizations which are under the umbrella of the Forest Stewardship Council. For example, clearcutting and the use of herbicides (and for some, plantations) are discouraged as activities that should not occur on private lands (Gordon 1996). Gordon (1996) suggested that forestland owners need to decide which procedural approach they will pursue. Either the International Organization of Standards (ISO) as a systems approach or the Forest Stewardship Council (FSC) as a performance-based approach might be more suitable for their system. However, Gordon (1996) concluded that both approaches are important for formal certification. There are elements of both approaches that are needed for the implementation of a successful forest certification.

The current format of data-intensive protocols is difficult for a private landowner to pursue. This is mainly due the large amount of information needed to satisfy all the values that these criteria and indicators demand. On the other hand, the lack of clarity in the criteria and indicators also makes it difficult for a landowner to know what information they need to have for their lands (see Chapter 4). The information collected for an assessment will ultimately be interpreted and determined by the expert individuals brought in to conduct the evaluation. This occurs since each protocol appears to require a large amount of data which gives the perception of detail and specificity. However, indicators end up being very subjective because the driving variables have not been a priori identified for that site. Therefore, if the evaluators are good in their field, the evaluation will be interpreted in a logical and consistent manner encompassing only the critical needs for that piece of land. However, if the expertise of the individuals is not as well developed, the potential for an erroneous evaluation is quite high. It is imperative that required documentation of certification does not result in the collection of information that can be interpreted differently at each site being evaluated, even though the data appear to be similar.

8.2   Opportunities and challenges identified by certifiers and certified owners

Kristiina A. Vogt, Bruce C. Larson, Daniel J. Vogt, John C. Gordon, Anna Fanzeres, Jennifer L. O'Hara, Peter A. Palmiotto

Other opportunities and challenges to certification have been shared by those who certify forestland owners and those who have been certified. Many of these opportunities and challenges must also be dealt with when examining the future of certification as a tool for assessing forest management practices and societal values for these lands. Dealing with challenges and taking advantage of the opportunities listed below, however, is an inadequate way to make certification one of the primary tools in natural resource management. Many of the issues presented in Section 8.1 need to be addressed for certification to become a generally accepted tool because of weaknesses that make protocols difficult to consistently implement. A summary of opportunities and challenges applicable to the United States is presented next.

   Northrup (1998) identified the following opportunities, needs, and challenges for certification:

Opportunities

- Boost for corporate responsibility and public relations
- Environmental communities received great public relations from their involvement with this issue
- Positive alternative for public lands and management
- A non-regulatory approach to remove bad interactions with the federal government
- Way for consumers to know which products are certifiable in the global markets
- Model for other sustainable resource uses

Challenges

- Need to reach private landowners
- Need new forestry training programs

- Need new ways to access capital and mobilize capital
- Need public policy tools for Sustainable Forest Management
- Tax policies need to change and need better incentives

Kane Hardwoods/Collins Pine in Kane, Pennsylvania has had their operations certified. Since they are a family-owned business that has existed for four generations, it was easier to become certified since there was social stability in the company (Puller 1998). Kane Hardwoods pursued certification because the family felt that it was a way to reestablish public trust in the company and they were interested in providing wood to the new markets being generated by discriminating customers interested in certified wood. Puller (1998) summarized the following advantages and disadvantages of the certification:

Advantages

- The first certification resulted in the forest managers believing that the assessment of their management made them have a better management system. This resulted from the internal audit that required justification of forest management practices and from the use of new tools (GIS, new inventory of lands).
- New market opportunities were created.
- There was improved morale in the company since the forest managers had verification that they were managing well.
- The evaluation showed that the company was too conservative in how much timber they were cutting and that they could cut more.

Disadvantages

- The initial cost of the certification and the requirement for annual reviews (~$7,000 each) (see 7.6)
- Internal economics became transparent
- Loss of management options (i.e., if an area was placed into an ecosystem reserve but contained high-quality timber)
- Had to hire additional personnel to cover areas not covered (e.g., wildlife biologist)

Puller (1998) sees threats to certification occurring from confusing labels that exist in the marketplace. When Kane Hardwoods pursued their certification, they used the SCS label, but then put on their own label. He feels that this results in too many labels being placed on a product and that eventually the furniture will be covered by labels (Puller 1998).

Puller (1998) stated that several extensive changes have occurred in the company's procedures and practices since being certified. There has been a greater need to document and collect more information and to conduct more analyses of decisions. This has made the process more formal for management. The gains achieved during the first certification process were mainly in the area of better management (Puller 1998). Since better management was achieved, there is a question about the value of being recertified when this initial certification is no longer valid. The expected economic returns have not occurred. For example, a niche market (expensive furniture, smaller manufacturers) is mainly buying the certified wood and it has been hard to advertise the fact that they have certified wood.

Jim Drescher of Windhorse Farms in Nova Scotia suggested that landowners participate in certification to (Dresher 1998):

Advantages

- Access the markets
- Improve management techniques
- Be recognized for good forest management

Owners such as Seven Islands in Maine pursued certification because they perceived the three main benefits from undergoing this evaluation to be:

Advantages

- An independent review of management practices
- A way to differentiate themselves from other large landowners and managers
- Market benefits from selling certified products

The first benefit was often the most important reason to pursue certification and Seven Islands thought this independent review would provide a thorough, unbiased estimate of their management. They felt that this review would be a great benefit to them as landowners and managers in deciding how to direct their future programs. When Seven Islands was certified in 1994, the market benefits were unclear since markets were just being developed and it was not clear if market advantages would accrue to them.

Paul Smith's College, a private four-year college owning 14,000 acres in the Adirondack Park in upstate New York, chose to initiate the process of certification in order to affirm the sustainability of their forest management plans. They were also interested in serving as a model for private colleges and universities that actively manage forestlands. The certification process confirmed for the college the areas to which they needed to pay more attention (i.e., identification of wildlife and endangered species habitat). The assessment showed that the decentralized structure of the management of the college's lands led to a confusing decision-making hierarchy. A redefinition of the forest manager's position, the addition of another staff forester, and specialized training were actions taken to address the issues raised. Clearer and more deliberate documentation of management plans and activities, along with the development of monitoring protocols also resulted from the certification process. The challenge for Paul Smith's is to implement these changes in a timely manner and to maintain them. This challenge is viewed as an opportunity because the college feels the changes are necessary, integral steps that will assist them in their educational mission to teach and promote sound forest stewardship.

Some of the advantages articulated by certifiers and those who are certified have been achieved by an independent review of management practices and independent verification that good forest management was being practiced. Some of the other advantages have been more difficult to achieve, such as accessing more markets or obtaining a higher premium on wood being sold (see 8.3). Other advantages which have not been consistently produced will probably not occur unless a better system of assessing forest resources is produced. The incentives for achieving these other advantages do not currently exist.

In the United States, the possibility exists that federal government lands may participate in certification (this is in contrast to other places around the world where government lands have already been certified, see 2.5.2). This would dramatically increase the amount of land that could be certified in the United States. This idea is receiving considerable discussion since several public forests have been certified (e.g., Minnesota, Massachusetts, and Pennsylvania). Some of the discussion for certifying federal lands is being generated at the community level and is being considered as a vehicle to restore potentially degraded forests and to revitalize communities dependent on forest resources (Donovan 1998b). However, certification of federal forestlands is being approached "cautiously" by some certifiers. Smartwood's executive director declared that "perhaps certification can play a part in this process" (Donovan 1998b).

Smartwood only responded to federal certification requests in 1998 that met the following requirements (Donovan 1998b):

1. Are strongly supported and/or sponsored by local environmental and community development groups
2. Are part of either a forest restoration or community economic development and revitalization program
3. Are considered pilot certification initiatives
4. Are open to input from the FSC national working group and/or representatives of regional FSC working groups

This means that for federal lands to be certified, the role of the public (e.g., the local community) will be strongly emphasized and used as part of the assessment protocols. According to Donovan (1998b), this is the only way that any certification assessment will have credibility. The public will have to initiate the certification process instead of it being an initiative generated at the level of the federal agencies. This begins to introduce that different types of certification protocols may be produced that will vary depending on what values are driving the initiation of this process. The approach being taken by Smartwood would suggest a greater emphasis would be placed on information on the social side of the equation for government lands. This does not exclude all the other information typically required to be collected for certification. It does suggest that it would be more difficult for government lands to pursue certification because of the greater role of the public in the process.

8.3 Factors affecting future use of certification as an assessment tool

Kristiina A. Vogt, J. Scott Estey, Andrew Hiegel

When one examines the statistics for forestlands certified in the early 1990s by organizations that are currently FSC accredited, 19.6% of the forestlands are no longer certified (Table 2.2). Most of this reduction in certified forestland area is due to one large tract of land located in Indonesia. There are many plausible explanations as to why six of the previously certified forests are no longer on the list. In one case, withdrawal of the certification label was justified by the certifier due to changes in land ownership (Amacol Ltda, Brazil). When certified areas have not complied with the baseline commitments of the protocols, protests from environmental NGOs have also caused certification to be withdrawn. Another reason for no longer being certified is that some companies have decided not to pursue renewal of certification after the expiration their initial certification.

This non-pursuit of certification can be explained by a company's inability to satisfy the certification requirements. In some cases, the advantages of certification were not apparent and did not justify the increased costs to the owner.

Three factors can potentially affect whether certification is maintained as one of the dominant tools in natural resource management:

- Costs of certification
- Imbalance between supply and markets
- Lack of participation of mainstream wood industry organizations

These will be briefly discussed in sections 8.31 through 8.3.3.

8.3.1   Costs of certification

Some forest owners are not pursuing certification because of economic factors. The suppliers may not be able to pass on the costs of certification to customers in the form of higher prices for certified wood products. When higher prices are not obtained in the marketplaces because the prices for certified wood products have not increased, (Mater 1998), other benefits must exist in order for certification to succeed. If certification resulted in increased access to other markets or specialty markets, certification could be financially beneficial to a forest owner. Some companies may pursue certification simply because maintaining sustainable forests (e.g., Seven Islands, Collins Pine) satisfies the owner's environmental values.

Certification increases a forestland owner's operating and management costs. These increased costs can discourage a forestland owner from pursuing certification when his primary purpose in owning forestland is not to manage timber. This is especially relevant for many non-industrial forestland owners who do not primarily manage their land for natural resource extraction (see 7.4). The costs to acquire the information required for certification can be insurmountable for landowners who do not wish to manage their forestland in order increase its value in the marketplace (see 7.5). Since they own more than half of the total forestland in the United States, private landowners may be excluded from even considering certification. It is difficult for this group to document that they are practicing good forest management. The result is that they cannot be suppliers of certified wood.

When considering the costs and benefits of certification, the downside of obtaining higher prices for certified wood must be considered. One major economic disadvantage of certification is that it raises the prices of wood products enough to place them in direct competition with some of the other building materials available today. These alternative building products might have higher environmental costs, but not necessarily bad images with the public (i.e., cement, steel, and brick) (Koch 1991, Moffet 1993). This trend is occurring in the United States as wood prices rise (Perez-Garcia 1995).

8.3.2   Balancing wood supply and lack of markets

In addition to cost factors, other issues related to the supply of certified wood and the ability to sell it may impede the use of certification as a general assessment tool. In the United States, there appears to be an insufficient supply of certified wood in the marketplace, while at the same time, certified forest owners are unable to sell all their labeled wood. In concrete terms, the quantity of certified wood products available is still quite low. The amount of worldwide industrial roundwood currently supplied by companies certified under the FSC label is less than 0.6% (de Callejon et al. 1998). This is partially because of how few companies and land areas are currently certified (Section 3.3). According to de Callejon et al. (1998), most of the wood produced on certified lands is not distinguished from the uncertified wood sold in the marketplace. It is interesting to note that of the total volume of certified wood produced worldwide, according to FSC's list of certified areas (see Table 2.2–2.8), the wood coming from Poland accounts for almost half of the world markets (De Callejon et al. 1998). This means that wood sources are limited to a few locations.

Despite the low quantity of certified wood available in the marketplace, many companies are unable to sell their certified wood. One reason is that some companies do not market any of their products (e.g., Ston Forestal, Costa Rica), or only market some of their products (e.g., Seven Islands, Maine), using the certification label. In the instance of Kane Hardwood/Collins Pine, 14% of their sales are of certified wood and 9% of their wood volume is certified (Puller 1998). The inability to sell their certified wood is a problem for wood suppliers who are interested in producing certified products, but are unable to sell them as such (de Callejon et al. 1998). This is balanced by customers who want certification to ensure that forests are being adequately managed. However, customers are not producing a demand for certified products significant enough for certification to become a realistic option for suppliers. "This contradiction accounts for much of the skepticism about the future market for certified wood products" (de Callejon et al. 1998).

The contradiction in the supply vs. the selling of certified timber is demonstrated by Seven Islands Land Company in Maine. Seven Islands is a prime example of a company that has all the characteristics necessary to profit from selling certified wood. They own a large amount of forestland (975,000 acres) in northern Maine which means they can supply a large amount of certified wood; however, they have been unable to find markets for all their certified wood.

The following information characterizes Seven Islands Land Company: they were certified by Scientific Certification Systems in 1994, land ownership has been with the Pingree family since the 1840s, and there are currently more than 150 heirs to the land. The family's philosophy emphasizes land stewardship responsibilities for their forestlands. They do not have any wood-processing facilities (e.g., company sawmill), and they contract with loggers and truckers to cut and deliver timber to independent mills. Several of the mills to which Seven Island regularly sells its wood have become chain-of-custody certified. Wood is sold in log form and manufactured into several certified product lines, including cedar shingles (Maibec Industries); spruce-fir dimensional lumber (Material Blanchet Inc.); pallet stock and graded hardwood; lumber; veneers; and flooring (Becesco, Gilles Begin Lumber, Inc., YWI, Green River Lumber) (*Environmental Building News* 1994).

However, Seven Islands has been unable to sell all the certified wood they produce, which show there is a problem in implementing the benefits of certification to landowners. Although Seven Islands produces 90 million board feet of certified logs of softwood, they are only able to market 80 million board feet of certified lumber (*Environmental Building News* 1994). The problem that Seven Islands has faced has been the difficulty in finding a regional distributor for the softwood. Currently, about 6 million board feet of beech, maple, and birch lumber and flooring are being sold as certified by Seven Islands (*Environmental Building News* 1994). However, the total volume sold as certified hardwood lumber and flooring is approximately 30% of the hardwood production at Seven Islands (*Environmental Building News* 1994).

Despite not finding more markets for certified wood, certification has benefited Seven Islands and has satisfied the stewardship values of the family. Seven Islands Land Company has made money through its certification and has obtained a better management plan for their lands. The president of Seven Islands was quoted on the front page of the

Wall Street Journal, saying they had made money through certification and, through increased market prices, had recouped the costs of certification within 18 months (Ulman 1997). Apparently because Seven Islands is a vertically structured company, the use of a label works for them.

Producing certified wood has given an advantage to some of Seven Islands' products because of increased access to markets, rather than because of increased premium prices. They have been unable to compete with uncertified wood selling at lower prices (MacAlpine 1996). However, the incentive for landowners to practice sustainable forestry has been based on their ability to obtain price premiums for certified wood, which that they are unable to achieve. This inability obtain price premiums will probably have an effect on whether some forestland owners will pursue certification.

8.3.3   Lack of participation by mainstream wood industry organizations

In the United States, another problem that certification faces is the mainstream wood industry's lack of attraction to and participation in its activities. In this country, certifying organizations under the FSC label have been more successful in attracting small to mid-size companies to certification (see sections 3.3, 3.5–3.7). The lack of participation by the large companies means that the impact of certification on sustainable natural and social systems in United States will be decreased. For example, despite the fact that other countries have certified areas potentially focused on pulp and paper production, this segment in the United States has not been properly involved in certification. There are currently no certified pulp mills in the United States. Since total production of wood pulp is approximately 60 million metric tons in the United States (Jaakko Pöyry Consulting 1996), its impact on forest management is too significant to ignore. Wood consumption for all pulps accounts for over half of the total annual harvest of industrial wood in the United States.

This means that large-scale, industrial forestry is not well represented in certification in the United States. When examining wood companies that have been certified under the FSC label, the lack of participation by the mainstream forest industry becomes apparent. As of March 1998, FSC reported a total of 30 forest management certificates in the United States, with a total certified acreage of slightly over 1.4 million ha (3.5 million ac) (FSC 1998c). This acreage represents less than 1% of the timberland in the United States (USDA Forest Service 1994).

Of the 30 management certificates existing in the United States, only one was awarded to a company that ranked among the top 40 companies in terms of timberland ownership (*Pulp & Paper North American Fact Book* 1998). It was Seven Islands Land Company in Maine, with 975,000 acres of forestland (FSC 1998c) — a company that is non-integrated to wood products or to pulp and paper production. More than 0.7 million ha (1.8 million ac) of the FSC-certified forests were public lands (FSC 1998c), which are also non-integrated to forest product production facilities. Of the total land area certified by FSC, 121,457 ha (300,000 ac) are owned by companies that are integrated with forest products production assets. However, most of that land is owned by one company (Collins Pine, 113,360 ha or 280,000 ac) that owns and operates several sawmills.

In order for certification programs to be implemented at broad scales in the United States, it will be critical for it to be accepted among the large-scale, commodity-oriented producers of products (e.g., softwood dimensional lumber; structural and non-structural panels; and pulp and paper). Producers of commodity products of this type are, by far, the largest consumers of wood in the United States and are significant timberland owners. The manufacturers holding chain-of-custody certification tend to be small companies. Of the chain-of-custody-certified companies, none ranks within the top 40 United States forest-products companies, based on annual sales revenue (*Pulp & Paper* 1998).

FSC certification tends to be concentrated among the niche and/or higher value-added wood products such as veneers, furniture, moldings and millwork, and flooring. Within these general categories, FSC certification seems to be gaining a wider acceptance among hardwood-based wood products that have a relatively high degree of consumer visibility. However, those products account for a relatively small portion of the wood consumption in the United States.

Currently, the majority of the FSC chain-of-custody certified manufacturers are secondary manufacturers that are not integrated to primary production facilities. Without significant sources of certified wood from primary manufacturers — which are, more often than not, large-scale commodity producers — secondary manufacturers will have difficulty producing certified products.

There are many reasons large forest products companies do not pursue FSC-type certification. For example, companies may simply disagree with the FSC label's underlying principles and science that drive the performance requirements of the systems. They also may not wish to incur the fixed, up-front costs associated with the audit process. Additionally, they may be hesitant to open their records and management practices to unaffiliated third parties for fear of losing a competitive or strategic advantage. Recently, Champion International publicly stated that they will pursue third-party SFI accreditation for all their forestlands. In this case, the values inherent in the different certification approaches determine who a timber company will utilize for their certification.

8.4   Reasons for forestland owners not to become certified

Kristiina A. Vogt, Christie Potts, Heidi Kretser, Jennifer Heintz, Allyson Brownlee, J. Scott Estey

Several reasons can be given for a forestland owner not pursuing certification. The following factors have been implicated in not pursuing certification:

- Aversion to government or third-party groups/organizations in regulating or being involved in making decisions on how their land is managed
- Strong values supporting private property rights
- Limited capital to invest in management or in developing documentation verifying past and future management
- For small and large private, non-industrial landowners, the costs of certification cannot be redeemed, either because the land area is so small that financial return is more long-term or the product does not meet the needs of the high-value end market
- For small, private, non-industrial landowners, the need to hire specialists (e.g., wildlife, specialists plant biologists) to collect information required as part of the certification process that the owner has to pay for
- Requirement by the certification protocols for public participation in decision making on management of the forestlands

- Perception by owners that the land is not well managed or has a legacy of land use associated with the previous owner that the current owner does not want to take responsibility for

- Timber production is a secondary reason for the landowner to own the forestland and other values such as recreation and aesthetics are more important

- High turnover rate of ownership of forestlands (in the decade range) — partially due to federal and state death tax burdens

- Chain-of-custody requirement in some protocols

Many small, non-industrial forestland owners do not become certified for two main reasons: (1) they cannot supply much of the basic, detailed, data documentation required by most of the certification protocols (e.g., management plan) (see 7.5) and (2) the costs to be certified are prohibitive (see 7.5, 7.6). In most cases, they do not have detailed, complete surveys of plants and wildlife for their land, nor do they have detailed documentation of how the land has been, and will be, managed. In most cases, small private landowners do not have a management plan for their lands. For example, an estimated 5% of private landowners (comprising 30% of private forestlands) have a written management plan (Birch 1995). Statistics show that as the sizes of the land parcels increase, the numbers of owners having written plans increases (Birch 1995). There appears to be a direct relationship between the existence of a management plan and the size of forest ownership. The fact that so few small forestland owners have basic documentation (e.g., a management plan) for their lands cannot be entirely explained by their inability to pay for such a plan. For example, through the Forest Stewardship Incentive Program, enacted through the 1990 Farm Bill, the government previously paid up to 75% of the cost to develop a management plan (Argow 1996). (However, this program is no longer funded.) In addition, many large, private-industry owners have assistance programs that will develop these plans at little or no cost to the owner.

The size of the land ownership frequently reappears as a variable in determining whether or not forest owners will pursue certification on their private lands. When the land area is small, the costs to be certified are not redeemable by the landowner in a couple of years. For this reason, many certifying organizations do not believe it is worthwhile to spend time certifying small land ownerships. Although it is not explictly stated by most certifying organizations, there appears to be a lower limit at which certification is not worth pursuing (see 2.5, 7.5). Northrup (1998) stated that Smartwood required a landowner to have a minimum of 1,000 acres in order for them to be considered for certification; otherwise, it would be cost prohibitive. An examination of Tables 2.1–2.8 corroborates the dominance of large land areas being preferentially certified.

Values and the perception of the degree to which an outsider should determine how private property is managed also have a strong influence on whether or not small forestland owners decide to pursue certification (see 3.2.2, 7.5). Many small, private landowners, especially in the southern United States, do not feel that the public should participate in the management decisions being made on private lands. Private forestland owners in the South have larger parcel sizes and manage more of their acreage for timber production compared to their counterparts the northeastern United States, where parcels are smaller and more acreage is dedicated to passive uses (Birch 1995, Williams et al. 1996). Surveys of southern landowners noted their belief that "they have the right to use their land in any fashion without regulations...but believe in protection of the environment, while still being able to use their lands as they deem appropriate" (Williams et al. 1966). Most southern landowners (92%), a majority of whom are "timber oriented" and adverse to government involvement on their lands, stated that it was inappropriate for public opinion, demands, and/or desires to drive the development of their management plan (Jacobson et al. 1996). This value exists even though a survey in the Tennessee Valley revealed that 86% of private forestland owners felt private property rights were important only if the environment was not hurt and 76% felt that private property rights should be limited in order to protect the environment (Bliss 1993). This value expressed in the southern states is in contrast to that in the Northeast where forestland owners were less adverse to government involvement in private land management (Northern Forest Lands Council 1994). Many people living in the Northeast also had no difficulty incorporating public opinion in their management, even though they still strongly supported private property rights (Northern Forest Lands Council 1994).

Another major reason for landowners not to pursue certification is that large, industrial owners require a return on any investment and have not recouped the costs of certification within a reasonable time. These owners do not obtain a price premium on certified timber products because the market is geared to the high-value end products, such as furniture, roof shingles, and flooring, and not to 2x4s or pulp for paper. Many of the FSC-certified operations in the United States have been small to medium-size companies who have less intensive management and high-value end products, such as hardwood furniture. It appears to be easier for a certified supplier to receive a premium over a non-certified supplier on products that are visible and of high value. For example, customers are more willing to pay extra for a chair constructed from certified mahogany than for a certified, 2x4 spruce stud that will be hidden behind paint and wallboard. For this reason, detractors of FSC certification argue that it will always be limited to niche markets and to people who do not adopt it for its potential economic incentives.

Certification implies a certain quality of forest management to a consumer, so only managers who think they meet the standards will probably pursue certification. Most of the organizations that pursued certification (e.g., Collins Pine/Kane Hardwoods, Seven Islands Land Management) felt they had managed their lands well prior to pursuing certification. In fact, they used certification as a means to verify that their management practices were sustainable.

A common element among all the certification protocols (see 4.2.1) is the need to have surveys of wildlife habitat and of use of their lands. Wildlife planning is probably a common impediment for landowners when they are considering certification of their forestlands. Documented surveys of wildlife populations and habitats are not necessary for assessing whether good forest management is being practiced, so this type of information is not typically collected and documented by managers — these decisions have generally been left to the discretion of each forest manager. This is not the case in the southeastern United States, where hunting clubs pay private forest owners for the right to hunt on their lands. In these cases, there are probably good records of wildlife, but they are not necessarily available to the public. In fact, the National Survey of Fishing, Hunting, and Wildlife Watching found that in 1996, 77 million people participated in outdoor-wildlife recreational activities in the United States. Of the 63 million wildlife watchers, about 10 million participated on private lands and spent $106 million on private land-use fees ($67 per participant) (Waddington 1998).

When data is not available on wildlife for a particular management unit, certification is sometimes conditional on hiring a wildlife biologist. In the case of Seven Islands, assessment for wildlife was conducted by hiring wildlife consultants and the additional training of their managers who had some training in wildlife biology. If one is a small, private forestland owner, having to survey wildlife is quite unrealistic, as much of the wildlife is not confined within the boundaries of the management unit. In this case, being a small, private landowner means that it might be necessary to do wildlife surveys at a landscape level. However, if the landowners must consider all external factors in planning their land management, they would have a severe disincentive in pursuing certification.

The high turnover rate of forestland ownership also contributes to the problem of developing sustainable forest management plans for any given piece of land. Often, landowners must subdivide and sell their land in order to pay property taxes (see 5.3). "Federal and state tax burdens cause disruptions in management and demand may cause heirs to abandon timber production programs" (Peters et al. 1997). All across the United States, forests and agricultural lands are being converted to higher value uses, such as development. Certification does not deal with issues related to longevity of ownership of property, zoning, or future owners because of the high turnover rate of land ownership. By not considering small, non-industrial private landowners as customers of the certification process, perhaps certifiers may not consider these issues that must be dealt with. If half the timber supply in the United States is obtained from these non-industrial, private forestland owners (Birch 1996), these factors will have to be considered. They cannot be ignored in any program working toward sustainable management of forests for economic, ecological, and social values.

Because of the decreased supply of timber coming from federal lands, these non-federal forests are contributing a larger portion of the United States' wood fiber needs. This shift to a higher proportion of the timber supplies coming from non-federal lands places greater importance on facilitating sustainable forest management activities on these lands. However, this increased dependence on private forestlands for timber supply is coinciding with a period of rapid reduction in current forest landholdings to smaller acreages (NCR 1998, see 5.3, 7.5). The legacy of existing laws that regulate private property is resulting in lands being subdivided into smaller tracts and in shorter periods of land tenure, as people inheriting lands cannot afford to maintain them. The increasing prevalence of smaller land holdings is another reason to have certification protocols that can be used by small, private landowners. However, many of the existing protocols under the FSC label are not structured to deal with small landholdings. By not looking at a landscape level, many small landowners are excluded from managing — or even attempting to manage — for all the values required by certification. In many cases, landowners cannot afford the costs required for good forest management or are unable to utilize silvicultural intervention to achieve these goals since timber prices are often too low (Ashton 1996).

The small, non-industrial landowners can access much of the information required as part of certification if they know where to find it (see 7.5.4). For example, the state-level Cooperative Extension Service has offered a full range of advice and forest management assistance since 1914. Consulting foresters assist landowners at the local level in marketing and timber sales, wildlife habitat management, and offer many other on-the-ground types of services. Many universities offer educational classes for non-professionals. The American Forest & Paper Association, through their Sustainable Forestry Initiative, trains loggers in "sustainable" harvesting techniques in an effort to improve NIPF harvesting techniques. There are also national and regional landowner associations with various missions and agendas. Some serve as information clearinghouses through newsletters and correspondence, others act as lobbying groups for the rights of private landowners. Associations act as referral sources for consulting foresters, service foresters, and forestry extension education materials and courses. Finally, there are landowner-assistance programs sponsored by the forest industry (often with plantation-style operations in the South) to manage forestlands. It will require effort by a small, non-industrial landowner to find all of the information currently available regarding the management of forestlands.

Information alone will not be sufficient in helping small landowners to have their forestlands certified. One approach for dealing with the problems of small landowners is to form local landowners' stewardship associations (Potts et al. 1998). Since most NIPFs use their land as a secondary or supplemental source of income, there is little incentive for small private landowners to actively manage and produce timber on their land in an "environmentally friendly way" (Potts et al. 1998). A local land stewardship association could create incentives for small landowners to participate in certification. They could provide a means to access information, reduce overhead costs to implement the techniques, and decrease the costs to access markets by sending products from many properties to the same place at the same time.

The chain of custody is one of the factors that has been identified as an impediment to small landowners in becoming certified (see 7.7). Since small landowners can be an important source of wood in the United States, their inability to satisfy the requirements for certification is not a pattern that should be maintained. Protocols need to be developed which address the need for small, forestland owners to become part of the thriving community of wood-market suppliers.

8.5   The past and future goals of certification

Kristiina A. Vogt, Brett Furnas, Brian Rod, Bruce C. Larson, Daniel J. Vogt, John C. Gordon, Anna Fanzeres, Jennifer L. O'Hara, Manrique Rojas, Peter A.Palmiotto

For certification to be credible:

- It will need to avoid the traps of the sustainability paradigm debate.
- It will need to develop a working framework that has justifiable standards which are not exclusively value-laden.
- There is also a need for standards that clearly show the condition that is being aimed for as part of certification. These will incorporate the feedback loops between the ecological and socio-economic components of the system.
- It will need to identify how overarching and local-level mechanisms for promoting forest certification are precluding different segments of forestland owners from pursuing this initiative.
- There is a need for certifying organizations to seriously evaluate the procedures and standards they use to evaluate forest management operations.

8.5.1   Sustainability revisited

Presently, forest certification does not guarantee that a forest will be well managed or managed in a sustainable manner (Kiekens 1997). Does this mean that certification is not working in analyzing the sustainability of an ecosystem, or that the structures of certification protocols are not geared to assess good forest management? Or, do these values for the system condition drive certifiers towards the selection of certain types and amount of indicators? Do we need to change our thinking on forest management practices, or on just the assessment approaches?

Part of the difficulty being experienced by certification protocols in assessing sites is related to their goals. These goals have directed assessments to be based strongly on values, and several values simultaneously become the driving variables for how management is assessed.

Values in certification are being generated from its goal to produce or maintain sustainable social and natural systems (see 3.1.2). Integral to the sustainability paradigm is the need to link the sustainability of the social system (i.e., stable socio-economic systems, alleviate poverty, etc.) with the natural system (i.e., maintaining productivity and its health). However, the socio-economic components of sustainability deal mainly with human values and existing protocols of forest certification do not expose these values or establish the links between the socio-economic side of the equation and the ecological conditions. To understand and detect these interactions is fundamental in determining the carrying capacity of the natural system to resist or regenerate human practices such as forest management.

The difficulties encountered by pursuing the goal of sustainable systems have driven certifiers away from using sustainability, to adopting terms like good forest management and good forest stewardship (see 6.1.3.2). However, even these terms have been difficult for some people to understand or to define because of their unfamiliarity with the elements of good forest management (see 3.1.3). Although the goals have changed in what is being pursued with certification, the standards being used apparently still do not evaluate or guide even this more simple goal.

Sustainability by itself is not bad goal to aim for with certification, however, given the heavy value-laden burden on this paradigm, this goal should not be used to define criteria and indicators of forest management. A more appropriate set would have to take into consideration all the constraints that exist on the site. However, sustainability might help identify what general values are desired in any forest, and might force one to define whether the four elements (see 3.1.2) of sustainability are the desired goals for any assessment. So, sustainability by itself is not bad, but it is being used in the wrong manner to drive the design of certification protocols. This is the same problem in ecosystem management where the definitions are general by necessity and will therefore not explicitly define what parameters need to be monitored for its assessment (Vogt et al. 1997a).

Evaluation criteria must determine whether certification reflects the reality of the site and how well they link the goals of ecological, social, economic, and silvicultural sustainability. These criteria can be evaluated by measuring some of the following indicators:

- Volume of wood being certified
- Landowner satisfaction
- Revenue generated from certification
- Analysis of on-ground management practices
- Public profile/credibility
- Social/economic/ecological sustainability reached
- New management systems developed
- Mimicry of pre-existing system based on stand developmental principles that incorporate longer-term temporal scales as success criteria
- Success of assessments based on analyses conducted on realistic time scales for the forest of interest, and not the five-year period currently used
- Restoration of forestlands on previously degraded lands adjusted to reflect the legacy condition of the site

It is important to recognize that the main link of those supporting and promoting certification is striving toward the ideals of sustainability and not necessarily making claims of defining or assessing sustainability. It is important to recognize that "sustainability" is still poorly defined and our ability to measure it is limited. There are several underlying concepts that are important to achieve on a forestland. The following concepts are based on exceptional forest stewardship that demonstrates a commitment to the holistic, long-term conservation of a forest:

- A forest has many different values resulting from the many different perspectives by which forests are perceived.
- Individuals and societies perceive and value forests differently.
- A forest may be seen as a source of aesthetic value, recreational opportunities, economic benefit, and critical and renewable natural resources.
- A forest also holds many inherent values for wildlife, biodiversity, and the maintenance of ecological processes and environmental services, etc. Thus, good forest stewardship should consider the whole forest, and aim to manage the forest holistically.
- Not all forests are created equal.
- Recognize the importance of the long-term conservation of forest resources and values over short-term gains and profitability.

8.5.2 Justifiable principles, criteria, and indicators

In sections 3.2.2, 6.2.3, 7.4, and 7.5, the different values embedded in each protocol were discussed. Since these values also determine the indicators used to assess a site, they raise interesting questions on what is being certified. Are we really certifying good forest management or are we certifying the values of the organizations supporting and promoting forest certification?

Forest certification requires that forest management fulfills and focuses on multiple economic, environmental, and social objectives. Management for multiple objectives is not a recent concept — it has been promoted and practiced since the early 1900s. Gifford Pinchot who established the U.S. Forest Service, translated the concept of "wise use" into the management practices of "multiple use of resources" (i.e., combination of conservation, leisure, grazing, lumber industry, and water-demand interests) (Hays 1959, Winters 1974, Shabecoff 1993, Jordan 1995). The novelty of the multiple management in forest certification is that it includes a wide range of information from many disciplines. Moreover, it is a "trustable" process in the eyes of consumers. Certification is not led by governments, but is a voluntary compliance and a third-party inspection that is applicable to forests throughout the world. In other words, certification acts as a guarantee to the final consumer by linking the product with the specific site and management system it came from. However, these objectives also create problems because of the conflicting values that they introduce into forest certification.

The principles and criteria as presently articulated for by the Forest Stewardship Council (FSC 1996) are an attempt to "manage forests for everything." As Ashton (1996) stated, "One cannot aim to manage forests for everything." To counteract this management approach that tries to incorporate everything will require clear definition of the main objectives of management. It will also require the development of a list of those hoped for but potentially unattainable minor objectives. However, criteria for assessing the forest condition should be based on the main objectives.

The management goals will determine what is to be sustained. This will vary depending on the participants and the sites being evaluated and, therefore, will have an impact on the guidelines being used to certify a specific forest. Mostly, the goals and values embedded in any assessment become the major forces for determining how an assessment is orchestrated and what the requirements are for certification. In one case, loggers attempt to maximize the extraction of timber for a given land base in order to maximize the economic returns. The certifying organizations predominantly attempt to conserve the highest number of species and ecosystem functions possible, while allowing a low-level of timber extraction. These organizations also attempt to provide economic returns for local communities and, therefore, improve local people's well-being. The problem is finding a balance among the four aspects of sustainability mentioned in Section 3.1.2. At present, it appears that if resource extraction is maximized at a sustainable level for commercial species, then the viability of other components of the ecosystems could be jeopardized (De Leo and Levin 1997).

To sustain these values, certification is attempting to broker a balance between what appear to be conflicting values and abilities. However, given the characteristics of the organizations who participated in this process, or possibly due to the political/cultural characteristics of the period when protocols were developed, they are extremely value-based. This suggests the need to conduct an evaluation of the tools and mechanisms being used for forest certification. This necessary exercise must incorporate existing scientific knowledge and have realistic goals about desired end points. For example, one aspect being eagerly pursued — maintenance (or enhancement) of biodiversity — could be an unattainable objective. Ashton (1996) suggested that the conservation of biodiversity and timber harvesting is compatible and the quasi-romantic viewpoint that traditional communities are the ultimate protectors of forest biodiversity has to be faced in terms of their interest and efforts towards useable crops (Wilk 1991, Atran 1993, Orlove 1996, Zimmerer 1996).

8.5.3 Some suggestions on how to improve forest certification

For forest certification to be credible, third-party assessments made by independent organizations are crucial. Also, there is a clear need to have a universal accreditor of certifying organizations. "Given the proliferation of environmental claims and the lack of established standards for evaluating them, various interest groups have long called for the establishment of an accreditor for certifying organizations" (Cabarle et al. 1995). This led to formation of the Forest Stewardship Council in 1994. However, there are still many certifying organizations that are not accredited and the variety of organizations and claims cast a cloud over consumer confidence. A seal from an accreditor is a readily recognizable sign of quality that gives assurance to consumers that the claims have been verified by a trustworthy organization.

Credibility is a key factor in this entire process. Important questions for forest certification organizations are: how much and how often can the public hear about decertification and still believe in this mechanism? How well can a certifier explain that a forest has crossed a threshold of resistance and resilience, despite the assessment ensuring sustainability? There are many more questions forest certification will have to answer to build consensus. Maybe the alternatives will come from existing experiences, as with environmental marketing and third-party certification in the industrial arena. These initiatives might be road maps for improving forest certification. The industrial experience specifically suggests ways to incorporate the principles of selectivity, life cycle assessment, and criteria verification protocols into forest certification programs. The industrial experience also highlights the importance of specific, scientifically supportable criteria and a clear, well-worded message to the consumer.

Another possibility for forest certification is the establishment of more selective targets for their labels. The identification of a selectivity level should be a key item of discussion in the public process used to negotiate the development of new standards. This selectivity could define clearly for the public exactly what is being ensured for each product. One reason for the FSC to consider selectivity levels is to assure itself and its stakeholders that its standards are challenging enough. The standards must be capable of recognizing the leaders in managing forests with all the human values that are considered environmental. In addition, these standards need not be so restrictive that they fail to encourage others to strive for these higher environmental standards. On the other hand, SFI needs to consider selectivity to demonstrate that its standards do indeed raise the bar for the entire industry, and do not merely modify the current status quo. The ISO 14024 Series recommends that selectivity levels be narrow enough to distinguish the labeled products from that which is available in the rest of the market for a particular product category. Although not an eco-seal, SFI should identify a selectivity level (i.e., 80% or lower) that distinguishes the new standard from the status quo.

In setting standards that attain a desired selectivity level, forest certification could follow the example of standard setting used by regulatory agencies in the United States. For example, the California Air Resources Board collects survey information about each company's market share, its environmental performance prior to a new standard, and the estimated cost to comply with a new standard. The survey information is used to estimate the portion of the market that already meets a range of hypothetical standards. This type of analysis also looks at the portion of the market that could meet a range of standard levels by making changes that fall within specified cost parameters. The results of the analysis are used in the discussion with stakeholders about how stringent the proposed standard should be (CARB 1997). If confidentiality concerns could be surmounted, FSC could use this process with its stakeholders, and SFI with AF&PA members.

There are lessons to can be learned from some of the other approaches that have been developed to assess whether fraudulent claims are being made for practices or products to be designated as environmental. The approaches being discussed with ISO are more objective and specific in nature than are the protocols for FSC and SFI forest certification programs. In the case of FSC-accredited organizations, the overarching document of Principles and Criteria generated quite broad and subjective field assessment guidelines. Similar problems exist with SFI. For example, one of their criteria requires participants to "continually apply the knowledge gained through research, science, technology, and field experience for conserving biological diversity" (AF&PA 1995). Many of these issues have been discussed in detail in Chapter 4. A mechanism for determining what information must be used in evaluating the certification potential of a landowner was also introduced in Chapter 5 where there is a discussion of the minimum data needs of any evaluation.

Another way of making forest certification more objective is to specify transparent mechanisms for the evaluation of forest certification, as recommended in the ISO 14024 Series and implemented in many government certification programs (e.g., California Environmental Technology Certification Program and its emission reduction credits). The best examples of industrial certification programs spell out the process by which criteria are verified. The present emphasis on very detailed descriptions and discussions of how to define criteria and indicators fails to determine whether the appropriate standards have been selected for a given site. Forest certification programs need to develop protocols that outline how criteria will be measured in the field, and for quantitative criteria the sampling and testing methodologies should be agreed upon. For criteria of an unavoidably qualitative nature, a detailed evaluation procedure should be written that standardizes the judging process for the field evaluators. Some of the subjectivity inherent in these sorts of criteria could be minimized by asking the evaluators to rank or compare what they observe at the site being assessed, with a set scale of hypothetical observations. By taking steps to increase the conformity of the evaluation process to determine what variables need to assess for a site and how to weight that information, the subjectivity nature of these criteria can be eliminated or made more manageable.

One aspect of forest certification that must be considered by supporters and promoters of this idea is the need for more interaction with governmental initiatives or institutions. Forest certification programs can be faced with a problem known in government jargon as "underground regulation." In short, this danger exists when the agreed-upon intention of a regulation (or other programs developed based on input from a pool of stakeholders) is circumvented because of loosely defined program elements left open to the subjective interpretation of those who implement the program. The SFI is particularly susceptible to this jeopardy because of the close relationship between AF&PA and the companies it certifies. The legitimacy of SFI could be ameliorated if a protocol were developed that "farmed out" the certification process to an outside consultant firm with connections to both the commercial and environmental protection sides of forestry. This is, in fact, what SFI is currently beginning to do.

It is important to keep in mind that these certification protocols are young and continually evolving. It is also important that they begin to address some of the problems identified here (mainly related to the weighting of the information and using the minimum number of data that really drive the processes in that system). Much of the forest certification protocols are caught up in pushing the ecosystem management philosophy, without having an understanding of how to select among the huge amounts of information that can potentially comprise an ecosystem. The demand for incorporating the social and natural sciences in the data needs is another reflection of our understanding that humans are strong drivers controlling the functioning of ecosystems around the world. What is missing is the road map that shows people how to do this. Many ideas are presented in this book for how to facilitate and ensure that sustainability is really being assessed. It is important that the approach taken is not to get rid of forest certification because it has so many problems, but to make it more effective and representative of the values we demand from forestlands. These programs are young and, therefore, are changing as they evolve. The problem is that if it is perceived that they are not truly assessing the quality of management and that fraudulent claims are occurring, then all programs will be discredited. It is important to remember that the values must be maintained in the process of ensuring that we do not change the resilience of systems due to some activity. If these programs are managed properly, they will function as a hedge to deal with the uncertainty of our knowledge base and will not allow activities to occur that are not sustainable.

Forest certification assessments should use the best available knowledge and established techniques to evaluate forest stewardship. Fortunately, many standards of good forest stewardship are already well defined and can be accurately assessed with practical, well-tested techniques. Using the best current knowledge and techniques, foresters can make reasonable judgments about the long-term conservation outlook of forest resources under different management strategies. Although "holistic" or "exceptional" forest management is somewhat subjective, general agreement exists more about what represents exceptional and thoughtful forestry practices than about "sustainability." Existing tools also exist to assess economic viability, based on the owner's objectives and expectations; market trends; and timber and non-timber forest product yields. In addition, proven techniques exist for assessing non-timber forest values, (e.g., water quality, wetland protection, wildlife habitat, etc.). By using existing techniques, a forest can be certified now, rather than waiting indefinitely to develop the theories and tools necessary for evaluating "sustainability."

Certifying organizations have the flexibility to design assessment techniques within the limits of principles and criteria and outside review. Existing approaches that are voluntary and use independent, third-party assessments of forest management which promote forest productivity, forest ecosystems, and socio-economic stability offer the best models for forest certification (Heissenbuttel et al. 1995). Using a team of assessors with expertise in forest ecology, economics, and social sciences, as well as personal experience in the region, has many benefits. Assessors should be allowed to exercise their judgment within established standards and practices, as long as assessments are open to public comment and peer review.

The methods and techniques for making assessments must also be reasonable, cost-effective, and justifiable. Avoiding "sustainability" greatly reduces the need to rely on a wide range of surrogate indicators, and questionable predictions and assessments of large-scale ecological and social processes. Less data is needed and more of the necessary information is likely to be available. Landowners and forest managers will benefit by not having to produce or pay for excessive and often unreasonable data requirements and costs. Finally, certifying organizations and assessments should respond to advances in science and technology, and changes in the dynamic natural and social systems.

Appendix 1: The Forest Principles (1992)

Extracted from official United Nations documentation.

Non-legally binding authoritative statement of principles for a global consensus on the management, conservation and sustainable development of all types of forests

PREAMBLE:

(a) The subject of forests is related to the entire range of environmental and development issues and opportunities, including the right to socio-economic development on a sustainable basis.

(b) The guiding objective of these principles is to contribute to the management, conservation and sustainable development of forests and to provide for their multiple and complementary functions and uses.

(c) Forestry issues and opportunities should be examined in a holistic and balanced manner within the overall context of environment and development, taking into consideration the multiple functions and uses of forests, including traditional uses, and the likely economic and social stress when these uses are constrained or restricted, as well as the potential for development that sustainable forest management can offer.

(d) These principles reflect a first global consensus on forests. In committing themselves to the prompt implementation of these principles, countries also decide to keep them under assessment for their adequacy with regard to further international cooperation on forest issues.

(e) These principles should apply to all types of forests, both natural and planted, in all geographic regions and climatic zones, including austral, boreal, sub-temperate, temperate, subtropical and tropical.

(f) All types of forests embody complex and unique ecological processes which are the basis for their present and potential capacity to provide resources to satisfy human needs as well as environmental values, and as such their sound management and conservation is of concern to the Governments of the countries to which they belong and are of value to local communities and to the environment as a whole.

(g) Forests are essential to economic development and the maintenance of all forms of life.

(h) Recognizing that the responsibility for forest management, conservation and sustainable development is in many States allocated among federal/national, state/provincial and local levels of government, each State, in accordance with its constitution and/or national legislation, should pursue these principles at the appropriate level of government.

PRINCIPLES/ELEMENTS:

1. (a) "States have, in accordance with the Charter of the United Nations and the principles of international law, the sovereign right to exploit their own resources pursuant to their own environmental policies and have the responsibility to ensure that activities within their jurisdiction or control do not cause damage to the environment of other States or of areas beyond the limits of national jurisdiction."

   (b) The agreed full incremental cost of achieving benefits associated with forest conservation and sustainable development requires increased international cooperation and should be equitably shared by the international community.

2. (a) States have the sovereign and inalienable right to utilize, manage and develop their forests in accordance with their development needs and level of socio-economic development and on the basis of national policies consistent with sustainable development and legislation, including the conversion of such areas for other uses within the overall socio-economic development plan and based on rational land-use policies.

   (b) Forest resources and forestlands should be sustainably managed to meet the social, economic, ecological, cultural and spiritual human needs of present and future generations. These needs are for forest products and services, such as wood and wood products, water, food, fodder, medicine, fuel, shelter, employment, recreation, habitats for wildlife, landscape diversity, carbon sinks and reservoirs, and for other forest products. Appropriate measures should be taken to protect forests against harmful effects of pollution, including air-borne pollution, fires, pests and diseases in order to maintain their full multiple value.

   (c) The provision of timely, reliable and accurate information on forests and forest ecosystems is essential for public understanding and informed decision-making and should be ensured.

   (d) Governments should promote and provide opportunities for the participation of interested parties, including local communities and indigenous people, industries, labour, nongovernmental organizations and individuals, forest dwellers and women, in the development, implementation and planning of national forest policies.

3. (a) National policies and strategies should provide a framework for increased efforts, including the development and strengthening of institutions and programmes for the management, conservation and sustainable development of forests and forestlands.

   (b) International institutional arrangements, building on those organizations and mechanisms already in existence, as appropriate, should facilitate international cooperation in the field of forests.

   (c) All aspects of environmental protection and social and economic development as they relate to forests and forestlands should be integrated and comprehensive.

4.   The vital role of all types of forests in maintaining the ecological processes and balance at the local, national, regional and global levels through, inter alia, their role in protecting fragile ecosystems, watersheds and freshwater resources and as rich storehouses of biodiversity and biological resources and sources of genetic material for biotechnology products, as well as photosynthesis, should be recognized.

5. (a) National forest policies should recognize and duly support the identity, culture and the rights of indigenous people, their communities and other communities and forest dwellers. Appropriate conditions should be promoted for these groups to enable them to have an economic stake in forest use, perform economic activities, and achieve and maintain cultural identity and social organization, as well as adequate levels of livelihood and well-being, through, inter alia, those land tenure arrangements which serve as incentives for the sustainable management of forests.

   (b) The full participation of women in all aspects of the management, conservation and sustainable development of forests should be actively promoted.

6. (a) All types of forests play an important role in meeting energy requirements through the provision of a renewable source of bio-energy, particularly in developing countries, and the demands for fuelwood for household and industrial needs should be met through sustainable forest management, afforestation and reforestation. To this end, the potential contribution of plantations of both indigenous and introduced species for the provision of both fuel and industrial wood should be recognized.

   (b) National policies and programmes should take into account the relationship, where it exists, between the conservation, management and sustainable development of forests and all aspects related to the production, consumption, recycling and/or final disposal of forest products.

   (c) Decisions taken on the management, conservation and sustainable development of forest resources should benefit, to the extent practicable, from a comprehensive assessment of economic and non-economic values of forest goods and services and of the environmental costs and benefits. The development and improvement of methodologies for such evaluations should be promoted.

   (d) The role of planted forests and permanent agricultural crops as sustainable and environmentally sound sources of renewable energy and industrial raw material should be recognized, enhanced and promoted. Their contribution to the maintenance of ecological processes, to offsetting pressure on primary/old-growth forest and to providing regional employment and development with the adequate involvement of local inhabitants should be recognized and enhanced.

   (e) Natural forests also constitute a source of goods and services, and their conservation, sustainable management and use should be promoted.

7. (a) Efforts should be made to promote a supportive international economic climate conducive to sustained and environmentally sound development of forests in all countries, which include, inter alia, the promotion of sustainable patterns of production and consumption, the eradication of poverty and the promotion of food security.

   (b) Specific financial resources should be provided to developing countries with significant forest areas, which establish programmes for the conservation of forests including protected natural forest areas. These resources should be directed notably to economic sectors, which would stimulate economic and social substitution activities.

8. (a) Efforts should be undertaken towards the greening of the world. All countries, notably developed countries, should take positive and transparent action towards reforestation, afforestation and forest conservation, as appropriate.

(b) Efforts to maintain and increase forest cover and forest productivity should be undertaken in ecologically, economically and socially sound ways through the rehabilitation, reforestation and re-establishment of trees and forests on unproductive, degraded and deforested lands, as well as through the management of existing forest resources.

(c) The implementation of national policies and programmes aimed at forest management, conservation and sustainable development, particularly in developing countries, should be supported by international financial and technical cooperation, including through the private sector, where appropriate.

(d) Sustainable forest management and use should be carried out in accordance with national development policies and priorities and on the basis of environmentally sound national guidelines. In the formulation of such guidelines, account should be taken, as appropriate and if applicable, of relevant internationally agreed methodologies and criteria.

(e) Forest management should be integrated with management of adjacent areas so as to maintain ecological balance and sustainable productivity.

(f) National policies and/or legislation aimed at management, conservation and sustainable development of forests should include the protection of ecologically viable representative or unique examples of forests, including primary/old-growth forests, cultural, spiritual, historical, religious and other unique and valued forests of national importance.

(g) Access to biological resources, including genetic material, shall be with due regard to the sovereign rights of the countries where the forests are located and to the sharing on mutually agreed terms of technology and profits from biotechnology products that are derived from these resources.

(h) National policies should ensure that environmental impact assessments should be carried out where actions are likely to have significant adverse impacts on important forest resources, and where such actions are subject to a decision of a competent national authority.

9. (a) The efforts of developing countries to strengthen the management, conservation and sustainable development of their forest resources should be supported by the international community, taking into account the importance of redressing external indebtedness, particularly where aggravated by the net transfer of resources to developed countries, as well as the problem of achieving at least the replacement value of forests through improved market access for forest products, especially processed products. In this respect, special attention should also be given to the countries undergoing the process of transition to market economies.

(b) The problems that hinder efforts to attain the conservation and sustainable use of forest resources and that stem from the lack of alternative options available to local communities, in particular the urban poor and poor rural populations who are economically and socially dependent on forests and forest resources, should be addressed by Governments and the international community.

(c) National policy formulation with respect to all types of forests should take account of the pressures and demands imposed on forest ecosystems and resources from influencing factors outside the forest sector, and intersectoral means of dealing with these pressures and demands should be sought.

10.   New and additional financial resources should be provided to developing countries to enable them to sustainably manage, conserve and develop their forest resources, including through afforestation, reforestation and combating deforestation and forest and land degradation.

11.   In order to enable, in particular, developing countries to enhance their endogenous capacity and to better manage, conserve and develop their forest resources, the access to and transfer of environmentally sound technologies and corresponding know-how on favorable terms, including on concessional and preferential terms, as mutually agreed, in accordance with the relevant provisions of Agenda 21, should be promoted, facilitated and financed, as appropriate.

12. (a) Scientific research, forest inventories and assessments carried out by national institutions which take into account, where relevant, biological, physical, social and economic variables, as well as technological development and its application in the field of sustainable forest management, conservation and development, should be strengthened through effective modalities, including international cooperation. In this context, attention should also be given to research and development of sustainably harvested non-wood products.

   (b) National and, where appropriate, regional and international institutional capabilities in education, training, science, technology, economics, anthropology and social aspects of forests and forest management are essential to the conservation and sustainable development of forests and should be strengthened.

   (c) International exchange of information on the results of forest and forest management research and development should be enhanced and broadened, as appropriate, making full use of education and training institutions, including those in the private sector.

   (d) Appropriate indigenous capacity and local knowledge regarding the conservation and sustainable development of forests should, through institutional and financial support, and in collaboration with the people in local communities concerned, be recognized, respected, recorded, developed and, as appropriate, introduced in the implementation of programmes. Benefits arising from the utilization of indigenous knowledge should therefore be equitably shared with such people.

13. (a) Trade in forest products should be based on non-discriminatory and multilaterally agreed rules and procedures consistent with international trade law and practices. In this context, open and free international trade in forest products should be facilitated.

   (b) Reduction or removal of tariff barriers and impediments to the provision of better market access and better prices for higher value-added forest products and their local processing should be encouraged to enable producer countries to better conserve and manage their renewable forest resources.

   (c) Incorporation of environmental costs and benefits into market forces and mechanisms, in order to achieve forest conservation and sustainable development, should be encouraged both domestically and internationally.

   (d) Forest conservation and sustainable development policies should be integrated with economic, trade and other relevant policies.

   (e) Fiscal, trade, industrial, transportation and other policies and practices that may lead to forest degradation should be avoided. Adequate policies, aimed at management, conservation and sustainable development of forests, including where appropriate, incentives, should be encouraged.

14. Unilateral measures, incompatible with international obligations or agreements, to restrict and/or ban international trade in timber or other forest products should be removed or avoided, in order to attain long-term sustainable forest management.

15. Pollutants, particularly air-borne pollutants, including those responsible for acidic deposition, that are harmful to the health of forest ecosystems at the local, national, regional and global levels should be controlled.

Appendix 2: Summary of Chapter 11 of Agenda 21: Combating Deforestation (1992)

Extracted from official United Nations documentation

Forests are a source of timber, firewood and other goods. They also play an important role in soil and water conservation, maintaining a healthy atmosphere and maintaining biological diversity of plants and animals.

Forests are renewable and, when managed in a way that is compatible with environmental conservation, can produce goods and services to assist in development.

Now, forests worldwide are threatened by uncontrolled degradation and conversion to other uses because of increasing human pressure. There is agricultural expansion, overgrazing, unsustainable logging, inadequate fire control and damage from air pollution. Damage to and loss of forests causes soil erosion, reduces biological diversity and wildlife habitats, degrades watersheds and reduces the amount of fuel-wood, timber and other products available for human development. It also reduces the number of trees that can retain carbon dioxide, a greenhouse gas.

The survival of the forests depends on us recognizing and protecting their ecological, climate-control, social and economic values. These benefits should be included in the national economic accounting systems used to weight development options.

There is an urgent need to conserve and plant forests in developed and developing countries to maintain or restore the ecological balance, and to provide for human needs. National governments need to work with business, nongovernmental organizations, scientists, technologists, local community groups, indigenous people, local governments and the public to create long-term forest conservation and management policies for every forest region and watershed.

Better management will also require more information on the state of forests. In many cases, planners lack even basic information on the size and type of forests and on the amount of wood being harvested.

Governments should create national action programs for sustainable forestry development. This will require a broad range of actions, ranging from the use of satellite images of the forests to better logging equipment to government policies that encourage the most efficient use of the trees and other forest products.

Governments, along with business, nongovernmental and other groups can:

- Plant more forests to reduce pressure on primary and old growth forests. Plant valuable crops among the trees to further increase the value of managed forests.

- Breed trees that are more productive and resistant to environmental stress.

- Protect forests from fires, pests, poaching and mining and reduce pollutants that affect forests, including air pollution that flows across borders. Limit and aim to halt destructive shifting cultivation by addressing the underlying social and ecological causes.

- Use environmentally sound, more efficient and less polluting methods of forest harvesting and expand forest-based processing industries that use wood and other forest products.
- Minimize wood waste and find uses for tree species that are now discarded or ignored. Promote small-scale forest-based enterprises that support rural development and local entrepreneurship.
- Increase the amount of value-added secondary processing of forest products to increase the amount of employment and revenue for each tree harvested.
- Develop urban forestry for the greening of all places where people live.
- Promote the use of such forest products as medicinal plants, dyes, fibers, gums, resins, fodder, rattan, bamboo and works of local artisans.
- Encourage low-impact forest use, such as eco-tourism and the managed supply of genetic materials, such as those used to develop medicines.
- Reduce damage to forests by promoting sustainable management of areas adjacent to the trees.

In order to get more value from their forests, some countries will need international cooperation in the form of advice on modern technologies, and the use of fair terms of trade, without unilateral restrictions and bans on forest products.

In addition to encouraging sustainable use of forests, countries need to create or expand protected-area systems to preserve some forests. Such forests are needed to preserve ecological systems, biological diversity, landscapes, and wildlife habitat. Forests also need to be preserved for their social and spiritual values, including that of traditional habitats of indigenous people, forest dwellers and local communities.

Appendix 3: Some Intergovernmental Initiatives on Forest Issues after UNCED

Extracted from official United Nations documentation.

The Rome Statement on Forestry. On March 1995, the FAO called upon the first ministerial-level meeting to take place under the invitation of its Director-General. For two days, in Rome, 120 Forestry Ministers (or forestry-related governmental officials), gathered to discuss "the importance of forests to sustainable development at local, national, and international levels as well as the pursuit of an integrated and balanced approach between the environmental and developmental functions of forests" (Backiel, 1995). The final product of this meeting — The Rome Statement on Forestry — was a statement of principles and actions, submitted to the United Nations Committee on Sustainable Development during its 3rd meeting and to be implemented nationally and internationally. Backiel (1995) summarizes the scope of the proposed actions as follows:

- The need to consider other aspects in the development of approaches for forestry that address underlying economic and social issues, for example, poverty, agriculture, population, and education;
- The development and application of criteria and indicators to measure progress toward sustainable forest management;
- Enhance international cooperation;
- Promote non-discriminatory trade;
- Recognize the greater benefits of increased public participation and transparency in decision-making;
- Study and assess the role of certification of forest products and its effects on trade and consumption;
- Recommend that FAO advise and cooperate with involved countries to develop individual countries' approaches to sustainable capacity; and
- Agreement to move forward in a step-by-step process to build consensus on international forest issues even though there was no agreement to pursue a legally binding document.

The Helsinki Process. A pan-European discussion on the protection of forests happened prior to UNCED during the First Ministerial Conference that took place in Strasbourg in December 1990, when six resolutions were drawn: 1) European Network of Permanent Sample Plots for Monitoring of Forest Ecosystems; 2) Conservation of Forest Genetic Resources; 3) Decentralized European Data Bank on Forest Fires; 4) Adapting the Management of Mountain Forests to New Environmental Conditions; 5) Expansion of the EUROSILVA Network of Research on Tree Physiology; and 6) European Network for Research into Forest Ecosystems.

In June 1993, the Second Ministerial Conference was convened, this time aiming, more specifically, to respond to the resolutions signed by European countries, about forests, during UNCED, when four resolutions were adopted (Leal 1997): 1) General Guidelines for the Sustainable Management of Forests in Europe; 2) General Guidelines for the Conservation of the Biodiversity of European Forests; 3) Forestry Cooperation with Countries with Economies in Transition; and 4) Strategies for a Long-Term Adaptation of Forests in Europe to Climate Change.

The Helsinki Process involves 37 temperate countries, including Russia. The next step was the development of European Criteria and Indicators for Sustainable Forest Management involving both quantitative and descriptive (qualitative) indicators on issues such as forest cover and their contribution to the carbon cycle, maintenance of the health and vitality of forest ecosystems, balance between wood production and other non-timber forest resources, maintenance and enhancement of biological diversity, maintenance of forests' protective functions (notably soil and water), and maintenance of socio-economic functions and conditions.

These proposed indicators focus mainly on the existence or creation of a political/institutional framework and/or capability at this level to implement a sustainable approach to forest management. Nevertheless, there is wording — at what can be considered a field-implementation level — on data collection to provide quantification on changes in stock and the physical appearance of forests (e.g., diseases and burnt areas). Guidelines at the management-unit level (Management Unit Level Recommendations for Sustainable Forest Management) are being proposed to be adopted on a voluntary basis by forest owners (Crampton and Ozinga, 1997).

The Montreal Process. In September 1993, after the developments of the Helsinki Process, the Conference on Security and Cooperation in Europe (CSCE) sponsored an international seminar in Montreal, Canada on the Sustainable Development of Boreal and Temperate Forests. Both European and non-European CSCE members participated jointly with observers from tropical, temperate, and boreal countries. The International Institute for Environment and Development (IIED) presented a paper commissioned to Duncan Poore (1993) attempting to define the terms "sustainable development" and "sustainable management," as well their inter-relationships. In June 1994, a Working Group on Criteria and Indicators for the Conservation and Sustainable Management of Temperate and Boreal Forests was formed, which became known as the Montreal Process (Canadian Secretariat of the Montreal Process 1997). This initiative involves 12 countries. Ten countries joined the process in the beginning: Australia, Canada, Chile, China, Japan, Republic of Korea, Mexico, New Zealand, USA, and Russia (which being over two continents has joined this effort as well); Argentina and Uruguay were the last two countries to endorse the Santiago Declaration.

The Santiago Declaration, issued in February 1995, ratified a set of seven national-level criteria and 67 indicators for sustainable forest management which, despite the differences in numbers from the Helsinki Process, is considered highly comparable with the latter (Wijewardana et al. 1997). Six of the criteria and indicators relate to the condition, attributes, functions, or benefits of forests and the last one is regarding the overall policy and institutional framework to support sustainable forest management: 1) Conservation of biological diversity; 2) Maintenance of productive capacity of forest ecosystems; 3) Maintenance of forest ecosystems' health and vitality; 4) Conservation and maintenance of soil and water resources; 5) Maintenance of forest contribution to global carbon cycles; 6) Maintenance and enhancement of long-term multiple socio-economic benefits to meet the needs of societies; 7) Legal, institutional, and economic framework for forest conservation and sustainable management.

These indicators are seen as tools for assessing national trends and for describing, monitoring, and evaluating progress towards sustainability, although they are not seen as performance standards or are intended to assess directly sustainability at the forest management level (Canadian Secretariat of the Montreal Process, 1997). The individual countries' interpretations of specific indicators under each criterion will be the tools for guiding sustainable forest management on the ground. However, given the tremendous differences in technical capability, land ownership, population size, and economic development, to mention a few issues, a great effort will be needed for harmonizing the approaches. The ultimate goal is that the Montreal Process conceptual framework of criteria and indicators will reflect an ecosystem-based approach (Canadian Secretariat of the Montreal Process, 1997).

Presently, the member countries are in very different stages of data gathering and development of their own set of indicators. In the Report of Progress of Secretariat of the Montreal Process to the FAO XI World Forestry Congress (Turkey, 13–22 Oct 1997), there was references to only a few countries' status quo: Argentina lacks data for most indicators, but is conducting a country assessment with World Bank funds, relying on large, international companies (mostly North American) to purchase vast tracks of native forests in the country on which to conduct sustainable forest management and dictate directives in these practices. Australia, on the other hand, has data about most indicators, but has to conciliate the interests of three levels of government, plus private forest ownership in order to implement the Montreal Process framework in their temperate, tropical, and plantation forests. In Canada as well, data collection is in good order but also decentralized legislation, policies and regulations are a challenge for the Canadian Council of Forest Ministry (CCFM) responsible for developing the country's criteria and indicators. Chile, despite having forest products as the second most important export commodity, is still struggling to place the sustainable use of forests on the public agenda. Moreover, the means of data collection will have to be improved to match the demands of the Montreal Process. Japan has good control of forest data collection and has established two experimental sites for developing methods on ecosystem management and monitoring indicators. Korea has the monitoring capacity to inform on the productive capacity of forests using remote sensing techniques, but for the other indicators there has been great difficulty in defining their meaning and how data collection can take place; the U.S. is currently finalizing the development of specific indicators under each criterion through the United States Forest Service. Full implementation is expected by 1998.

The Amazonian Tarapoto Proposal. Eight countries encompassing Amazonia — Colombia, Bolivia, Brazil, Guyana (formerly English Guyana), Peru, Surinam (formerly Dutch Guyana) and Venezuela — signed in 1978 an Amazonian Cooperation Treaty (Tratado de Cooperação Amazonica — TCA). French Guyana, still being a colony, did not have the required sovereignty to participate in this treaty. The goals are to promote and coordinate the rational exploitation of natural resources in their respective parts of Amazonia. And as stated by Carazo (1997), given the objective of maintaining equilibrium between economic growth and conservation of the environment, this treaty was a "precursor of the concept of sustainable development." In 1995, in the need to respond to UNCED determinations and in face of the ongoing international movement towards developing criteria and indicators of sustainable forest management appropriate for specific ecosystems' types and/or political realities, the TCA convened a meeting in Peru to draft a set of forest-use guidelines. The Tarapoto Proposal has a total of 12 criteria and 77 indicators grouped as: national level — 7 criteria and 47 indicators; management unit level — 4 criteria and 23 indicators; and services at the global level — 1 criterion and 7 indicators (Carazo 1997).

According to Viana (1995), the author of a subsidiary document for this governmental exercise by the Amazonian countries, the main goal of the proposed criteria and indicators was to be simple, objective, and easy to implement, while reflecting the complexity of the social, environmental, and economic aspects of forest sustainability. The utilitarian purpose of the forest is stresses throughout the document as a means to alleviate poverty, which is seen as one of the biggest causes of Amazonia's devastation. Also, the sovereignty of each of the countries of the region to decide what is best and how it should be done, in terms of natural resources use and/or conservation, was a basic statement (Carazo, 1997).

Presently, national consultation meetings have been conducted in some of the member countries with the participation of all other members (Colombia, Ecuador, and Peru), and the basic proposed set is receiving proposals for modifying, eliminating, or including indicators (Carazo, 1997). The general approach of this Amazonian set is still centered in data generation and gathering such information as forest extent, economic return, level of investments, proportion between protected and non-protected areas, etc. However, this set is one of the few that, in the main document, make suggestions of criteria and indicators at the management-unit level, despite still not dealing with concrete thresholds of sustainability, but rather, promoting the approach of data collection, such as percentage of area under protection (e.g., forests and soils), level of production diversification, adoption — or not — of sound technology, and water protection systems.

The Central American Process of Lepaterique. The Central American Process on the development of guidelines for sustainable forest management has been named Lepaterique in recognition of a Honduran Lenca community whose inhabitants, through the sustainable use of forest resources, are demonstrating in practical terms how to achieve sustainable development (Zapata, 1997). This process in Central America is based on a regional integration agreement to commonly achieve peace, freedom, democracy and development — System of Central American Integration (SICA). In 1990, the Central American Commission for Environment and Development (CCAD) established the Secretariat's subdivision on environmental issues. The other subdivisions are economic, social, and educacional/cultural.

Post-UNCED, the countries in the region signed two agreements: 1) the Conservation of the Biodiversity and Protection of Priority Forest Areas in Central America, 1992; and 2) the Management and Conservation of Natural Forest Ecosystems and the Development of Forest Plantations, 1993. The following year, the presidents of all Central American countries signed the Alliance for Sustainable Development (ALIDES), a political accord whose purpose is to develop an integrated strategy within the framework of Agenda 21, with the governmental programs and activities abiding by seven principles: respect for life; the improvement of the quality of life; respect for, and the utilization of, the vitality and diversity of the land in a sustainable way; the promotion of peace and democracy; respect for a multicultural society and ethnic diversity; economic integration between the countries of the region and of those countries of the rest of the world; and intergenerational responsibility with sustainable development (Zapata, 1997).

Since 1995, the institutions listed above have been working on the development of criteria and indicators for sustainable forest management in Central America. This exercise has been based on the experiences gained with other processes, especially Helsinki, Montreal, and Tarapoto. The proposal presented before the 9th Meeting of the Central American Council for Forests "recommended that, when defining criteria and indicators for the sustainable management of forests, account should be taken not only of technical criteria but of those socio-cultural criteria associated with native communities and their representative organizations" (Zapata, 1997).

The Lepaterique process has seven principles: political responsibility; forest maintenance and health; contribution from forests regarding environmental services; maintenance of biological diversity; forest production; development of science and technology; and forests' role to fulfill present and future social needs. These principles have four criteria and forty indicators at the regional level and eight criteria and fifty-two indicators at the national level. The criteria proposed for all Central American countries are more in line with general political guidelines and include to: the existence of political, legal, institutional, technical, economic and social frameworks to conduct sustainable use of forest resources; the conservation and maintenance of environmental services; the maintenance of the production capacity; and the maintenance and improvement of multiple social, economic, and cultural benefits of forested ecosystems. On the other hand, the criteria being proposed at the national level, similar to the indicators under each of the regional criteria, state the need for quantification of bureaucratic mechanisms; forest extent; health and vitality; the existence — or not — of biological diversity; and the forests' production, social, economic, and cultural functions (Zapata, 1997).

The initiatives of the African Timber Organization (ATO) and for the Dry Zone in Africa. In the aftermath of UNCED, in an effort to respond to the agreements of Agenda 21, the 13 member countries of the African Timber Organization requisitioned two regional studies in 1993–94 to create awareness among local governments about the issue of criteria and indicators for sustainable forest management, and to conduct consultations with major logging companies on the feasibility of implementing field tests in their sites. "… the basic reason for ATO to adopt this initiative was not only because of its linkages to conservation of natural resources but to assist in providing policy guidelines to a market oriented mechanism that would improve the competitiveness of African Timber on the world market in many ways, among which through timber certification" (Garba, 1997).

ATO's approach to the development of criteria and indicators for the humid African tropical forests was to conduct field-intensive tests in cooperation with the Center for International Forestry Research (CIFOR). Six ATO countries have been selected for these experiments and, so far, two tests have been conducted in Côte d'Ivoire (1995) and in Cameroon (1996). Based on the results of these tests, the first draft of ATO's criteria and indicators has been launched with five principles, two sub-principles, 28 criteria, and 60 indicators covering the following aspects (Garba, 1997): 1) Sustainability of the forest and its multiple functions are high political priorities; 2) Areas devoted to forestry activities or the permanent forest state are not declining; 3) Forests are adequately managed and developed irrespective of their role (sustainable timber production [in quantity and quality] is guaranteed, and sustainable production of non-timber forest products is ensured); 4) The main ecological functions of the forest are maintained; 5) The rights and duties of stakeholders should be clearly defined, perceived, and accepted by all.

In those countries and regions that have not yet been involved in the discussion and development of criteria and indicators for sustainable forest management, FAO, in collaboration with other organizations, such as ITTO, UNDP, UNEP, and IUCN, have been taking the lead to promote action. The set of criteria and indicators for the countries of the sub-Saharan dry-zone Africa was a product of a FAO/UNEP Expert Meeting organized in Nairobi, Kenya and held from November 21–24, 1995. The outcome of this meeting was reported to the 10th Session of the African and Wildlife Commission held in South Africa from November 27–December 1, 1995. The FAO/UNEP effort was endorsed, but it was recommended that there be further discussion and improvement and/or adaptation of the criteria and indicators at a national level, and that these two leading UN agencies continue to monitor and support the process (Taal, 1997). The indicators agreed upon at the meeting generic in character and are products of three case studies covering the situation, trends, and needs of forested ecosystems in western, eastern and southern dry-zone Africa. There are a total of seven criteria and 47 indicators with some of them specifically classified as: ecosystem indicators, species indicators, genetic indicators, indicators of economic benefit, and indicators of the distribution of benefits (Taal, 1997): 1) Maintenance and improvement of forest resources, including their contribution to global carbon cycles; 2) conservation and enhancement of biological diversity in forest ecosystems; 3) maintenance of forest ecosystem health, vitality, and integrity; 4) maintenance and enhancement of production functions of forests and other wooded lands; 5) maintenance and improvement of protection functions in forest management; 6) maintenance and enhancement of socio-economic benefits; 7) adequacy of legal, institutional, and policy frameworks for sustainable forest management.

The Near-East Process. This process also was promoted by a FAO/UNEP joint initiative, and the start-up meeting took place in Cairo, Egypt (October 15–17, 1997). Its outcomes, presented at the 12th Session of the Near East Forestry Commission, received the following recommendations: nomination of representatives in each country (national focal points) to further the work, creation of training opportunities for the implementation of criteria and indicators of sustainable forest management through regional seminars, and continuing assistance by FAO on identifying sources of funding for this process (El-Lakany, 1997).

Different from other forested regions, the Near East region has a series of unique characteristics listed by El-Lakany (1997) which are worth mentioning: vegetation types are determined by the combined effects of climate and topography and are largely influenced by demography and land-use systems; aridity is predominant; forested areas are diminishing at a rapid rate; there is rapid and unplanned urban expansion; there are uncontrolled tourist activities; the region is a net importer of wood products; fuelwood and non-wood products are the major uses of forest resources; man-made plantation forests represent the bulk of forest resources; endemic species of wildlife are facing serious threats of extinction; forest areas are in their majority under governmental ownership, but there are communal, corporate, and private properties and/or use of forest resources (e.g., agro-pastoralism and collection of non-wood forest products); and NGOs are increasing their role in the region, pushing for environmental and social awareness, but policymakers and the civil society (both urban and rural) still do not fully reflect the incorporation of such concepts.

Similar, to the Dry-Zone Africa Process, also promoted by FAO/UNEP, the Near East Process classified their indicators as ecosystem indicators, species indicators, genetic indicators, indicators of economic benefit, and indicators of the distribution of benefits, but added a few more categories, such as: external influence indicators, forest vitality indicators, anthropogenic influence indicators of site degradation, and indicators of participation among forest stakeholders. The seven criteria proposed in this process are: extent of forest resources; conservation of biological diversity; health, vitality and integrity; productive capacity and functions; protective and environmental functions; maintenance and development of socio-economic functions and conditions; and the legal and institutional frameworks (El-Lakany, 1997).

References

Abdulhadi, R., Kartawinata, K. and Sukardjo, S., Effects of mechanized logging in the lowland Dipterocarp forest at Lempake, East Kalimantan, *Malaysian Journal of Forestry*, 44, 407, 1981.

Abt Associates, *Status report on the use of environmental labels worldwide*, prepared for the United States Environmental Protection Agency, Office of Pollution Prevention and Toxics, EPA Contract, 1993.

Adamowicz, W. L., Boxall, P. C., Luckert, M. K., Philips, W. E. and White, W. A., *Forestry, economics and the environment*, CAB International, Wallingford, U.K., 1996.

Adams, P. W., Flint, A. L. and Fredriksen, R. L., Long-term patterns in soil moisture and revegetation after a clearcut of a Douglas fir forest in Oregon, *Forest Ecology and Management*, 41, 249, 1991.

AF&PA (American Forest & Paper Association), *Sustainable forestry: principles and implementation guidelines*, American Forest & Paper Association, Washington, D.C., 1995.

Agee, J. K., *Eastside forest ecosystem health assessment*, USDA, Forest Service, Pacific Northwest Research Station, Corvallis, OR, 359, 1993.

Aley, J., Burch, W. R., Conover, B. and Field, D., *Ecosystem management: adaptive strategies for natural resources organizations in the twenty-first century*, Taylor & Francis, Philadelphia, PA, London, U.K., 1998.

Alonso-Martinez, M. and Sanchez-Urbina, M., *Proyecto piloto para la creación de una "reserva extractiva de madera caida" en la Peninsula de Osa, Costa Rica*, Programa de Conservacion y Manejo Forestal. Fundación TUVA, Puerto Jimenez, Osa, Costa Rica, 1994.

Alpizar, E., Watson, V. and Alonso, M., *Proyecto piloto: "reservas extractivas de madera caida" (REMAC)*, Fundación TUVA y Centro Cientifíco Tropical, San Jose, Costa Rica, 1996.

Amaral, P., Verissimo, A., Barreto, P. and Vidal, E., *Floresta para sempre: um manual para produção de madeira na Amazônia*, Imazon, WWF and USAID, Belém, Pará, Brazil, 1998.

Apps, M. T. and Kurz, W. A., Assessing the role of Canadian forests and forest sector activities in the global carbon balance, *World Resource Review*, 3, 333, 1991.

Argow, K. A., This land is their land: the potential and diversity of nonindustrial private forests, *Journal of Forestry*, 94(2), 30, 1996.

Argow, K. A., Interview with Keith Argow, president of the National Woodland Owners Association, 1998.

Armitage, D., An integrative methodological framework for sustainable environmental planning and management, *Environmental Management*, 19(4), 469, 1995.

Arnt, R. A. and Schwartzman, S., *Um artifício orgânico: transição na Amazônia e ambientalismo (1985–1990)*, Rocco, Rio de Janeiro, Brazil, 1992.

Arts, B., *The political influence of global NGOs: case studies on the climate and biodiversity conventions*, International Books, Utrecht, The Netherlands, 1998.

Ashton, P. S., Stand-level concepts and indicators for certification of forest management, *The Conference on Economic, Social, and Political Issues in Certification of Forest Management*, Malaysia, 1996.

Atran, S., Itza Maya tropical agroforestry, *Current Anthropology*, 14(5), 633, 1993.

Backiel, A., forests and sustainable development: the Rome statement on forestry, *Journal of Forestry*, 93(10), 13, 1995.

Baharuddin, H. J. G., Timber certification: An overview, *Unasylva*, 46(183), 1995.

Baker, H. G., Patterns of plant invasion in North America, *Ecology of Biological Invasions of North America and Hawaii*, Mooney, H. A. and Drake, J. A., Springer-Verlag, New York, 44, 1986.

Barnard, J. E., Lucier, A. A., Johnson, A. H., Brooks, R. T., Karnosky, D. F. and Richter, D. D., *Changes in forest health and productivity in the United States and Canada*, National Atmospheric Precipitation Program, Washington, D.C., NAPAP State of Science and Technology Report, 1990.

Barrett, R., *Quality manager's complete guide to ISO 9000*, Prentice-Hall, Englewood Cliffs, New Jersey, 1993.

Bartelmus, P., *Accounting for sustainable development*, United Nations Department of International Economic and Social Affairs, New York, U.S.A., working paper n° 8, November, 1987.

Barthod, C., Criteria and indicators for sustainable temperate forest management: 1992–1996, *Unasylva*, 49(192), 53, 1998.

Bassman, J. H., Influence of two site preparation treatments on ecophysiology of planted *Picea engelmannii* x *glauca* seedlings, *Canadian Journal of Forest Research*, 19, 1359, 1989.

Bastl, M., Kocár, P., Prach, K. and Pysek, P., The effect of successional age and disturbance on the establishment of alien plants in man-made sites: an experimental approach, *Plant invasions: studies from North America and Europe*, Brock, J. H., Wade, M., Pysek, P. and Green, D., Backhuys Publishers, Leiden, 191, 1997.

Batzli, G. O., White, R. G., S. F. Maclean, J., Pitelka, F. A. and Collier, B. D., The herbivore-based trophic system, *An Arctic ecosystem: the coastal tundra at Barrow*, Brown, J., Miller, P. C., Tieszen, L. L. and Bunnel, F. L., Hutchinson & Ross, 335, 1980.

Bazzaz, F. A., Life history of colonizing plants: Some demographic, genetic, and physiological features, *Ecology of biological invasions of North America and Hawaii*, Mooney, H. A. and Drake, J. A., Springer-Verlag, New York, 96, 1986.

Bazzaz, F. A. and Parrish, J. A. D., Organization of grassland communities, *Grasses and grasslands: systematics and ecology*, Estes, J. R., Tyrl, R. J. and Brunken, J. N., University of Oklahoma Press, Norman, Oklahoma, 233, 1982.

Beanlands, G., Cumulative effects and sustainable development, *Defining and measuring sustainability: the biogeophysical foundations*, Munasinghe, M. and Shearer, W., World Bank, Washington, D.C., 77, 1995.

Beek, R. a. d., CATIE's contribution to sustainable forest management in the humid tropical forests of Central America, *Forest codes of practice*, Dykstra, D. P. and Heinrich, R., FAO, Rome, 15, 1996.

Belsky, M. H., Implementing the ecosystem management approach: Optimism or fantasy?, *Ecosystem Health*, 1(4), 214, 1995.

Benchmark Environmental Consulting, *ISO 14000: An uncommon perspective — Five questions for proponents of ISO 14000 Series*, The European Environmental Bureau, October, 1995.

Bengtsson, J., Jones, H. and Setala, H., The value of biodiversity, *Trends in Ecology and Evolution (TREE)*, 12(9), 334, 1997.

Berish, C. W. and Ewel, J. J., Root development in simple and complex tropical successional ecosystems, *Plant Soil*, 106, 73, 1988.

Berry, J., Brewer, G. D., Gordon, J. C. and Patton, D. R., Closing the gap between ecosystem management and ecosystem research, *Policy Sciences*, 31(1), 55, 1998.

Billing, W. D., *Plants and Ecosystems*, Wadsworth Publishing Co., Belmont, 1978.

Binkley, D., Does forest removal increase rates of decomposition and nitrogen release?, *Forest Ecology and Management*, 8, 229, 1984.

Binkley, D., Richter, D., David, M. B. and Caldwell, B., Soil chemistry in a loblolly longleaf pine forest with interval burning, *Ecological Applications*, 2(2), 157, 1992.

Birch, T. W., *The private forestland owners of the United States: preliminary findings. Report to the USDA Forest Service*, State and Private Forestry, Northeast Forest Experiment Station, Radnor, PA, 1995.

Blank, R. R. and Young, J. A., *Lepidium latifolium*: Influences on soil properties, rate of spread, and competitive stature, *Plant Invasions: studies from North America and Europe*, Brock, J. H., Wade, M., Pysek, P. and Green, D., Backhuys Publishers, Leiden, The Netherlands, 69, 1997.

Bliss, J. C., *Alabama's non-industrial private forest owners: snapshots from a family album*, Auburn University, Alabama Cooperative Extension Service Center, 1993.

Blossey, B. and Notzold, R., Evolution of increased competitive ability in invasive non-indigenous plants: a hypothesis, *Journal of Ecology*, 83, 887, 1995.

Boerner, R. E. J., DeMars, B. G. and Leicht, P. N., Spatial patterns of mycorrhizal ineffectiveness of soils along a successional chronosequence, *Mycorrhiza*, 6(2), 79, 1996.

Bond, W. J., Keystone species, *Biodiversity and ecosystem function*, Schulze, E. D. and Mooney, H. A., Springer-Verlag, Berlin, 237, 1993.

Borgström, B. E., The way the world is going: the society-nature dichotomy in development rhetorics, *Nature and society in historical context*, Teich, M., Porter, R. and Gustafsson, B., Cambridge University Press, Cambridge, UK; New York; Melbourne, Australia, 332, 1997.

Bormann, B. T., Brookes, M. H., Ford, E. D., Kiester, A. R., Oliver, C. D. and Weigand, J. F., A broad, strategic framework for sustainable ecosystem management, *Eastside forest ecosystem health assessment*, USDA, Forest Service, 62, 1993.

Bormann, F. H. and Likens, G. E., *Pattern and Process in a Forested Ecosystem*, Springer-Verlag, New York, 253, 1979.

Bormann, F. H., Likens, G. E., Fisher, D. W. and Pierce, R. S., Nutrient loss accelerated by clearcutting of a forest ecosystem, *Science*, 159(3817), 882, 1968.

Bowden, R. D., Biodiversity and ecosystem function: using natural attributes of islands, *Islands: biological diversity and ecosystem function*, Vitousek, P. M., Loope, L. L. and Aderson, H., Springer-Verlag, Berlin, 221, 1995.

Bowen, B., Trading fairly, *The Courier*, http://www.oneworld.org/euforic/courier/166e_bow.htm, ACP-EU. #166 (Nov.-Dec.), 72, 1997.

Bradbury, D., "Green" forest products gain marketing niche, but environmentally certified lumber still has a long way to go with the average consumer, *Portland Press Herald*, Portland, ME, 1B, 1997.

Brady, N. C., *The nature and properties of soils*, 10th Ed., Macmillan Publishing, New York, 621, 1990.

Brady, N. C. and Weil, R. R., *The nature and properties of soils*, 11th Edition, Prentice Hall, Upper Saddle River, New Jersey, 740 pages, 1996.

Brais, S., Camire, C. and Pare, D., Impacts of whole-tree harvesting and winter wind-throwing on soil pH and Base status of clayey sites of northwestern Quebec, *Canadian Journal of Forest Research-Revue Canadienne de Recherche Forestiere*, 25(6), 997, 1995.

Bray, D. B., Carreon, M., Merino, L. and Santos, V., On the road to sustainable forestry: the Mayan of Quintana Roo are striving to combine economic efficiency, ecological sustainability and a democratic society, *Culture Survival Quarterly*, Spring, 38, 1993.

Brewer, G. D., The challenges of interdisciplinary research, *The theory and practice of interdisciplinary work*, Bergendal, Sweden, 1998.

Brockmann, K. L., Hemmelskamp, J. and Hohmeyer, O., *Certified tropical timber and consumer behavior: the impact of a certification scheme for tropical timber from sustainable forest management*, Physica-Verlag, ZEW, Heidelberg, Mannheim, Germany, 178 pages, 1996.

Brothers, T. S. and Spingarn, A., Forest fragmentation and alien plant invasion of central Indiana, *Conservation Biology*, 6(1), 91, 1992.

Brown, J. H., Patterns, modes and extents of invasions by invertebrates, *Biological invasions: a global perspective*, Drake, J. A., Mooney, H. A., Castri, F. d., Groves, R. H., Kruger, F. J., Rejmanek, M. and Williamson, M., Wiley & Sons, New York, 1989.

Bruce, R. A., The comparison of the FSC forest certification and ISO environmental management schemes and their impact on small retail business, *Management School*, University of Edinburgh, Scotland, Edinburgh, dissertation presented for the Master of Business Administration degree, 44 pages, 1998.

Brunson, M. and Shelby, B., Assessing recreational and scenic quality: how does new forestry rate, *Journal of Forestry*, 90(7), 37, 1992.

Brunson, M. W., Human dimensions in silviculture, *Natural resource management: the human dimension*, Ewert, A. W., Westview Press, Boulder, CO, 1996.

Brynn, D., Statement representing the Vermont Family Forests, Addison County Forester, VT, Dept. of Forests and Parks, 1998.

Bueren, E. M. L. v. and Blom, E. M., *Hierarchical framework for the formulation of sustainable forest management standards*, Tropenbos Foundation, 1997.

Burch, W. R., Jr. and Deluca, D. R., *Measuring the Social Impact of Natural Resource Policies*, University of New Mexico Press, Albuquerque, NM, 1984.

Burks, D. C., *Place of the Wild*, Island Press/Shearwater Books, Washington, D.C.; Covelo, CA, 1994.

Cabarle, B., The timber trade and the search for "good wood," *Timber certification: implications for tropical forest management*, O'Hara, J., Endara, M., Wong, T., Hopkins, C. and Maykish, P., Yale School of Forestry and Environmental Studies, New Haven, CT, U.S.A., 5, 5–6 February, 1994.

Cabarle, B. and Heiner, H., The role of nongovernmental organizations in forestry, *Journal of Forestry*, 92(6), 8, 1994.

Cabarle, B., Hrubes, R. J., Elliot, C. and Synnot, T., Certification accreditation: the need for credible claims, *Journal of Forestry*, 93(4), 12, 1995.

California Air Resources Board, *The California state implementation plan for ozone*, The Air Resources Board's Mobile Source and Consumer Products Elements, Sacramento, CA, 1994.

Callicott, J. B., *In defense of the land ethic*, State University of New York Press, Albany, NY, U.S.A., 1989.

Calow, P., Can ecosystems be healthy? Critical considerations of concepts, *Journal of Aquatic Ecosystem Health*, 1, 1, 1992.

Camara, L., *Certification, small landowners and comparison of a Mayan community forestry project in Mexico*, Yale University School of Environmental Studies, New Haven, CT, Unpublished paper for FES 508b: Sustainable Forest Management and Certification: Field Testing Current Approaches., 1998.

Canadian Forest Service, *The Montreal Process: Criteria and indicators for the conservation and sustainable management of temperate and boreal forests*, Quebec, Canada, 1995.

Canadian International Development Agency (CIDA), *Forest issues: non-wood forest products*, Hull, Quebec, 1992.

Canadian Secretariat of the Montreal Process, Progress on implementation of the Montreal Process on criteria and indicators for the conservation and sustainable management of temperate and boreal forests, *FAO XI World Forestry Congress*, Antalya, Turkey, 13–22 October, 1997.

Cannon, C. H., Peart, D. R., Leighton, M. and Kartawinata, K., The structure of lowland rain forest after selective logging in West Kalimantan, Indonesia, *Forest Ecology and Management*, 67(1–3), 49, 1994.

Carazo, V. R., Analysis and prospects of the Tarapoto proposal: criteria and indicators for the sustainability of the Amazonian forets, *FAO XI World Forestry Congress*, Antalya, Turkey, 13–22 October, 1997.

Cargill, S. M. and Jefferies, R. L., The effects of grazing by lesser snow geese on the vegetation of a sub-Arctic salt marsh, *Journal of Applied Ecology*, 21(2), 669, 1984a.

Cargill, S. M. and Jefferies, R. L., Nutrient limitation of primary production in a sub-Arctic salt marsh, *Journal of Applied Ecology*, 21(2), 657, 1984b.

Carlquist, S., *Island Biology: A Natural History of the Islands of the World*, Natural History Press, Garden City, 1965.

Carpenter, S. R., *Complex Interactions in Lake Communities*, Springer-Verlag, New York, 1988.

Carpenter, S. R., Frost, T. M., Kitchell, J. F. and Kratz, T. K., Species dynamics and global environmental change: a perspective from ecosystem experiments, *Biotic interactions and global change*, Kareiva, P. M., Kingsolver, J. G. and Huey, R. B., Sinauer Associates, Inc., Sunderland, 1993.

Carreon, M. M., Objetivos y obstaculos principales para ejecutar el manejo sostenible en las selvas de Quintana Roo, *Madera, Chicle, Caza y Milpa: contribuciones al manejo integral de las selvas de Quintana Roo, Mexico*, Snook, L. K. and Barrera de Jorgeson, A., PROAFT, INFAP, USAID, WWF-US, 1994.

Carrere, R. and Lohmann, L., *Pulping the south: industrial tree plantations and the world paper economy*, Zed Books Ltd., London and New Jersey, 1996.

Carson, R., *Silent Spring*, Penguin Books in association with Hamish Hamilton, Harmondsworth, Middlesex, U.K., 1965.

Case, T., Invasion resistance arises in strongly interacting species-rich model competition communities, *Proceedings of the National Academy of Science*, 87, 9610, 1990.

Chapin, F. S. I., Sala, O. E., Burke, I. C., Grime, J. P., Hooper, D. U., Laurenroth, W. K., Lombard, A., Mooney, H. A., Mosier, A. R., Naeem, S., Pacala, S. W., Roy, J., Steffen, W. L. and Tilman, D., Ecosystem consequences of changing biodiversity, *BioScience*, 48(1), 45, 1998.

Chen, J. Q., Franklin, J. F. and Spies, T. A., Vegetation responses to edge environments in old-growth Douglas-fir forests, *Ecological Applications*, 2(4), 387, 1992.

Chen, J. Q., Franklin, J. F. and Spies, T. A., Growing-season microclimatic gradients from clear-cut edges into old-growth Douglas-fir forests, *Ecological Applications*, 5(1), 74, 1995.

Chernov, Y. I., *The Living Tundra*, Cambridge University Press, Cambridge, 1985.

Chew, R. M., Consumers as regulators of ecosystems: an alternative to energetics, *Ohio Journal of Science*, 74, 359, 1974.

Chichilnisky, G., Sustainable development and north-south trade, *Protection of global biodiversity: converging strategies*, Guruswamy, L. D. and McNeely, J. A., Duke University Press, Durham, USA; London, U.K., 33, 1998.

Chilvers, G. A. and Burdon, J. J., Further studies on a native Australian eucalyptus forest invaded by native pines, *Oecologia*, 59, 239, 1983.

Christensen, N. L., Implementing ecosystem management: where do we go from here?, *Ecosystem Management*, Boyce, M. S. and Hanley, A., Yale University Press, New Haven, CT, 1997.

Clark, C. W., Economic biases against sustainable development, *Ecological economics: the science and management of sustainability*, Constanza, R., Columbia University Press, New York, 1991.

Clark, M. E., Rethinking ecological and economic education: A gestalt shift, *Ecological economics: the science of sustainability*, Constanza, R., Columbia University Press, New York, USA; Chichester, West Sussex, U.K., 1991.

Clark, R. N., Stankey, G. H. and Krueger, L. E., From new perspectives to ecosystem management: A social science perspective on forest management, *Ecosystem management: adaptive strategies for natural resources organizations in the twenty-first century*, Aley, J., Burch, W. R., Conover, B. and Field, D., Taylor & Francis, Philadelphia, PA; London, U.K., 1998.

Clark, T. W. and Minta, S. C., *Greater Yellowstone's future*, Homestead Publisher, Moose, WY, 1994.

Clayton, J. L. and Kennedy, D. A., Nutrient losses from timber harvest in the Idaho Batholith, *Soil Science Society of America Journal*, 49(4), 1041, 1985.

Colchester, M. and Lohmann, L., *The Tropical Forestry Action Plan: what progress?*, World Rainforest Movement and The Ecologist, Penang, Malaysia and Dorset, England, 1990.

Colfer, C. J. P., *Who counts most in sustainable forest management?*, Center for International Forest Research (CIFOR), Bogor, Indonesia, Working Paper 7, 16, October, 1995.

Colfer, C. J. P., Prabhu, R. and Wollenberg, E., *Principles, criteria and indicators: applying Ockham's Razor to the people-forest link*, Center for International Forestry Research (CIFOR), Bogor, Indonesia, Working Paper no. 8, 16, October, 1995.

Colfer, C. J. P., Woefel, J., Wadley, R. L. and Harwell, E., *Assessing people's perceptions of forests in Danau Sentarum Wildlife Reserve*, Center for International Forestry Research (CIFOR), Bogor, Indonesia, Working Paper 13, 23, October, 1996.

Conford, P., *The Organic Tradition: An Anthology of Writings on Organic Farming 1900–1950*, Green Books, Devon, England, 1988.

Connell, J. H., Diversity in tropical rain forests and coral reefs: high diversity of trees and corals is maintained only in a non-equilibrium state, *Science*, 199, 1302, 1978.

Constanza, R., Assuring sustainability of ecological economic systems, *Ecological economics: the science and management of sustainability*, Constanza, R., Columbia University Press, New York, 1991.

Constanza, R., Ecological economic issues and considerations in indicator development, selection, and use: toward an operational definition of system health, *Ecological indicators*, McKenzie, D. H., Hyatt, D. E. and McDonald, V. J., Chapman & Hall, 1491, 1992.

Constanza, R., Ecological economics: reintegrating the study of humans and nature, *Ecological Applications*, 6(4), 978, 1996.

Constanza, R., Daly, H. E. and Bartholomew, J. A., Goals, agenda, and policy recommendations for ecological economics, *Ecological economics: the science and management of sustainability*, Constanza, R., Columbia University Press, New York, 1991.

Constanza, R., Norton, B. G. and Haskell, B. D., *Ecosystem Health: New Goals for Environmental Management*, Island Press, 1992.

Covington, W. W., Changes in forest floor organic matter and nutrient content following clear cutting in northern hardwoods, *Ecology*, 62, 41, 1981.

Cowling, R. M., Pierce, S. M. and Moll, E. J., Conservation and utilization of South Coast Renosterveld: an endangered South African vegetation type, *Biological Conservation*, 37, 363, 1986.

Crampton, A. and Ozinga, S., *EU Forest Watch Newsletter*, World Rainforest Movement, Brussels, Belgium, 1997.

Crawley, M. J., What makes a community invasible?, *Colonization, succession, and stability*, Gray, A. J., Crawley, M. J. and Edwards, P. J., Blackwell, Oxford, 1987.

Crawley, M. J., Chance and timing in biological invasions, *Biological invasions: a global perspective*, Drake, J. A., Mooney, H. A., Castri, F. d., Groves, R. H., Kruger, F. J., Rejmánek, M. and Williamson, M., John Wiley & Sons, Chichester, 315, 1989.

Cronk, Q. C. B. and Fuller, J. L., *Plant Invaders*, Chapman & Hall, New York, 1995.

Crossley, R., A review of global forest management certification initiatives: political and institutional aspects, *Conference on Economic, Social and Political Issues in Certification of Forest Management*, Malaysia, 12–16 May, May 12–16, 1996.

Crossley, R., Braga, C. A. P. and Varangis, P. N., *Is there a commercial case for tropical timber certification?*, The World Bank, Washington, D.C., June, 1994.

Crump, A., *Dictionary of environment and development, people, places, ideas, and organizations*, Earthscan Publications, London, 1991.

CSA, *A sustainable forest management system: guidance document*, Canadian Standards Association, October, 1996a.

CSA, *A sustainable forest management system: specifications document*, Canadian Standards Association, October, 1996b.

Cubbage, F. W., Public regulation of private forestry: proactive policy responses, *Journal of Forestry*, 89(12), 31, 1991.

Cubbage, F. W., Regulation of private forest practices: what rights, which policies?, *Journal of Forestry*, 93(6), 14, 1995.

Cubbage, F. W., O'Laughlin, J. and C.S. Bullock III, *Forest resource policy*, Wiley & Sons, New York, 1993.

Cuomo, C. J., *Feminism and Ecological Communities: An Ethic of Flourishing*, Routledge, London and New York, 1998.

Cushman, J. H., Ecosystem-level consequences of species additions and deletions on islands, *Islands: biological diversity and ecosystem function*, Vitousek, P. M., Loope, L. L. and Aderson, H., Springer-Verlag, Berlin, 135, 1995.

D'Antonio, C. M. and Dudley, T. L., Biological invasions as agents of change on islands versus mainlands, *Ecological Studies*, Vitousek, Springer-Verlag, Berlin, 1995.

Dahl, A. L., *The Eco-Principle: Ecology and Economics in Symbiosis*, George Ronald/Zed Books Ltd., Oxford, London, New Jersey, 1996.

Daly, H. E., *Economics, Ecology, Ethics: Essays Toward a Steady-State Economy*, W. H. Freeman and Company, San Francisco, 1980.

Dana, S. T. and Fairfax, S. K., *Forest and Range Policy*, McGraw-Hill Publishing Co., New York, 1980.

Davis, M. B., Holocene vegetational history of the eastern United States, *Late Quaternary Environments of the United States, Vol. 2*, Wright, H. E. J., University of Minnesota Press, Minneapolis, 166, 1983.

de Callejon, D. P., Lent, T., Skelly, M. and Crossley, R., Marketing products from sustainably managed forests: An emerging opportunity, *The business of sustainable forestry: case studies -- a project of the Sustainable Forestry Working Group*, The John D. and Catherine T. MacArthur Foundation, 3, 1998.

De Leo, G. A. and Levin, S., The multifaceted aspects of ecosystem integrity, http://www.consec-ol.org/vol1/iss1/art3, 1(1:3), 1997.

DeAngelis, D. L., Stability and connectance in food web models, *Ecology*, 56, 238, 1975.

Decker, D., Brown, T. L. and Knuth, B. A., Human dimensions research: its importance in natural resource management, *Natural resource management: the human dimension*, Ewert, A. W., Westview Press, United Kingdom, 29, 1996.

DeFerrari, C. M. and Naiman, R. J., A multi-scale assessment of the occurrence of exotic plants on the Olympic Peninsula, Washington, *Journal of Vegetation Science*, 5, 247, 1994.

Denslow, J. S., Patterns of plant species diversity during succession under different disturbance regimes, *Oecologia*, 46, 18, 1980.

di Castri, F., On invading species and invaded ecosystems: the interplay of historical chance and biological necessity, *Biological Invasions in Europe and the Mediterranean Basin*, di Castri, F., Hansen, A. J. and Debussche, M., Kluwer Academic Publishers, Dordrecht, 3, 1990.

di Castri, F., Hansen, A. J. and Debussche, M., *Biological Invasions in Europe and the Mediterranean Basin*, Kluwer, Dordrecht, 1990.

Diener, B. J., Strategic decisions at Portico S.A. — 1982–1997., *The business of sustainable forestry: Case Studies — a project of the Sustainable Forestry Working Group*, The John D. and Catherine T. MacArthur Foundation, Chicago, IL, 12, 1998.

Doak, D. F., Bigger, D., Harding, E. K., Marvier, M. A., O'Malley, R. E. and Thomson, D., The statistical inevitability of stability-diversity relationships in community ecology, *American Naturalist*, 151(3), 264, 1998.

Dobson, A., *Green Political Thought: An Introduction*, Unwin Hyman Ltd., London, 224, 1990.

Dodd, M., Silvertown, J., McConway, K., Potts, J. and Crawley, M., Community stability: a 60-year record of trends and outbreaks in the ocurrence of species in the Park Grass Experiment, *Ecology*, 83, 277, 1995.

Donovan, R., Lecture at Yale School of Forestry and Environmental Studies: FES 508b, Sustainable Forest Management and Certification: Field Testing Current Approaches, Director Smartwood, 1998b.

Donovan, R. Z., Certification of Federal Lands, FAX, 1998a.

Doolittle, M. L., Future availability of non-industrial private lands in the South, *Non-Industrial Private Forests in the 1990s*, Auburn University Press, 56, 1993.

Doorman, F., *Global Development: Problems, Solutions, Strategy*, International Books, Utrecht, The Netherlands, 1998.

Doran, J. W. and Parkin, T. B., Quantitative indicators of soil quality: A minimum data set, *Methods for Assessing Soil Quality*, Doran, J. W. and Jones, A. J., Soil Society of America, Madison, WI, 25, 1996.

DPCSD (United Nations Department for Policy Coordination and Sustainable Development), *Scientific research, forest assessment and development of criteria and indicators for sustainable forest management*, Commission on Sustainable Development — Ad-Hoc Intergovernmental Panel on Forests, New York, Secretary-General's Category III: Report, Third Session, 19–20 September, 1996.

Drake, J. A., Mooney, H. A., Di Castri, F., Groves, R. H., Kruger, F. J., Rejmánek, M. and Williamson, M., *Biological Invasions: A Global Perspective*, John Wiley & Sons, Chichester, 1989.

Drengson, A. and Taylor, D., *Ecoforestry: The Art and Science of Sustainable Forest Use*, New Society Publishers, Gabriola Island, B.C., Canada, 1997.

Drescher, J., Measuring sustainability. Lecture at Yale School of Forestry and Environmental Studies: FES 508b, Sustainable Forest Management and Certification: Field Testing Current Approaches, Forest ecologist and small woodlot owner, Windhorse Farm, Nova Scotia, 1998.

Driver, B. L., Manning, C. J. and Peterson, G. L., Toward better integration of the social and biophysical components of ecosystems management, *Natural Resource Management: The Human Dimension*, Ewert, A. W., Westview Press, Boulder, CO, 1996.

Dube, Y. C., Strategic planning for the conservation and development of forests: guidelines for improvement, unpublished manuscript, 1996.

During, H. J. and Willems, J. H., The impoverishment of the bryophyte and lichen flora of the Dutch chalk grasslands in the thirty years 1953–1983, *Biological Conservation*, 36, 143, 1986.

Durwael, S., The development of fair trade in the Netherlands, http://www.web.net/fair-trade/fair645.html, 1998.

Dykstra, D. P. and Heinrich, R., *Forest codes of practice*, FAO, Rome, 132 pages, 1996.

Eder, K., The institutionalisation of environmentalism: ecological discourse and the second transformation of the public sphere, *Risk, Environment and Modernity: Towards a New Ecology*, Lash, S., Szerszynski, B. and Wynne, B., Sage Publications, London, Thousand Oaks, New Delhi, 1996.

Edmonds, R. L. and McColl, J. G., Effects of forest management on soil nitrogen in Pinus radiata stands in the Australian Capital Territory, *Forest Ecology and Management*, 29, 199, 1989.

Edwards, N. T. and Ross-Todd, B. M., The effects of stem girdling on biogeochemical cycles within a mixed deciduous forest in eastern Tennessee. I. Soil solution chemistry, soil respiration, litterfall, and root biomass studies, *Oecologia*, 40, 247, 1979.

Ehrlich, P. R., *The Population Bomb*, Ballantine Books, New York, 223, 1968.

Ehrlich, P. R., Attributes of invaders and the invading processes: vertebrates, *Biological Invasions: A Global Perspective*, Drake, J. A., Mooney, H. A., Castri, F. d., Groves, R. H., Kruger, F. J., Rejmánek, M. and Williamson, M., John Wiley & Sons, Chichester, 315, 1989.

El-Lakany, M. H., Criteria and indicators for sustainable forest management in the Near East, *FAO XI World Forestry Congress*, Antalya, Turkey, 13–22 October, 1997.

Ellefson, P. V. and Cheng, A. S., State forest practices programs: regulation of private forestry comes of age, *Journal of Forestry*, 92(5), 34, 1994.

Elliot, W. J., Page-Dumroese, D. and Robichaud, P. R., The effects of forest management on erosion and soil productivity, *Soil Quality and Erosion*, 195, 1999.

Elliott, C., Paradigms of forest conservation, *Unasylva*, 47(187), 3, 1996.

Elliott, C., *WWF Guide to Forest Certification*, World Wildlife Fund -UK, London, U.K., 43, 1997.

Ellis, R. C., Lowry, R. K. and Davies, S. K., The effect of regeneration burning upon the nutrient status of soil in two forest types in southern Tasmania, *Plant and Soil*, 65(2), 171, 1982.

Elton, C. S., *The Ecology of Invasions by Animals and Plants*, Methuen, London, 181, 1958.

Enrenfeld, D. W., The conservation of non-resources, *American Scientist*, 64(6), 648, 1976.

Entry, J. A., Stark, N. M. and Loewenstein, H., Effect of timber harvesting on extractable nutrients in a northern Rocky Mountain forest soil, *Canadian Journal of Forest Research — Revue Canadienne de Recherche Forestiere*, 17(7), 735, 1987.

Environmental Building News, certified wood from the Seven Islands forests, Vol. 3 No. 6, November/December 1994.

Environmental Law Institute, The evolution of national wildlife law, Council on Environmental Quality, Washington, D.C., 1977.

Ervin, J., The consultive process, *Certification of Forest Products: Issues and Perspectives*, Viana, V. M., Island Press, Washington, D.C., 1996.

Ervin, J., Panel Discussion at Yale School of Forestry and Environmental Studies: FES 508b, Sustainable Forest Management and Certification: Field Testing Current Approaches., U.S. contact, Forest Stewardship Council, 1998.

Ervin, J., Elliot, C., Cabarle, B. and Synnott, T., The accreditation process, *Certification of Forest Products: Issues and Perspectives*, Viana, V. M., Ervin, J., Donovan, R. Z., Elliot, C. and Gholz, H., Island Press, Washington, D.C., 1996.

Estes, J. A. and Palmisano, J. F., Sea otters: their role in structuring near-shore communities, *Science*, 185, 1058, 1974.

Evans, B., Technical and scientific elements of forest management certification programs, *Conference on Economic, Social and Political Issues in Certification of Forest Management*, Malaysia, May 12–16, May 12–16, 1996.

Ewel, J., Berish, C., Brown, B., Price, N. and Raich, J., Slash and burn impacts on a Costa Rican wet forest site, *Ecology*, 62(3), 816, 1981.

Ewel, J. J., Designing agricultural ecosystems for the humid tropics, *Ann. Rev. Ecol. Syst.*, 17, 245, 1986.

Ewel, J. J., Mazzarino, M. J. and Berish, C. W., Tropical soil fertility: changes under monocultures and successional communities of different structure, *Ecological Applications*, 1, 289, 1991.

Ewert, A. W., Human dimensions research and natural resource management, *Natural resource management: the human dimension*, Ewert, A. W., Westview Press, United Kingdom, 5, 1996.

Fabos, A., Soil compaction and its effect on growth potential of ponderosa pine stands in central Oregon, *School of Forestry and Environmental Studies*, Yale University, New Haven, Master's Thesis, 1997.

FAO, *Forestry for local community development*, United Nations Food and Agricultural Organization, Rome, Italy, 1978.

FAO, *FAO Yearbook*, United Nations Food and Agriculture Organization, Rome, Italy, 1983.

FAO, *Small-scale forest-based processing enterprises*, United Nations Food and Agricultural Organization, Rome, Italy, 1987a.

FAO, *Restoring the balance: women and forest resources*, United Nations Food and Agriculture Organization, Rome, Italy, 1987b.

FAO, *Women in community forestry*, United Nations Food and Agriculture Organization, Rome, Italy, 1989.

FAO, *Assessing forestry project impacts: issues and strategies*, United Nations Food and Agriculture Organization, Rome, Italy, 1993a.

FAO, *Management and conservation of closed forests in tropical America*, United Nations Food and Agriculture Organization, Rome, Italy, 1993b.

FAO, *Forest development and policy dilemmas*, United Nations Food and Agriculture Organization, Rome, Italy, 1994.

FAO, *State of the world's forests*, United Nations Food and Agriculture Organization, Rome, Italy, 1995a.

FAO, *Harmonization of criteria and indicators for sustainable forest management*, United Nations Food and Agriculture Organization, Rome, Italy, 1995b.

FAO, *The state of the world's forests 1997 (SOFO 1997)*, United Nations Food and Agriculture Organization, Rome, 1997.

Farber, S., Local and global incentives for sustainability: failures in economic systems, *Ecological economics: the science and management of sustainability*, Constanza, R., Columbia University Press, New York, 1991.

Farnsworth, N. R., Screening plants for new medicines, *Biodiversity*, Wilson, E. O., National Academy Press, Washington, D.C., 83, 1988.

Federal Trade Commission, *Guides for the use of environmental marketing claims*, publications of revised guidelines, 10 pages, 1996.

Feeny, D., Berkes, F., McCay, B. J. and Acheson, J. M., The tragedy of the commons: twenty-two years later, *Human Ecology*, 18(1), 1, 1990.

Fletcher, R. A., McAlexander, J. and Hansen, E., STORA: The road to certification, *The business of sustainable forestry: case studies — a project of the Sustainable Forestry Working Group*, The John D. and Catherine T. MacArthur Foundation, Chicago, IL, 14, 1998.

Foil, R. R. and Ralston, C. W., The establishment and growth of loblolly pine seedlings on compacted soils, *Soil Science Society of America Proceedings*, 31, 565, 1967.

Forman, R. T. T. and Godron, M., *Landscape ecology*, Wiley & Sons, New York, 1986.

Fownes, J. H., Effects of diversity on productivity: quantitative distribution of traits, *Islands: biological diversity and ecosystem function*, Vitousek, P. M., Loope, L. L. and Aderson, H., Springer-Verlag, Berlin, 177, 1995.

Fox, M. D. and Fox, B. J., The susceptibility of natural communities to invasions, *Ecology of biological invasions*, Groves, R. H. and Burdon, J. J., Cambridge University Press, Cambridge, 57, 1986.

Frank, D. A. and McNaughton, S. J., Stability increases with diveristy in plant communities: empirical evidence from 1988 Yellowstone drought, *Oikos*, 62(3), 360, 1991.

Franklin, J. F., Towards a new forestry, *Journal of American Foresters*, 95, 37, 1989.

Franklin, J. F., A kinder, gentler forestry, *Evergreen*, July 2–4, 1990.

Franklin, J. F., Preserving biodiversity: species, ecosystems, or landscapes, *Ecological Applications*, 3(2), 202, 1993.

Frazer, D. W., McColl, J. G. and Powers, R. F., Soil nitrogen mineralization in a clearcutting chronosequence in a northern California conifer forest, *Soil Science Society of America Journal*, 54, 1145, 1990.

Friedland, A. J., Craig, B. W., Miller, E. K., Herrick, G. T., Siccama, T. G. and Johnson, A. H., Decreasing lead levels in the forest floor of the northeastern U.S.A., *Ambio*, 21, 400, 1992.

Froehlich, H. A., Soil compaction from logging equipment: effects on growth of young ponderosa pine, *Journal of Soil and Water Conservation*, Nov-Dec, 276, 1979.

Froehlich, H. A., Aulerich, D. E. and Curtis, R., *Designing skid trail systems to reduce soil impacts from tractive logging machines*, Oreg. State Univ. Sch. For., For. Res. Lab, Corvalis, OR, Res. pap., 1981.

FSC, *FSC process guidelines for developing regional certification standards*, Forest Stewardship Council, Oaxaca, Mexico, 5 pages, 1994.

FSC, *FSC Statutes*, Forest Stewardship Council, Oaxaca, Mexico, 16 pages, (Ratified September 1994, editorial revision October 1996), 1996.

FSC, *Group certification: guidelines for certification bodies (Draft 1.1)*, Forest Stewardship Council, Oaxaca, Mexico, 10 pages, 7 April, 1998a.

FSC, *Status of FSC-endorsed independent forest certification in the United States*, Forest Stewsrdship Council — U.S.A., http://www.fscoax.org/html/1–2.html., 1998b.

FSC, *Forests certified by FSC-accredited certification bodies*, Forest Stewsrdship Council, Oaxaca, Mexico, 7 pages, 13th November, 1998c.

FSC, *FSC principles and criteria*, Forest Stewardship Council, Oaxaca, Mexico, 8 pages, January, 1999a.

FSC, *FSC policy on percentage-based claims*, Forest Stewardship Council, oaxaca, Mexico, 7 pages, February, 1999b.

Furniss, R. L. and Carolin, V. M., *Western forest insects*, USDA, Forest Service, Miscellaneous Publication, 1977.

Gabus, A., Seeking a sustainable solution, *Timber Trade Journal*, 5th June, 1993.

Gangloff, D., Forests: A new view — new knowledge and an evolving conservation ethic reflect a promising future for our nation's trees., *American Forests*, Autumn, 19, 1998.

Garba, M. L., The initiatives of the African Timber Organization on criteria and indicators for sustainable forest management, *FAO XI World Forestry Congress*, Antalya, Turkey, 13–22 October, 1997.

GFPP (Global Forest Policy Project), *ISO's significant limitations regarding forest certification*, IPF background paper for Programme Element IV, 1996.

Ghazali, B. H. and Simula, M., *Certification schemes for all timber and timber products*, International Tropical Timber Organization (ITTO), Yokohama, Japan, 1994.

Gholz, H. L., Fisher, R. F. and Pritchett, W. L., Nutrient dynamics in slash pine plantation ecosystems, *Ecology*, 66, 647, 1985.

Gilbert, J. M., *Hunting and hunting reserves in medieval Scotland*, John Donald Publishers Ltd., Edinburgh, 1979.

Givinish, T. J., Does diversity beget stability?, *Nature*, 371, 113, 1994.

Glass, R. and Muth, R., The changing role of subsistence in rural Alaska, *Fifty-fourth North American Wildlife and Natural Resources Conference*, Wildlife Management Institute, Washington, D.C., 1989.

Godoy, R. A. and Bawa, K. S., The economic value and sustainable harvest of plants and animals from tropical forest: assumptions, hypothesis, and methods, *Economic Botany*, 47(3), 215, 1993.

Goldsmith, E., *Blueprint for Survival*, Houghton Mifflin, Boston, 189 pages, 1972.

Goldsmith, E., Development as colonialism, *The Ecologist*, 69, 1997.

Golley, F. B., Energy flux in the ecosystem, *Ecosystem structure and function*, Wiens, J. A., Oregon State University, Corvallis, 68, 1971.

Golley, F. B., *A History of the Ecosystem Concept in Ecology*, Yale University Press, New Haven, CT, 1995.

Good Housekeeping Institute, Internet website, http://www.goodhousekeeping.com, 1998.

Goodland, R. J. A., Asibey, E. O. A., Post, J. C. and Dyson, M. B., Tropical moist forest management: the urgency of transition to sustainability, *Ecological economics: the science and management of sustainability*, Constanza, R., Columbia University Press, New York, 1991.

Gordon, J. C., Ecosystem management: idiosyncratic overview, *Defining Sustainable Forestry*, Aplet, G. H., Island Press, Washington, D.C., 240, 1994.

Gordon, J. C., The principles of sustainable forest management: order or chaos?, *International Conference and Labelling of Products from Sustainably Managed Forests*, 21, 1996.

Gosz, J. R., Ecotone hierarchies, *Ecological Applications*, 3(3), 369, 1993.

Gotelli, N. J., *A Primer for Ecology*, Sinauer Associates, Inc., Sunderland, 1995.

Gottlieb, R., *Forcing the Spring: The Transformation of the American Environmental Movement*, Island Press, Washington, D.C.; Covelo, California, 413 pages, 1993.

Gradwohl, J. and Greenberg, R., *Saving the Tropical Forests*, Island Press, Washington, D.C., 1988.

Granholm, H., Vähänen, T. and Sahlberg, S., *Intergovernmnetal Seminar on Criteria and Indicators for Sustainable Forest Management*, Ministry of Agriculture and Forestry, Helsink, Finland, Background Document, August 19–22, 1996.

Graveland, J. and van der Wal, R., Decline in snail abundance due to soil acidification causes eggshell defects in forest passarines, *Oecologia*, 105, 351, 1996.

Graveland, J., van der Wal, R., van Balen, J. H. and van Noordwijk, A. J., Poor reproduction in forest passerines from decline of snail abundance on acidified soils, *Nature*, 368, 446, 1994.

Green, D. M. and Baker, M. G., Species composition along a gradient of urbanization in the lower Sonoran Desert, Arizona, U.S.A., *Plant Invasions: Studies from North America and Europe*, Brock, J. H., Wade, M., Pysek, P. and Green, D., Backhuys Publishers, Leiden, 37, 1997.

Green Seal, Internet Website, http://www.greenseal.org, 1999.

Grocery Manufacturers of America, GMA calls on U.S. Government and WTO to advance trade: innovative consumer benefits in environmental decision making, Press Release, 1996.

Grove, J. M. and William R. Burch JR., A social ecology approach and implications of urban ecosystem and landscape analyses: a case study of Baltimore, Maryland, *Urban Ecosystems*, 1, 259, 1997.

Grove, R., The island and the history of environmentalism: the case of St. Vincent, *Nature and Society in Historical Context*, Teich, M., Porter, R. and Gustafsson, B., Cambridge University Press, Cambridge, UK; New York; Melbourne, Australia, 148, 1997.

Grumbine, E. R., What is ecosystem management, *Conservation Biology*, 8(1), 27, 1994.

Guariguata, M. R. and Dupuy, J. M., Forest regeneration in abandoned logging roads in lowland Costa Rica, *Biotropica*, 29(1), 15, 1997.

Guariguata, M. R., Rheingaus, R. and Montagnini, F., Early woody invasion under tree plantations in Costa Rica: implications for forest restoration, *Restoration Ecology*, 3(4), 252, 1995.

Gunderson, L. H., Holling, C. S. and Light, S. S., *Barriers and bridges to the renewal of ecosystems and institutions*, Columbia University Press, New York, 593 pages, 1995.

Gupta, M., Singh, R. and Jha, M., Soil chemical properties of silver fir and spruce forest under different systems of silviculture: changes in soil nitrogen, *Indian Forester*, 4, 235, 1989a.

Gupta, M. K., Singh, R. P. and Jha, M. N., Soil chemical properties of silver fir and spruce forest under different systems of silviculture: changes in soil phosphorus, *Indian Forester*, 115(11), 802, 1989b.

Guruswamy, L. D. and McNeely, J. A., *Protection of global biodiversity: converging strategies*, Duke University Press, Durham, USA; London, U.K., 425 pages, 1998.

Hahn-Schilling, B., Heuveldop, J. and Palmer, J., *A comparative study of evaluation systems for sustainable forest management (including principles, criteria, and indicators)*, Cartagena, Colombia, Supplement to Consultants Report Commissioned to Respond to Decision 6 (XV) of the ITTC, 1994.

Hammond, A., Adriaanse, A., Rodenburg, E., Bryant, D. and Woodward, R., *Environmental indicators: a systematic approach to measuring and reporting on environmental policy performance in the context of sustainable development*, World Resources Institute, Washington, D.C., 43 pages, May, 1995.

Hammond, H., *Seeing the Forest Among the Trees*, Polestar Press Ltd., Vancouver, B.C., Canada, 1992.

Hansen, E., Forest certification and its role in marketing strategy, *Forest Products Journal*, 47(3), 16, 1997.

Hanski, I. and Gilpin, M., Metapopulation dynamics: brief history and conceptual domain, *Biological Journal of the Linnean Society*, 42, 3, 1991.

Hardin, G., Paramount positions in ecological economics, *Ecological Economics: The Science and Management of Sustainability*, Constanza, R., Columbia University Press, New York, 1991.

Harris, L. D., *The fragmented forest: Island Biogeography Theory and the preservation of biotic diversity*, The University of Chicago Press, Chicago, 1984.

Haskell, B. D., Norton, B. G. and Constanza, R., What is ecosystem health and why should we worry about it?, *Ecosystem Health: New Goals for Environmental Management*, Constanza, R., Norton, B. G. and Haskell, B. D., Island Press, 1992.

Hauselmann, P., *ISO inside out*, WWF International, Discussion Paper, January, 1997.

Haynes, R. W., *Analysis of the timber situation in the United States: 1989–2040*, USDA Forest Service — Rocky Mountain Forest and Range Experimental Station, Fort Collins, Colorado, General Technical Report, 1990.

Hays, S. P., *Conservation and the Gospel of Efficiency: The Progressive Conservation Movement, 1890–1920*, 1979 Edition; original by Harvard University Press, Atheneum, New York, 1959.

Hayward, J. and Vertinsky, I., High expectations, unexpected benefits: what managers and owners think of certification, *Journal of Forestry*, 97 (February, Number 2), 13, 1999.

Heaton, K., *Perspectives on certification from the Smartwood certification program*, Rainforest Alliance, New York, April, 1994.

Hedin, L. O., Granat, L., Likens, G. E., Buishand, T. A., Galloway, J. N., Butler, T. J. and Rodhe, H., Steep declines in atmospheric base cations in regions of Europe and North America, *Nature*, 367, 351, 1994.

Heissenbuttel, J., Cabarle, B., Cashwell, J., Coulombe, M., Mater, J., Stuart, W., Winterhalter, D. and Hill, L., Forest certification, *Journal of Forestry*, 93(4), 6, 1995.

Henderson, H., *Building a Win-Win World: Life Beyond Global Economic Warfare*, Berret-Koehler, San Francisco, 1996.

Hendricks, B., Transformative possibilities: reinventing the convention on biological diversity, *Protection of global biodiversity: converging strategies*, Guruswamy, L. D. and McNeely, J. A., Duke University Press, Durham, USA; London, U.K., 360, 1998.

Hendrickson, O. Q., Chatarpaul, L. and Burgess, D., Nutrient cycling following whole-tree and conventional harvest in northern mixed forest, *Canadian Journal of Forest Research-Revue Canadienne de Recherche Forestiere*, 19(6), 725, 1989.

Henry, J. D. and Swan, J. M. A., Reconstructing forest history from live and dead plant material: an approach to the study of forest succession in southwest New Hampshire, *Ecology*, 55, 772, 1974.

Herbold, B. and Moyle, P. B., Introduced species and vacant niches, *The American Naturalist*, 128(5), 751, 1986.

Hester, A. J. and Hobbs, R. J., Influence of fire and soil nutrients on native and non-native annuals at remnant vegetation edges in the western Australian wheat belt, *Journal of Vegetation Science*, 3, 101, 1992.

Heuveldop, J., *Assessment of sustainable forest management*, Federal Research Centre for Forestry and Forest Products (BFH)/Institute for World Forestry, Hamburg, 165, September 1994, 1994.

Heywood, V. H., Patterns, extents and modes of invasions by terrestrial plants, *Biological invasions: a global perspective*, Drake, J. A., Mooney, H. A., Castri, F. d., Groves, R. H., Kruger, F. J., Rejmánek, M. and Williamson, M., John Wiley & Sons, Chichester, 31, 1989.

Higgins, S. I. and Richardson, D. M., A review of models of alien plant spread, *Ecological Modelling*, 87, 249, 1996.

Hobbs, R. J. and Atkins, L., Effect of disturbance and nutrient addition on native and introduced annuals in plant communities in the western Australian wheat belt, *Australian Journal of Ecology*, 13(2), 171, 1988.

Hobbs, R. J. and Huenneke, L. F., Disturbance, diversity, and invasion: implications for conservation, *Conservation Biology*, 6(3), 324, 1992.

Hobbs, R. J. and Mooney, H. A., Community and population dynamics of serpentine grassland annuals in relation to gopher disturbances, *Oecologia (Berlin)*, 67, 342, 1985.

Holdren, J. P., Daily, G. C. and Ehrlich, P. R., The meaning of sustainability: biogeophysical aspects, *Defining and Measuring Sustainability: The Biogeophysical Foundations*, Munasinghe, M. and Shearer, W., World Bank, Washington, D.C., 1995.

Hollenhorst, S. J., Brock, S. M., Freimund, W. A. and Twery, M. J., Effects of gypsy moth infestation on near-view aesthetic preferences and recreation behavior intentions, Eighth Central Hardwood Conference, USDA, Forest Service Northeastern Forest Experiment Station, Radnor, PA, 23, 1991.

Hooper, D. U., The role of complementary and competition in ecosystem responses to variation in plant diversity, *Ecology*, In press.

Hooper, D. U. and Vitousek, P. M., Effects of plant composition and diversity on nutrient cycling, *Ecological Monographs*, In press.

Hornbeck, J. W. and Kropelin, W., Nutrient removal and leaching from a whole-tree harvest of northern hardwoods, *Journal of Environmental Quality*, 11(2), 309, 1982.

Horsch, R., Lecture at Yale School of Forestry and Environmental Studies, co-president of the sustainable development sector and general manager of the Monsanto Company, 1999.

Hughes, R. F., Vitousek, P. M. and Tunison, J. T., Effects of invasion by fire-enhancing C_4 grasses on native shrublands in Hawaii Volcanoes National Park, *Ecology*, 72, 743, 1991.

Hulbert, S. H., The non-concept of species diversity: a critique and alternative parameters, *Ecology*, 52, 577, 1971.

Hunter, M. L., *Wildlife, Forests and Forestry: Principles of Managing Forests for Biological Diversity*, Prentice Hall, New Jersey, 1990.

Huntly, N. R. and Inouye, R. S., Pocket gophers and ecosystems: patterns and mechanisms, *BioScience*, 38, 786, 1988.

Hurd, L. E., Mellinger, M. V., Wolf, L. L. and Mcnaughton, S. j., Stability and diversity at three trophic levels in terrestrial successional ecosystems, *Science*, 173, 1134, 1971.

Huston, M. A., *Biological Diversity: The Coexistence of Species on Changing Landscapes*, Cambridge University Press, Cambridge, U.K., 1994.

Huston, M. A., Hidden treatments in ecological experiments: re-evaluating the ecosystem function of biodiversity, *Oecologia*, 110, 449, 1997.

IBGE (Instituto Brasileiro de Geografia e Estatística), SIDRA — Censo Agropecuario, http://www.ibge.gov.br, 1997.

IISD (International Institute for Sustainable Development), Summary of the Nineteenth United Nations General Assembly Special Session to Review Implementation of Agenda 21, Carpenter, C., Doran, P., Gupta, A. and Wagner, L., 23–27 July, 1997.

Indian Forest Management Assessment Team, *An assessment of Indian forests and forest management in the United States*, Report to the Indian Tribal Timber Council, Portland, OR, 1993.

Isenmann, P., Some recent bird invasions in Europe and the Mediterranean Basin, *Biological invasions in Europe and the Mediterranean Basin*, di Castri, F., Hansen, A. J. and Debussche, M., Kluwer Academic Publishers, Dordrecht, 245, 1990.

ISO, *Environmental labelling: guiding principles, practices and criteria for multiple criteria-based practitioner programmes. Guide for Certification Procedures.*, International Standardization Organization, Geneva, Switzerland, Working Draft, 1995.

ISO, *Draft 09 Technical Report ISO/WD 14061 — TC207 N197 — Informative Reference Material to assist forestry organizations in the use of ISO 14001 and ISO 14004 Environmental Management System Standards*, 1997.

ITTO (International Tropical Timber Organization), *Status and potential of non-timber products in the sustainable development of tropical forests*, International Tropical Timber Organization, Yokohama, Japan, 1990.

ITTO (International Tropical Timber Organization), ITTO process in the development of criteria and indicators for the measurement of sustainable tropical forest management, *FAO XI World Forestry Congress*, Antalya, Turkey, 13–22 October, 1997.

IUFRO, Sustainable management of small-scale forestry, *IUFRO Symposium on Sustainable Management of Small Scale Forestry*, Murashima, Y., IUFRO, Kyoto, Japan, 8–13 September, 1997.

Jaakko Pöyry Consulting, *Is a Global Long-Term Fibre Shortage Ahead? Multi-Client Study*, Helsinki, Finland. Tarrytown, NY. Stockholm, Sweden, 1996.

Jaakko Pöyry Consulting, *File Data Regarding Trend Product Pricing for Northern Bleached Softwood Kraft Market Pulp*, Tarrytown, NY, 1998.

Jack, J., Panel Discussion at Yale School of Forestry and Environmental Studies: FES 508b, Sustainable Forest Management and Certification: Field Testing Current Approaches., American Forest & Paper Association, Washigton, D.C., 1998.

Jacob, M., Sustainable development and deep ecology: an analysis of competing traditions, *Environmental Management*, 18(4), 477, 1994.

Jacobson, M., Jones, E. and Cubbage, F., Landowner attitudes toward landscape-level management, *Symposium on Non-industrial Private Forests: Learning from the Past, Prospects for the Future.*, Minnesota Extension Service, Washington, D.C., 417, 1996.

Jagannathan, N. V., *Poverty, public policies and the environment*, World Bank, Washington, D.C., 1989.

Jamison, A., The shaping of the global environmental agenda: the role of nongovernmental organizations, *Risk, environment and modernity: towards a new ecology*, Lash, S., Szerszynski, B. and Wynne, B., Sage Publications, London, Thousand Oaks, New Delhi, 1996.

Janik, P., McDougle, J., Hubbard, J. and Payne, L., *Forest Service action strategy for state and private forestry services*, USDA Forest Service, Washington, D.C., Executive Summary, 1998.

Janzen, D. H., The external threat, *Conservation biology: the science of scarcity and diversity*, Soule, M. E., Sinauer, Sunderland, MA, 286, 1986.

Jenkins, M. B., *Sustaining forests and profits: the business of sustainable forestry*, The John D. and Catherine T. MacArthur Foundation, New York, 1997.

Johnson, C. E., Johnson, A. H., Huntington, T. G. and Siccama, T. G., Whole-tree clearcutting effects on soil horizons and organic matter pools, *Soil Science Society of America Journal*, 55(2), 497, 1991a.

Johnson, C. E., Johnson, A. H. and Siccama, T. G., Whole-tree clearcutting effects on exchangeable cations and soil acidity, *Soil Science Society of America Journal*, 55(2), 502, 1991b.

Johnson, D. W., Effects of forest management on soil carbon storage, *Water, Air, and Soil Pollution*, 64, 83, 1992.

Johnson, K. H., Vogt, K. A., Clark, H. J., Schmitz, O. J. and Vogt, D. J., Biodiversity and the productivity and stability of ecosystems, *Trends in Ecology and Evolution — TREE*, 11(9), 372, 1996.

Johnson, N. and Cabarle, B., *Surviving the Cut: Natural Forest Management in the Humid Tropics*, World Resources Institute, Washington, D.C., 1993.

Johnson, N. C. and Wedin, D. A., Soil carbon, nutrients, and mycorrhizae during conversion of dry tropical forest to grassland, *Ecological Applications*, 7(1), 171, 1997.

Johnson, R. L., Alig, R. J., Moore, E. and Mouton, R. J., NIPF landowners' view of regulation, *Journal of Forestry*, 95(1), 23, 1997.

Jones, C. G. and Hanson, H. C., *Mineral Licks, Geophagy, and Biogeochemistry of North American Ungulates*, Iowa State University Press, Ames, Iowa, 1985.

Jones, C. G., Lawton, J. H. and Shackak, M., Organisms as ecosystem engineers, *Oikos*, 69, 373, 1994.

Jones, C. G., Lawton, J. H. and Shackak, M., Positive and negative effects of organisms as physical ecosystem engineers, *Ecology*, 78(7), 1946, 1997.

Jonsson, A., Eriksson, T., Eriksonn, G., Kahr, M., Lundkvist, K. and Norell, L., Interfamily variation in nitrogen productivity of Pinus sylvestris seedlings, *Scandinavian Journal of Forest Research*, 12(1), 1, 1997.

Jordan, C. F., *Conservation: Replacing Quantity with Quality as a Goal for Global Management*, John Wiley & Sons, Inc., 1995.

Jusoff, K. and Mustafa, N. M. S. N., Guidelines on logging practices for the hill forest of peninsular Malaysia, *Forest codes of practice*, Dykstra, D. P. and Heinrich, R., FAO, Rome, 89, 1996.

Kareiva, P., Diversity begets productivity, *Nature*, 368, 686, 1994.

Kaufmann, M. R., An ecological basis for ecosystem management, USDA Forest Service, Rocky Mountain Forest and Range Experiment Station, Ft. Collins, CO, General Technical Report, 1994.

Kavanagh, R. P. and Lambert, M. J., Food selection by Greater Gilder, *Petauroides volans*: Is foliar nitrogen a determinant of habitat quality?, *Australian Wildlife Research*, 17, 285, 1990.

Kay, J. J. and Schneider, E., Embracing complexity: the challenge of the ecosystem approach, *Alternatives*, 20, 33, 1994.

Keating, M., *The Earth's Summit Agenda for Change*, Centre for Our Common Future, Geneva, Switzerland, 1993.

Keddy, P. A., Assembly and response rules: two goals for predictive community ecology, *Journal of Vegetation Science*, 3, 157, 1992.

Kellert, S., *The Value of Life*, Island Press, Washington, D.C., 1996.

Kellert, S., *Kinship to Mastery*, Island Press, Washington, D.C., 1997.

Kessler, W. B., Salwasser, H., Cartwright, C. W. J. and Caplan, J. A., New perspectives for sustainable natural resources management, *Ecological Applications*, 2(3), 221, 1992.

Keynes, J. M., *Activities 1940–1946 — Shaping the Postwar World: Breton Woods and Reparations*, Macmillan; Cambridge University Press for the Royal Economic Society, London; New York, 453 pages, 1980a.

Keynes, J. M., *Activities 1940–1946 — Shaping the Postwar World: Employement and Communities*, Macmillan; Cambridge University Press for the Royal Economic Society, London; New York, 539 pages, 1980b.

Kiekens, J.-P., *Certification: international trends and forestry and trade implications*, Environmental Strategies Europe, Bruxelles, November, 1997.

Kimmins, J. P., *Forest Ecology*, Macmillan, New York, 1987.

King, A. W. and Pimm, S. L., Complexity, diversity and stability: a reconciliation of theoretical and empirical results, *American Naturalist*, 122, 229, 1983.

Knight, D. H., The Yellowstone fire controversy, *The Greater Yellowstone Ecosystem. Redefining America's Wilderness Heritage*, Keiter, R. B. and Boyce, M. S., Yale University Press, New Haven, CT, 1991.

Knight, R. and Bates, S., *A New Century for Natural Resources Management*, Island Press, Washington, D.C, 1995.

Knoepp, J. D. and Swank, W. T., Long-term effects of commercial sawlog harvest on soil cation concentrations, *Forest Ecology and Management*, 93(1–2), 1, 1997.

Koch, P., *Wood vs. non-wood materials in U.S. residential construction: some energy-related international implications*, University of Washington, Center for International Trade in Forest Products (CINTRAFOR), Seattle, WA, Working Paper 36, 1991.

Kornas, J., Plant invasions in central Europe: historical and ecological aspects, *Biological Invasions in Europe and the Mediterranean Basin*, Castri, F. d., Hansen, A. J. and Debussche, M., Kluwer Academic Publishers, Dordrecht, 19, 1990.

Krag, R., Higginbotham, K. and Rothwell, R., Logging and soil disturbance in southeast British Columbia, *Canadian Journal of Forest Research*, 16(1345–1354), 1986.

Krebs, C. H., *Ecology: The Experimental Analysis of Distribution and Abundance*, Harper & Row, New York, 694 pages, 1972.

Kusuma, I. D., *Sustainability values inside the "traditional wisdom" to manage the forests. Case studies of the indigenous people in Kalimantan, Indonesia*, Yale University School of Environmental Studies, Unpublished paper for FES 508b: Sustainable Forest Management and Certification: Field Testing Current Approaches., 1998.

Kuusipalo, J., Adjers, G., Jafarsidik, Y., Otsamo, A., Tuomela, K. and Vuokko, R., Restoration of natural vegetation in degraded *Imperata cylindrica* grassland: understory development in forest plantations, *Journal of Vegetation Science*, 6(2), 205, 1995.

Lagerlof, J. and Wallin, H., Cropping systems, field margins, and invertebrate fauna in Swedish agriculture, *INTECOL Bulletin*, 16, 55, 1988.

Lampman, S., The World Bank's forest policy under review, *Journal of Forestry*, 97(2), 5, 1999a.

Lampman, S., World Bank and WWF alliance, *Journal of Forestry*, 97(2), 7, 1999b.

Landerfield, J. S. and Hines, J. R., National accounting for non-renewable natural resource in the mining industry, *The Review of Income and Wealth*, 1(March), 1985.

Landis, J., Personal communication, forester of Smartwood in Richmond, VT, 1998.

Landres, P. B., Verner, J. and Thomas, J. W., Ecological uses of vertebrate indicator species: a critique, *Conservation Biology*, 2(4), 316, 1988.

Langeland, K., *Exotic woody plant control*, Florida Cooperative Extension Service, Ft. Lauderdale, FL, 1990.

LaPointe, G., Panel Discussion at Yale School of Forestry and Environmental Studies: FES 508b, Sustainable Forest Management and Certification: Field Testing Current Approaches, Sustainable Forestry Certification Coalition. Montreal, Canada, 1998.

Larson, D. R., Shifley, S. R., Thompson, F. R., Brokkshire, B. L., Dey, D. C., Kurzejeski, E. W. and England, K., Ten guidelines for ecosystem researchers: lessons from Missouri, *Journal of Forestry*, 95(4), 4, 1997.

Lash, W. H., *Green showdown at the WTO*, Center for the Study of American Business, Policy Brief #174, 1996.

Laurance, W. F., Edge effects in tropical forest fragments: application of a model for the design of nature reserves, *Biological Conservation*, 57, 205, 1991.

Laurenroth, W. K., Grassland primary productivity, *Grassland Primary Production: North American Grasslands in Perspective*, French, N. R., Springer-Verlag, New York, 3, 1979.

Laurenroth, W. K., Coffin, D. P., Burke, I. C. and Virginia, R. A., Interactions between demographic and ecosystem processs in a semi-arid and an arid grassland: a challenge for plant functional types, *Plant functional types: their relevance to ecosystem properties and global change*, Smith, T. M., Shugart, H. H. and Woodward, F. I., Cambridge University Press, Cambridge, 1997.

Lawton, J. H., Non-competitive populations, non-convergent communities, and vacant niches: the herbivores of bracken, *Ecological Communities: Conceptual Issues and the Evidence*, Strong, D. R., Simberloff, D., Abele, L. G. and Thistle, A. B., Princeton University Press, Princeton, NJ, 67, 1984.

Lawton, J. H. and Brown, K. C., The population and community ecology of invading insects, *Philosophical Transactions of the Royal Society*, B, 314, 606, 1986.

Leal, L. C., The pan-European process — Helsinki Process, *FAO XI World Forestry Congress*, Antalya, Turkey, 13–22 October, 1997.

Leidy, R. A. and Fiedler, P. L., Human disturbance and patterns of fish species diversity in the San Francisco Bay drainage, California, *Biological Conservation*, 33, 247, 1985.

Leopold, A., *Game Management*, Scribners, New York, 1933.

Lever, C., *Naturalized Mammals of the World*, Longman, London, 1985.

Levesque, C. A., Northern Forest Lands Council: a planning model use of regional natural resource land, *Journal of Forestry*, 93(6), 36, 1995.

Levin, S. A., Scale and sustainability, *Defining and measuring sustainability: the biogeophysical foundations*, Munasinghe, M. and Shearer, W., World Bank, Washington, D.C., 1995.

Likens, G. E., Bormann, F. H. and Johnson, N. M., Nitrification: importance to nutrient losses from a cutover forested ecosystem, *Science*, 163, 1205, 1969.

Linnard, W., *Welsh Woods and Forests: History and Utilization.*, Amgueddfa Genedlaethol Cymru-National Museum of Wales, Caerdydd/Cardiff, Wales, 1982.

Lodge, D. M., Biological invasions: lessons for ecology, *Trends in Ecology and Evolution*, 8(4), 133, 1993.

Lohmann, L., Belief, imagination and desire in cost-benefit analysis: Some thoughts for further thinking, unpublished manuscript, 1997.

Long, J. L., *Introduced Birds of the World*, David & Charles, London, 1981.

Loomis, J. B., *Integrated Public Lands Management: Principles and Applications to National Forests, Parks, Wildlife Refuges, and BLM Lands*, Columbia University Press, New York, 1993.

Loope, L. L., Hammann, O. and Stone, C. P., Comparitive conservation biology of oceanic archipelagoes, *BioScience*, 38(4), 272, 1988.

Loope, L. L. and Mueller-Dombois, D., Characteristics of invaded islands with special reference to Hawaii, *Biological Invasions: A Global Perspective*, Drake, J. A., Mooney, H. A. and castri, F. d., Wiley, New York, 257, 1989.

Lovejoy, T. E., Bierregaard, R. O., Rylands, A. B., Malcolm, J. R., Quintela, C. E., Harper, L. H., Brown, K. S., Powell, A. H., Powell, G. V. N., Schubar, H. O. R. and Hays, M. B., Edge and other effects of isolation on Amazon South America forest fragments, *Conservation biology: the science and scarcity and diversity*, Soule, M. E., Sinauer, Sunderland, MA, 256, 1986.

Ludwig, D., Hilborn, R. and Walters, C., Uncertainty, resource exploitation, and conservation: lessons from history, *Science*, 260(2 April), 17, 1993.

Luloff, A. E., Jacob, S., Bourke, L. and Finley, J., Achieving better forest management through a forestland preservation program: a Pennsylvania perspective, *Ecosystem Health*, 2(1), 69, 1996.

Lyke, J., Forest product certification revisited: an update, *Journal of Forestry*, 10, 16, 1996.

MacAlpine, D., Certified doors, floors, and furniture: corporate America takes the field. Vol. 6, No. 1, http://www.web/net/goodwood/understory/new/macalpine.html, 1996.

MacArthur, R. H., Fluctuations of animal populations and a measure of community stability, *Ecology*, 36, 533, 1955.

MacArthur, R. H. and Wilson, E. O., An equilibrium theory of insular zoogeography, *Evolution*, 373, 1963.

MacArthur, R. H. and Wilson, E. O., *The Theory of Island Biogeography*, Princeton University Press, Princeton, New Jersey, 1967.

MaCarthy, T. S., Ellery, W. N. and Bloem, A., Some observations on the geomorphological impact of hippopotamus (*Hippopotamus amphibius* L.) in the Okavango Delta, Botswana, *African Journal of Ecology*, 36, 44, 1998.

MacDonald, I. A. W. and Cooper, J., Insular lesson for global biodiversity conservation with particular reference to alien invasions, *Islands: Biological Diversity and Ecosystem Function*, Vitousek, P. M., Loope, L. L. and Aderson, H., Springer-Verlag, Berlin, 189, 1995.

MacDonald, I. A. W., Kruger, F. J. and Ferrar, A. A., The ecology and management of biological invasions in southern Africa, *National Synthesis Symposium on the Ecology of Biological Invasions*, Oxford University Press, Cape Town, 1986.

MacDonald, M., *Agendas for Sustainability: Environment and Development into the Twenty-First Century*, Routledge, London and New York, 1998.

Machlis, G. E., Force, J. E. and Burch, W. R., The human ecosystem. part 1. The human ecosystem as an organizing concept in ecosystem management., *Society and Natural Resources*, 10, 347, 1997.

Machlis, G. E. and McKendry, J. E., Maps and models for natural resource management: powerful tools from the social sciences., *Natural Resource Management: The Human Dimension*, Ewert, A. W., Westview Press, Boulder, CO, 1996.

Mack, R. N., Temperate grasslands vulnerable to plant invasions: characteristics and consequences, *Biological Invasions: A Global Perspective*, Drake, J. A., H. A, M., Castri, F. d., Groves, R. H., Kruger, F. J., Rejmánek, M. and Williamson, M., John Wiley & Sons, Chichester, 155, 1989.

Majer, J. D., Recher, H. F. and Ganeshanandam, S., Variation in foliar nutrients and its relation to arthropod communities on eucalyptus trees in New South Wales and western Australia, *Australian Journal of Ecology*, 13, 65, 1992.

Malmer, A. and Grip, H., Soil disturbance and loss of infiltrability cause by mechanized and manual extraction of tropical rainforest in Sabah, Malaysia, *Forest Ecology and Management*, 38(1–12), 1990.

Mankin, W., Lecture at Yale School of Forestry and Environmental Studies: FES 508b, Sustainable Forest Management and Certification: Field Testing Current Approaches, 1998.

Mann, L. K., Johnson, D. W., West, D. C., Cole, D. W., Hornbeck, J. W., Martin, C. W., Riekerk, H., Smith, C. T., Swank, W. T., Tritton, L. M. and Vanlear, D. H., Effects of whole-tree and stem-only clearcutting on postharvest hydrologic losses, nutrient capital and regrowth, *Forest Science*, 34(2), 412, 1988.

Margalef, R., On certain unifying principles in ecology, *American Naturalist*, 97, 357, 1963.

Margules, C. R., Landscape-level concepts and indicators for the conservation of forest biodiversity and sustainable forest management, *Conference on economic, social and political issues in certification of forest management*, Malaysia, May 12–16, 1996.

Marinelli, J., Introduction: Redefining the weed, *Invasive Plants: Weeds of the Global Garden*, Randall, J. M. and Marinelli, J., Brooklyn Botanic Garden, Brooklyn, 4, 1996.

Markham, D., l, *Willapa Indicators for a Sustainable Community*, Willapa Alliance and Ecotrust, 1995.

Marks, D. K., *Women and Grass Roots Democracy in the Americas: Sustaining the Initiative*, 2nd Edition, North-South Center Press, University of Miami, Coral Gables, Florida, 1996.

Marsh, A. S. and Siccama, T. G., Use of formerly plowed land in New England to monitor the vertical distribution of lead, zinc and copper in mineral soil, *Water Air and Soil Pollution*, 95(1–4), 75, 1997.

Martensen, G. D., Cushman, J. H. and Whitham, T. G., Impact of gopher disturbance on plant species diversity in a shortgrass prairie community, *Oecologia*, 83, 132, 1990.

Martin, C. W. and Harr, R. D., Logging of mature Douglas fir in western Oregon has little effect on nutrient output budgets, *Canadian Journal of Forest Research*, 19, 35, 1989.

Martus, C. E., Haney, H. J. and Siegel, W. C., Local forest regulatory ordinances: trends in the Eastern U.S., *Journal of Forestry*, 93(6), 27, 1995.

Mater, C., Lecture at Yale School of Forestry and Environmental Studies: FES 508b, Sustainable Forest Management and Certification: Field Testing Current Approaches, vice president of Mater Engineering, Ltd., Corvallis, Oregon, 1998.

Mater, J., *Reinventing the Forest Industry*, Green Tree Press, Wilsonville, Oregon, 1997.

Mathias, A., Key implementation issues for developing countries of certification and labeling, *International conference on certification and labeling of products from sustainably managed forests*, Jan. 25, 1996, 1996.

Matson, P. A., Vitousek, P. M., Ewel, J. J. and Mazzarino, M. J., Nitrogen transformations following tropical forest felling and burning on a volcanic soil, *Ecology*, 68(3), 491, 1987.

Mattson, W. J. and Addy, N. D., Phytophagous insects as regulators of forest primary production, *Science*, 190, 515, 1975.

May, P. H., Building institutions and markets for non-wood forest products from the Brazilian Amazon, *Unasylva*, 165(42), 9, 1991.

May, R. M., Will a large complex system be stable?, *Nature*, 238, 414, 1972.

May, R. M., *Stability and Complexity in Model Ecosystems*, Princeton University Press, Princeton, 1973.

McCay, B. J. and Acheson, J. M., *The question of the commons: the culture and ecology of communal resources*, The University of Arizona Press, Tucson, Arizona, U.S.A., 1987.

McCormack, R. J., A review of forest practice codes in Australia, *Forest codes of practice*, Dykstra, D. P. and Heinrich, R., FAO, Rome, 105, 1996.

McCormick, J., *Reclaiming Paradise: The Global Environmental Movement*, Indiana University Press, Bloomington, xv, 1989.

McIntyre, S. and Lavorel, S., Predicting richness of native, rare, and exotic plants in response to habitat and disturbance variables across a variegated landscape, *Conservation Biology*, 8(2), 521, 1994.

McKenzie, D. H., Hyatt, D. E. and McDonald, V. J., *Ecological Indicators, 1 and 2*, Elsevier/Chapman & Hall, London and New York, 1992.

McNabb, K. L., Miller, M. S., Lockaby, B. G., Stokes, B. J., Clawson, R. G., Stanturf, J. A. and Silva, J. N. M., Selection harvests in Amazonian rainforests: long-term impacts on soil properties, *Forest Ecology and Management*, 93, 153, 1997.

McNaughton, S. J., Structure and function in California grasslands, *Ecology*, 49, 962, 1968.

McNaughton, S. J., Diversity and stability of ecological communities: a comment on the role of empiricism in ecology, *American Naturalist*, 111, 515, 1977.

McNaughton, S. J., Serengeti ungulates: feeding selectivity influences the effectiveness of plant defense guilds, *Science*, 199, 806, 1978.

McNaughton, S. J., Ecology of a grazing ecosystem: the Serengeti, *Ecological Monographs*, 53, 259, 1985.

McNaughton, S. J., Biodiversity and function of grazing ecosystems, *Biodiversity and ecosystem function*, Schulze, E. and Mooney, H. A., Springer-Verlag, Berlin, 361, 1993.

McNaughton, S. J., Coughenour, M. b. and Wallace, L. L., Interactive process in grassland ecosystems, *Grassess and Grasslands: Systematics and Ecology*, Estes, J. R., Tyrl, R. J. and Brunken, J. N., University of Oklahoma Press, Norman, Oklahoma, 167, 1982.

McNaughton, S. J., Osterheld, M., Frank, D. A. and Williams, K. J., Ecosystem level patterns of primary productivity and herbivory in terrestrial habitats, *Nature*, 341, 142, 1989.

Medeiros, A. C., Loope, L. L. and Cole, F. R., Distribution of ants and their effects on endemic biota of Haleakala and Hawaii Volcanoes National Park: a preliminary assessment, *Proc. 6th Conf. Nat. Sci.*, Hawaii Volcanoes National Park, 39, 1986.

Meffe, G. K. and Carroll, C. R., *Principles of Conservation Biology*, Sinauer Associates, Sunderland, MA, 1994.

Mellinger, M. V. and McNaughton, S. J., Structure and function of successional vascular plant communities in central New York, *Ecological Monographs*, 45, 161, 1975.

Mendez, G. J., Manejo del bosque natural en la region Huetar Norte de Costa Rica, *Revista Forestal Centroamericana*, 2, 42, 1993.

Merino, L. P., Analysis of social elements in forestry certification, *Conference on economic, social and political issues in certification of forest management*, Malaysia, 12–16 May, 1996.

Meyerson, L. A., Vogt, K. A., Dunning, G. W. and Gordon, J. C., *Invasive alien species*, Yale School of Forestry and Environmental Studies, New Haven, CT, Yale Forest Forum Series Publication, 1998.

Mikola, P., Mycorrhizae and feeder root diseases, *Ectomycorrhizae. Their Ecology and Physiology*, Marks, G. C. and Kozlowski, T. T., Academic Press, New York, 1973.

Miller, R. B., Interactions and collaboration in global change across the social and natural sciences, *Ambio*, 23(1), 19, 1994.

Moffet, J., *A comparison of product diffusion and distributed lag models for estimating wood/non-wood substitution in the U.S. window market.*, University of Washington, Center for International Trade in Forest Products (CINTRAFOR), Seattle, WA, working paper, 1993.

Mooney, H. A. and Drake, J. A., *Ecology of Biological Invasions of North America and Hawaii*, Springer-Verlag, Berlin, 1986.

Morris, L. A., Pritchett, W. L. and Swindel, B. F., Displacement of nutrients into windrows during site preparation of a flatwood forest, *Soil Science Society of America Journal*, 47, 591, 1983.

Motzkin, G., Foster, D., Allen, A., Harrod, J. and Boone, R., Controlling site to evaluate history: vegetation patterns of a New England sand plain, *Ecological Applications*, 66, 345, 1996.

Mou, P. U., Fahey, T. J. and Hughes, J. W., Effects of soil disturbance on vegetation recovery and nutrient accumulation following whole-tree harvest of a northern hardwood ecosystem, *Journal of Applied Ecology*, 30, 661, 1993.

MPCG (Ministério do Planejamento e Coordenação Geral), *O desafio brasileiro e o programa estratégico*, Ministério do Planejamento e Coordenação Geral, Rio de Janeiro, 74 p. ill.; 25 cm., 1968.

Mroz, G. D., Jurgensen, M. F. and Frederick, D. J., Soil nutrient changes following whole-tree harvesting on three northern hardwood sites, *Soil Science Society of America Journal*, 49(6), 1552, 1985.

Mueller-Dombois, D., A non-adapted vegetation interferes with water removal in a tropical rain forest area in Hawaii, *Tropical Ecology*, 14, 1, 1973.

Mueller-Dombois, D., Island ecosystems: what is unique about their ecology, *Island Ecosystems: Biological Organization in Selected Hawaiian Communities*, Mueller-Dombois, D., Bridges, K. W. and Carson, H. L., Hutchinson Ross, Stroudsburg, 485, 1981.

Mueller-Dombois, D., Impoverishment in Pacific island forests, *The Earth in Transition: Patterns and Processes of Biotic Impoverishment*, Woodwell, G. M., Cambridge University Press, Cambridge, 199, 1990.

Mueller-Dombois, D., Biological diversity and disturbance regimes in island ecosystems, *Islands: Biological Diversity and Ecosystem Function*, Vitousek, P. M., Loope, L. L. and Aderson, H., Springer-Verlag, Berlin, 163, 1995.

Munasinghe, M. and Shearer, W., Defining and measuring sustainability: the biogeophysical foundations, *International Conference on the Definition and Measurement of Sustainability: The Biophysical (Biogeophysical) Foundations*, World Bank, Washington, D.C., 440 pages, 1995.

Munson, A. D., Margolis, H. A. and Brand, D. G., Intensive silvicultural treatment: impacts on soil fertility and planted conifer response, *Soil Science Society of America Journal*, 57, 246, 1993.

Murawski, D., Gunatilleke, I. and Bawa, K. S., The effects of selective logging on inbreeding in *Shorea megistophylla* (Dipterocarpaceae) from Sri Lanka, *Conservation Biology 8*, 4, 997, 1994.

Murcia, C., Edge effects in fragmented forests: implications for conservation, *Trends in Ecology and Evolution*, 10(2), 58, 1995.

Murrieta, R. S. S., *Relatório final do teste de critérios e indicadores sociais de sustentabilidade de manejo florestal comunitário — Amazônia, Brazil*, Department of Anthropology, Hale Building, University of Colorado at Boulder, Boulder, Colorado, U.S.A., 40, unpublished manuscript, 1998.

Myers, N., *The Primary Source: Tropical Forests and Our Future*, 1st ed., Norton, New York, 399 pages, 1984.

Myers, N., The world's forests: need for a policy appraisal, *Science*, 268(12th May), 823, 1995.

Naeem, S., Thompson, L. J., Lawler, S. P., Lawton, J. H. and Woodfin, R. M., Declining biodiversity can alter the performance of ecosystems, *Nature*, 368, 734, 1994.

Naiman, R. J., Beechie, T. J., Benda, L. E., Bisson, P. A., McDonald, L. H., Conner, M. D., Olson, P. L. and A., S. E., Fundamental elements of ecologically healthy watersheds in the Pacific Northwest coastal ecoregions, *Watershed Management: Balancing Sustainability and Environmental Change*, Naiman, R. J., Springer Verlag, New York, 1992.

Naiman, R. J., Melilo, J. M. and Hobbie, J. E., Ecosystem alteration of boreal forest streams by beavers (*Castor canadensis*), *Ecology*, 67(5), 1254, 1986.

NAPAP (National Atmospheric Precipitation Assessment Program), *1992 Report to Congress*, Washington, D.C., 37, 1993.

Neeson, E., *A History of Irish Forestry*, The Liliput Press Ltd., Dublin, Ireland, 1991.

Neyland, M. G. and Brown, M. J., Disturbance of cool temperate rainforest patches in eastern Tasmania, *Australian Forestry*, 57, 1, 1994.

Noordik, H. and Hoben, M., *The TUVA Fallen Timber Monitoring System*, Fundación TUVA y Centro Científico Tropical, San Jose, Costa Rica, 1996.

Norberg-Hodge, H., *Shifting Direction: From Global Dependence to Local Interdependence, The Case Against Globalization and a Turn Toward the Local*, Mander, J. and Goldsmith, E., Sierra Club Books, San Francisco, 1996.

Northern Forest Lands Council, *Technical appendix: a compendium of technical research and forum proceedings from the Northern Forest Lands Council*, February, 1994.

Northrup, M., Lecture at Yale School of Forestry and Environmental Studies: FES 508b, Sustainable Forest Management and Certification: Field Testing Current Approaches., Rockefeller Brothers Fund, New York, 1998.

Norton, B., Ecological integrity and social values: at what scale?, *Ecosystem Health*, 1(4), 228, 1995.

Norton, B. G., On the inherent danger of undervaluing species, *The preservation of species*, Norton, B. G., Princeton University Press, Princeton, 110, 1986.

Norton, B. G., *Why Preserve Natural Variety?*, Princeton University Press, Princeton, 1987.

Norton, B. G., Ecological health and sustainable resource management, *Ecological Economics: The Science and Management of Sustainability*, Constanza, R., Columbia University Press, New York, 1991.

Norton, B. G., A new paradigm for environmental management, *Ecosystem Health: New Goals for Environmental Health*, Constanza, R., Norton, B. G. and Haskell, B. D., Island Press, 1992.

Noss, R. F., A regional landscape approach to maintain diversity, *BioScience*, 3(11), 70, 1983.

Noss, R. F., Corridors in real landscapes: a reply to Simberloff and Cox, *Conservation biology*, 1, 159, 1987.

NRC (National Research Council Board of Agriculture), *Soil and Water Quality: An Agenda for Agriculture*, National Academy Press, Washington, D.C., 516, 1993.

NRC (National Research Council), *Forested Landscapes in Perspective: Prospects and Opportunities for Sustainable Management of America's Nonfederal Forests*, National Academy Press, Washington, D.C., 1998.

O'Laughlin, J., Forest ecosystem health assessment issues: definition, measurement, and management implications, *Ecosystem Health*, 2(1), 19, 1996.

O'Laughlin, J. O., What the law is and what it means, *Journal of Forestry*, 90(8), 6, 1992.

Odum, E. P., *Ecology*, Holt, Rinehart & Winston, New York, 152 pages, 1953.

Odum, E. P., *Fundamentals of Ecology*, W.B. Saunders Company, Philadelphia, Pennsylvania, 1959.

Odum, E. P., The strategy of ecosystem development, *Science*, 164, 262, 1969.

Oliver, C. D. and Larson, B. C., *Forest and Stand Dynamics*, John Wiley and Sons, Inc., New York, 1996.

Olmstead, C. J., Duryea, M. L. and Harrell, J. B., *Forestry Assistance for Florida Landowners*, Florida Cooperative Extension Service, Circular, 13 pages, 1989.

Orians, G. H., Site characteristics favoring invasions, *Ecology of Biological Invasions of North America and Hawaii*, Mooney, H. A. and Drake, J. A., Springer-Verlag, New York, 133, 1986.

Orlove, B. S. and Brush, S. B., Anthropology and the conservation of biodiversity, *Annual Review of Anthropology*, 25, 329, 1996.

Ortiz, R. A., Sociedad inca en el momento de la conquista, *Anthropologica, Departamento de Ciencias Sociales, Lima*, 10, 95, 1992.

Osberg, M. and Murphy, B. J., British Columbia forest practices code, *Forest codes of practice*, Dykstra, D. P. and Heinrich, R., FAO, Rome, 57, 1996.

Oswald, E. T. and Brown, B. N., Vegetation development on skid trails and burned sites in southeastern British Columbia, *The Forestry Chronicle*, 69(1), 75, 1993.

Otavio, C. and Mendes, V., O enxame que devora a Amazônia, *O Globo*, Rio de Janeiro, Brazil, seção O Pais, 1998.

Otavo, R. E., *Wood extraction with oxen and agricultural tractors*, Food and Agriculture Organization, Rome, Italy, 1986.

Ozanne, L. and Vlosky, R., Willingness to pay for environmentally certified wood products: a consumer perspective, *Forest Products Journal*, 47(6), 39, 1997.

Ozanne, L. K. V. and Smith, P. M., Strategies and perspectives of influencial environmental organizations towards tropical deforestation, *Forest Products Journal*, 43(4), 39, 1993.

Packham, J. M., Elliott, H. J. and Kile, G. A., Myrtle wilt in Tasmanian rainforests, *Monash Publications in Geography*, 41, 1992.

Padoch, C. and Peluso, N. L., *Borneo in Transition. People, Forest, Conservation and Development*, Oxford University Press, New York, 1996.

Padoch, C. and Peters, C., Managed forest garden in West Kalimantan, Indonesia, *Perspectives on Biodiversity: Case Studies of Genetic Resource Conservation and Development.*, Potter, C. C., American Association for the Advancement of Science, Washington, D.C., 1993.

Paine, R. T., Food web complexity and species diversity, *American Naturalist*, 100, 65, 1966.

Palmer, J. R., Curtin, D. and Graham, C., Monitoring forest practices, *Conference on economic, social and political issues in certification of forest management*, Malaysia, 12–16 May, 1996.

Palmiotto, P. A., The role of specialization in nutrient-use efficiency as a mechanism driving species diversity in a tropical rain forest, *Department of Forestry and Environmental Studies*, Yale University, New Haven, CT, unpublished dissertation, 282, 1998.

Palmiotto, P. A., Vogt, K. A., Ashton, P. M. S., Ashton, P. S., Vogt, D. J., Semui, H. and Seng, L. H., Soils and gaps: The influence of habitat heterogeneity on diversity in a tropical rain forest, Sarawak, Malaysia, *Forest Diversity and Dynamism*, University of Chicago Press, 1999.

Panetta, F. D., A system of assessing proposed plant introductions for weed potential, *Plant Protection Quarterly*, 8, 10, 1993.

Pastor, J., Naiman, R. J., Dewey, B. and McInnes, P., Moose, microbes, and the boreal forest, *BioScience*, 38, 770, 1988.

Patel-Weynand, T., *Forests and poverty: a view towards sustainable development*, United Nations Development Programme, New York, 105 pages, 1997a.

Patel-Weynand, T., Sukhomajri and Nanda: Managing common property resources at the village level, *Natural resource economics: theory and applications in India*, Kerr, J. M., Oxford University Press, New Delhi, India, 363, 1997b.

Pearce, D. and Moran, D., *The Economic Value of Biodiversity*, Earthscan Ltd., London, 1994.

Pearce, D. W. and Warford, J. J., *World Without End: Economics, Environment and Sustainable Development*, Oxford University Press, New York, 1993.

Pearce, F., Timber trade stripped of its green veneer, *New Scientist*, April(4), 1992.

Peart, D. R. and Foin, T. C., Analysis and prediction of population and community change: a grassland case study, *The Population Structure of Vegetation*, White, J., W. Junk, The Hague, 313, 1985.

Peet, R. K., The measurement of species diversity, *Annual Review of Ecology and Systematics*, 5, 285, 1974.

Pendleton, L. H., Sustainability from three perspectives: neoclassical economics, ecological economics and biology, Doctoral Qualifying Exam at Yale School of Forestry & Environmental Studies, 1994.

Perez-Garcia, J., Global economic and land-use consequences, *Journal of Forestry*, 93(7), 34, 1995.

Perry, D. A., *Forest Ecosystems*, The Johns Hopkins University Press, Baltimore, MD, 1994.

Perry, D. A., Borchers, J. G. and Borchers, S. L., Species migrations and ecosystem stability during climate change: the belowground connection, *Conservation Biology*, 4, 266, 1990.

Peskin, H. M., *Accounting for natural resource depletion and degradation in developing countries*, World Bank Environmental Department, Washington, D.C., Working Paper No.13, January, 1989.

Pierce, A. R., The challenges of certifying non-timber forest products, *Journal of Forestry*, 97(2), 1999.

Pimentel, D., Biological invasions of plants and animals in agriculture and forestry, *Ecology of Biological Invasions of North America and Hawaii*, Mooney, H. A. and Drake, J. A., Springer-Verlag, New York, 322, 1986.

Pimentel, D., McNair, M., Buck, L., Pimentel, M. and Kamil, J., The value of forests to world food security, *Human Ecology*, 25(1), 91, 1997.

Pimm, S. L., Complexity and stability: another look at MacArthur's original hypothesis, *Oikos*, 33, 351, 1979.

Pimm, S. L., The complexity and stability of ecosystems, *Nature*, 307, 321, 1984.

Pimm, S. L., *The Balance of Nature? Ecological Issues in the Conservation of Species and Communities*, The University of Chicago Press, Chicago, Illinois, 1991.

Pimm, S. L., Jones, H. L. and Diamond, J., On the risk of extinction, *American Naturalist*, 132, 757, 1988.

Pinedo-Vasquez, M., Human impact on várzea ecosystems in the Napo-Amazon, Peru, *Department of Forestry and Environmental Studies*, Yale University, New Haven, CT, unpublished dissertation, 1995.

Pomeroy, K. B., The germination and initial establishment of loblolly pine under various surface soil conditions, *Journal of Forestry*, 68, 166, 1949.

Poore, D., *Criteria for the sustainable development of temperate and boreal forests*, International Institute for Environment and Development, London, U.K., 1993.

Pope, C., *Forging an environmentally responsible trade policy*, Washington, D.C., Testimony before the House Ways and Means Trade Subcommittee, 1997.

Posey, D. and Dutfield, G., *Indigenous peoples and sustainability: cases and actions*, International Books, Utrecht, Holland, 1997.

Posey, D. A. and Balée, W. L., *Resource Management in Amazônia: Indigenous and Folk Strategies*, Vol. 7, New York Botanical Garden, Bronx, N.Y., U.S.A., 287, 1989.

Potts, C., Krester, H., Heintz, J. and Brownlee, A., *Landowner stewardship associations: a holistic approach to good management for small non-industrial private landowners*, Yale University School of Environmental Studies, New Haven, CT, unpublished paper for FES 508b: Sustainable Forest Management and Certification: Field Testing Current Approaches., 1998.

Powell, D. S., Faulkner, J. L., Darr, D. R., Shu, Z. and MacCleery, D. W., Forest resources of the United States, USDA, Forest Service, Rocky Mountain Forest and Range Experiment Station, Fort Collins, CO, General Technical Report, 132, 1993.

Power, M. E., Tilman, D., Estes, J. A., Menge, B. A., Bond, W. J., Mills, L. S., Daily, G., Castilla, J. C., Lubchenco, J. and Paine, R. T., Challenges in the quest for keystones, *BioScience*, 46(8), 609, 1996.

Prabhu, R., Colfer, C. J. P., Venkateswarlu, P., Tan, L. C., Soekmadi, R. and Wollenberg, E., *Testing criteria and indicators for the sustainable management of forests: phase 1 — final report*, CIFOR (Center for International Forestry Research), 1996.

Prescott, C. E., Effects of clearcutting and alternative silvicultural systems on rates of decomposition and nitrogen mineralization in a coastal montane coniferous forest, *Forest Ecology and Management*, 95, 253, 1997.

Price, P. W., Communities of specialists: Vacant niches in ecological and evolutionary time, *Ecological Communities: Conceptual Issues and the Evidence*, Strong, D. R., Simberloff, D., Abele, L. G. and Thistle, A. B., Princeton University Press, Princeton, NJ, 510, 1984.

Price, P. W., The resource-based organization of communities, *Biotropica*, 24, 273, 1992.

Primack, R. B., *Essentials of Conservation Biology*, Sinauer Associates Inc., Sunderland, Massachusetts, 1993.

Princen, T. and Finger, M., *Environmental NGOs in world politics: linking the local and the global*, Routledge, London, New York, 1994.

Puller, B., Lecture at Yale School of Forestry and Environmental Studies: FES 508b, Sustainable Forest Management and Certification: Field Testing Current Approaches, forest manager, Kane Hardwood/Collins Pine, Kane, Pennsylvania, 1998.

Pulp & Paper, *Pulp & Paper North American Fact Book*, Miller Freeman, Inc., San Francisco, CA, 1998.

Raffles, H., Igarape Guariba: Nature, locality, and the logic of Amazonian anthropogenesis, *School of Forestry and Environmental Studies*, Yale University, New Haven, CT, unpublished Ph.D dissertation, 1998.

Ramakrishnan, P. S., *Shifting Agriculture and Sustainable Development: An Interdisciplinary Study from North-Eastern India*, UNESCO, Paris, 1992.

Ramakrishnan, P. S. and Vitousek, P. M., Ecosystem-level processes and the consequences of biological invasions, *Biological Invasions: A Global Perspective*, Drake, J. A., H. A, M., Castri, F. d., Groves, R. H., Kruger, F. J., Rejmánek, M. and Williamson, M., John Wiley & Sons, Chichester, 281, 1989.

Ramney, J. W., Bruner, M. C. and Levenson, J. B., *Forest Island Dynamics in Man-Dominated Landscapes*, Springer, New York, 1981.

Randall, J. M., Plant invaders: how non-native species invade and degrade natural areas, *Invasive Plants: Weeds of the Global Garden*, Randall, J. M. and Marinelli, J., Brooklyn Botanic Garden, Brooklyn, 7, 1996.

Rapport, D. J., Historical changes in Ponderosa pine forests since Euro-American settlement, *Ecosystem Health*, 3, 171, 1997.

Raven, P. H. and McNeely, J. A., Biological extinction: its scope and meaning for us, *Protection of Global Biodiversity: Converging Strategies*, Guruswamy, L. D. and McNeely, J. A., Duke University Press, Durham, NC, London, U.K., 13, 1998.

Rawes, M. and Welch, D., Trials to recreate floristically rich vegetation by plant introduction in the Northern Penines, England, *Biological Conservation*, 4, 135, 1972.

Raynor, G. S., Wind and temperature structure in a coniferous forest and a contiguous field, *Forest Science*, 17, 351, 1971.

Read, M., *Truth or trickery*, World Wildlife Fund, U.K., London, 1994.

Reagan, D. P. and Waide, B., *The Food Web of a Tropical Rain Forest*, University of Chicago Press, Chicago, Illinois; London, U.K., 1996.

Reichard, S. H. and Hamilton, C. W., Predicting invasions of woods plants introduced into North America, *Conservation Biology*, 11(1), 193, 1997.

Reid, W. V., Halting the loss of biodiversity: international institutional measures, *Protection of Global Biodiversity: Converging Strategies*, Guruswamy, L. D. and McNeely, J. A., Duke University Press, Durham, USA; London, U.K., 168, 1998.

Reifsnyder, W. E., Wind profiles in a small isolated forest stand, *Forest Science*, 1, 289, 1955.

Reimer, N. J., Distribution and impact of alien ants in vulnerable Hawaiian ecosystems, *Exotic Ants: Biology, Impact, and Control of Introduced Species*, Williams, D. F., Westview Press, Boulder, CO, 11, 1994.

Rejmánek, M., Invasibility of plant communities, *Biological Invasions: A Global Perspective*, Drake, J. A., H. A, M., Castri, F. d., Groves, R. H., Kruger, F. J., Rejmánek, M. and Williamson, M., John Wiley & Sons, Chichester, 369, 1989.

Rejmánek, M. and Richardson, D. M., What attributes make some plant species more invasive?, *Ecology*, 77(6), 1655, 1996.

Repetto, R., *Natural resource accounting*, World Resource Institute, Washington, D.C., 1987.

Rice, D. S., *Eighth-century physical geography, environment, and natural resources in the Maya lowlands. Lowland Maya Civilization in the Eighth Century A.D.*, Dumbarton Oaks, Washington, D.C., 1993.

Richards, M., Stabilizing the Amazon frontier: Technology, institutions and policies, *Natural Resources Perspective*, www.oneworld.org/odi/nrp/10.html, 10 (July), 1996.

Richardson, D. M., Forestry trees as invasive aliens, *Conservation Biology*, 12(1), 18, 1998.

Richardson, D. M., Macdonald, I. A. W., Holmes, P. M. and Cowling, R. M., Plant and animal invasions, *The Ecology of Fynbos*, Cowling, R. M., Oxford University Press, Cape Town, 271, 1992.

Richter, D. D., Wells, C. G., Allen, H. L., April, R., Heine, P. R. and Urrego, B., Soil chemical change during three decades in an old-field loblolly pine (*Pinus taeda* L.) ecosystem, *Ecology*, 75, 1463, 1994.

Rickenbach, M. G., Kitteredge, D. B., Dennis, D. and Stevens, T., Ecosystem management: capturing the concept for woodland owners, *Journal of Forestry*, 96(4), 18, 1998.

Robertson, D., Memorandum issued to regional foresters and research-station directors, Forest Service, Washington, D.C., 1992.

Robertson, G. P. and Tiedje, J. M., Deforestation alters denitrification in a lowland tropical rain forest, *Nature*, 336, 756, 1988.

Rocha, G. E. P. and Barrera, M. S., *Programa de Monitoreo Ecologico y Socioeconomic de los Bosques del Pacifico Medio y Sur Colombiano*, Convenio BID-MINAMBIENTE-CONIF, Santafe de Bogotá, DC, Colombia, April, 1997.

Rodale, M., *Maria Rodale's Organic Gardening*, Rodale Press, Emmaus, PA, 1999.

Rolston, H. I., *Philosophy Gone Wild*, Prometheus Books, Buffalo, NY, 1986.

Romanowicz, R. B., Driscoll, C. T., Fahey, T. J., Johnson, C. E., Likens, G. E. and Siccama, T. G., Changes in the biogeochemistry of potassium following a whole-tree harvest, *Soil Science Society of America Journal*, 60(6), 1664, 1996.

Romme, W. H., Knight, D. h. and Yavitt, J. B., Mountain pine beetle outbreaks in the central Rocky Mountains: effects on primary production, *American Naturalist*, 127, 484, 1986.

Rosenberg, D. K., Noon, B. R. and Meslow, E. C., Biological corridors: form, function and efficiency, *BioScience*, 47, 677, 1997.

Roy, J., In search of the characteristics of plant invaders, *Biological Invasions in Europe and the Mediterranean Basin*, Castri, F. d., Hansen, A. J. and Debussche, M., Kluwer Academic Publishers, Dordrecht, Netherlands, 335, 1990.

Sackett, S., Haase, S. and Harrington, M. G., Restoration of southwestern Ponderosa pine ecosystems with fire, *Sustainable Ecological Systems: Implementing an Ecological Approach to Land Management*, Covington, W. W. and DeBano, L. F. t. C., USDA Forest Service, Rocky Mountain Forest Experiment Staion, Fort Collins, CO, General Technical Report, 1993.

Sagoff, M., Animals as inventions: biotechnology and intellectual property rights, *Protection of Global Biodiversity: Converging Strategies*, Guruswamy, L. D. and McNeely, J. A., Duke University Press, Durham, NC; London, U.K., 331, 1998.

Sala, O. E., Laurenroth, W. K. and McNaughton, S. J., Temperate grasslands, *Global Biodiversity Assessment*, Heywood, V. h., Cambridge University Press, Cambridge, 1995.

Salwasser, H., Gaining perspective: forestry for the future, *Journal of Forestry*, 88(11), 32, 1990.

Salzman, J., Informing the green consumer: the debate over the use and abuse of environmental labels, *Journal of Industrial Ecology*, 1(2), 11, 1997.

Sample, A. and Aplet, G. H., *Defining Sustainable Forestry*, Island Press, Washington, D.C., 1993.

Sample, A. V., Building partnership for ecosystem management on mixed ownership landscapes, *Journal of Forestry*, 92(8), 41, 1994.

Sampson, N., Private forests: more owners, fewer acres, *American Forests*, Autumn, 23, 1998.

Sanchez, P. A., Villachica, J. H. and Bandy, D. E., Soil fertility dynamics after clearing a tropical rainforest in Peru, *Soil Science Society of America Journal*, 47(6), 1171, 1983.

Sandoval, P. C., How certification and labelling may affect the capacity of forest usage by indigenous peoples, to enable them to satisfy their traditional needs, *International Conference on Certification and Labelling of Products from Sustainably Managed Forests. Session 5. DPIE*, Australian Department of Primary Industries & Energy, 1996.

Saunders, D. A. and Hobbs, R. J., Biological consequences of ecosystem fragmentation: a review, *Conservation Biology*, 5(1), 18, 1991.

Schaeffer, D. J. and Cox, D. K., Establishing ecosystem threshold criteria, *Ecosystem health: new goals for environmental management*, Constanza, R., Norton, B. G. and Haskell, B. D., Island Press, 1992.

Schenck, N. C. and Kinlock, R. A., Incidence of mycorrhizal fungi on six field crops in monoculture on a newly cleared woodland site, *Mycologia*, 72(445–456), 1980.

Schild, G., *Bretton Woods and Dumbarton Oaks: American economic and political postwar planning in the summer of 1944*, St. Martin's Press, New York, xiii + 254 pp, 1995.

Schlesinger, W. H., Changes in soil carbon storage and associated properties with disturbance and recovery, *The Changing Carbon Cycle: A Global Analysis*, Trabalk, J. R. and Reichle, D. E., Springer-Verlag, New York, 1986.

Schmidt, M. G., Macdonald, S. E. and Rothwell, R. L., Impacts of harvesting and mechanical site preparation on soil chemical properties of mixed-wood boreal forest sites in Alberta, *Canadian Journal of Soil Science*, 76(4), 531, 1996.

Schmidt, R., Berry, J. and Gordon, J., *Creating Integrated Forest Strategies*, United Nations Development Programme, New York, 1995.

Schneider, E. D., Monitoring for ecological integrity: the state of the art, *Ecological Indicators*, MCkenzie, D. H., Hyatt, D. E. and McDonald, V. J., Elsevier/Chapman & Hall, 14031419, 1992.

SCS (Scientific Certification Systems), *The Forest Conservation Program: Program Description and Operations Manual*, Scientific Certification Systems, Oakland, California, October, 1995.

SCS (Scientific Certification Systems), *Certification Standards for the Forest Conservation Program*, Scientific Certification Systems, 1998.

Seale, D. B., Physical factors influencing oviposition by wood frog, Rana sylvatica, in Pennsylvania, *Copeia*, 1982(3), 627, 1982.

Seastedt, T. R. and Crossley, D. A., Jr., Microarthropod response following cable logging and clearcutting in the southern Appalachians, *Ecology*, 62, 126, 1981.

Seip, H. K., *Forestry for Human Development: A Global Imperative*, Scandinavian University Press, Copenhagen, Oxford, Boston, 1996.

Seventh American Forest Congress, *Vision and Principles*, Yale Forest Forum. Yale School of Forestry and Environmental Studies, New Haven, CT, 1996.

SF&CW (Sustainable Forestry & Certification Watch), Key Actors on Forest Certification, Sustainable Forestry & Certification Watch, www.sfcw.org/keyactors.html, 1999.

SFF (Silva Forest Foundation), *Silva Forest Foundation Certification Program: Evaluation and Scoring Checklist (Draft I)*, October 16th, 1997.

SFF (Silva Forest Foundation), *Summary of Silva Forest Foundation Standards for Ecologically Responsible Timber Management*, Silva Forest Foundation, Slocan Park, B.C., Canada, 12 pages, July, silvafor@netidea.com, 1998.

SGS (Silviconsult Limited), *Tropical Forest Management: Position Paper on Certification*, University of Oxford, Forestry Research Programme (FRP), Oxford, U.K., 89, June 1994, 1994.

Shabecoff, P., *A Fierce Green Fire: the American Environmental Movement*, 1st ed., Hill & Wang, New York, 352 pages, 1993.

Shiva, V., Anderson, H., Schucking, H., Gray, A., Lohmann, L. and Cooper, D., *Biodiversity: Social and Ecological Perspectives*, Zed Books Ltd., London, UK; New Jersey, 1991.

Silvertown, J., Dodd, M. E., McConway, K., Potts, J. and Crawley, M., Rainfall, biomass variation and community composition in the Park Grass Experiment, *Ecology*, 75(8), 2430, 1994.

Simberloff, D., Which insect introductions succeed and which fail?, *Biological Invasions: A Global Perspective*, Drake, J. A., Mooney, H. A., Castri, F. d., Groves, R. H., Kruger, F. J., Rejmánek, M. and Williamson, M., John Wiley & Sons, Chichester, 61, 1989.

Simberloff, D. S., Farr, J. A., Cox, J. and Mehlman, D. W., Movement corridors: conservation bargains or poor investments?, *Conservation Biology*, 6, 493, 1992.

Simonis, U. E., *Beyond Growth: Elements of Sustainable Development*, Sigma, Berlim, Germany, 1990.

Simpson, B., *Lecture on sustainable forestry and certification for FES 508b. Sustainable Forest Management and Certification: Field Testing Current Approaches. Yale School of Forestry and Environmental Studies*, director, American Tree Farm System, Washington, D.C., New Haven, CT, 1998.

Simpson, B. B. and Ogorzaly, M. C., *Economic Botany: Plants in our World*, McGraw-Hill, New York, 1986.

Simula, M., Economics of certification, *Certification of Forest Products. Issues and Perspectives*, Viana, V. M., Ervin, J., Donovan, R. Z., Elliott, C. and Gholz, H., Island Press, Washington, D.C, 1996.

Singh, J. S. and Misra, R., Diversity, dominance, stability and net production in the grasslands at Varanasi, India, *Canadian Journal of Botany*, 47, 425, 1968.

Smartwood, *Generic Guidelines for Assessing Natural Forest Management*, Rainforest Alliance, Richmond, Vermont, November, 1998.

Smedes, G. W. and Hurd, L. E., An empirical test of community stability: resistance of a fouling community to biological patch-forming disturbance, *Ecology*, 62(6), 1561, 1981.

Smethurst, P. J. and Nambiar, E. K. S., Distribution of carbon and nutrients and fluxes of mineral nitrogen after clear-felling a Pinus radiata plantation, *Canadian Journal of Forest Research*, 20, 1490, 1990a.

Smethurst, P. J. and Nambiar, E. K. S., Effects of slash and litter management on fluxes of nitrogen and tree growth in a young Pinus radiata plantation, *Canadian Journal of Forest Research*, 20, 1498, 1990b.

Smith, C. L. and Voinov, A., Resource management: can it sustain Pacific Northwest fishery and forest systems?, *Ecosytem Health*, 2(2), 145, 1996.

Smith, C. T., McCormack, M. L., Hornbeck, J. W. and Martin, C. W., Nutrient and biomass removals from a red spruce — balsam fir whole-tree harvest, *Canadian Journal of Forest Research — Revue Canadienne de Recherche Forestiere*, 16(2), 381, 1986.

Smith, D. M., Larson, B. C., Kelty, M. J. and Ashton, P. M. S., *The practice of silviculture: applied forest ecology*, 9th Edition, John Wiley & Sons, Inc., New York, 537 pages, 1997.

Smith, E., *The quest for a certified 2x4. A case study of Seven Islands Land Company.*, Yale University School of Environmental Studies, New Haven, CT, Unpublished paper for FES 508b: Sustainable Forest Management and Certification: Field Testing Current Approaches., 18, 1998.

Smith, M., *The real dirt: farmers tell about organic and low-input practices in the Northeast*, Northeast Region Sustainable Agriculture Research and Education program, Burlington, VT, 1994.

Smith, P. B., Okoye, S. E., Wilde, J. d. and Deshingkar, P., *The world at the crossroads: towards a sustainable, equitable and liveable world — a report to the Pugwash Council*, Earthscan Publications Ltd., London, 1994.

Snook, L. K., Sustaining harvest of mahogany Swietenia macrophylla *King* from Mexico's Yucatan forest: past, present and future, *Timber, Tourist and Temples: Conservation and Development in the Mayan Forest of Belize, Guatemala and Mexico*, Primack, R., Bray, D., Galleti, H. and Ponciano, I., Island Press, Washington, D.C., 1998.

Soil Association, Internet Website, http://www.soilassociation.org, 1999.

Soil Science Society of America, Glossary of Soil Science Terms, http://www.soils.org, 1999.

Soil Survey Staff, *National Soil Survey Handbook . Title 430-VI*, Soil Conservation Service. U.S. Government Printing Office, Washington, D.C., 1993.

Soulé, M. E., *Conservation Biology: The Science of Scarcity and Diversity*, Sinauer, Sunderland, MA, 1986.

Southgate, D. and Runge, C. F., *The institutional origins of deforestation in Latin America*, University of Minnesota, St. Paul, Staff Paper, 7, www.ciesin.org/docs/002-407/002-407.html, 1990.

Specht, R. L., Changes in eucalyptus forests of Australia as a result of human disturbance, *The Earth in Transition. Patterns and Processes of Biotic Impoverishment*, Woodwell, G. M., Cambridge University Press, Cambridge, 1991.

Squire, R. O., Review of second rotation silviculture of *Pinus radiata* plantations in southern Australia: establishment practices and expectations, *Australian Forest*, 46, 83, 1983.

Steele, J. H., Marine functional diversity, *BioScience*, 47(7), 470, 1991.

Steinberg, E. K. and Kareiva, P., Challenges and opportunities for empirical evaluation of "spatial theory," *Spatial Ecology*, Tilman, D. and Kareiva, P., Princeton University Press, Princeton, 1997.

Stern, P., Young, O. and Druckman, D., *Global Environmental Change: Understanding the Human Dimensions*, National Academy Press, Washington, D.C., 1992.

Stewart, M. M. and Woolbright, L. L., Amphibians, *The food web of a tropical rain forest*, Reagan, D. P. and Waide, B., University of Chicago Press, Chicago, Illinois; London, U.K., 363, 1996.

Stone, C. P., Alien animals in Hawaii's native ecosystems: towards controlling the adverse effects of introduced vertebrates, *Hawaii's Terrestrial Ecosystems: Preservation and management*, Stone, C. P. and Scott, J. M., Cooperative National Park Resources Study Unit, University of Hawaii, Honolulu, 251, 1985.

Stone, E. L. and Kalisz, P. J., On the maximum extent of tree roots, *Forest Ecology and Management*, 46, 59, 1991.

Stott, D. E. and Kennedy, C. A., Impact of soil organic matter on soil structure, *Soil Quality and Erosion*, 57, 1999.

Strehlke, B., Fiji national code of logging practices, *Forest codes of practice*, Dykstra, D. P. and Heinrich, R., FAO, Rome, 43, 1996.

Sultan, S. E., Evolutionary implications of phenotypic plasticity in plants, *Evolutionary Biology*, Hecht, M. K., Wallace, B. and Prance, G. T., Plenum Press, New York, 1987.

Swift, M. J. and Anderson, J. M., Biodiversity and ecosystem function in agricultural systems, *Biodiversity and Ecosystem Function*, Schulze, E. D. and Mooney, H. A., Springer-Verlag, Berlin, 15, 1993.

Taal, B. M., Criteria and indicators for sustainable forest management in dry-zone Africa, *FAO XI World Forestry Congress*, Antalya, Turkey, 13–22 October, 1997.

Tansley, A. G., The use and abuse of vegetational concepts and terms, *Ecology*, 42, 237, 1935.

Teskey, R. O., Hinckley, T. M. and Grier, C. C., Temperature induced change in the water relations of *Abies amabilis* (Dougl.) Forbes, *Plant Physiology*, 74, 77, 1984.

Tew, D. T., Morris, L. A., Allen, H. L. and Wells, C. G., Estimates of nutrient removal, displacement and loss resulting from harvest and site preparation of a Pinus-Taeda plantation in the Piedmonts of North Carolina U.S.A., *Forest Ecology & Management*, 15(4), 257, 1986.

Thiollay, J. M., Disturbance, selective logging and bird diversity: a neotropical forest study, *Biodiversity and Conservation*, 6(8), 1155, 1997.

Tibor, T., *ISO 14,000: A Guide to the New Environmental Management Standards*, Irwin Professional Publishing, Chicago, Illinois, 1996.

Tietenberg, T. H., *Environmental and Natural Resource Economics*, 4th ed., Harper Collins College, New York, NY, xxvi, 1996.

Tilman, D., *Plant strategies and the Dynamics and Structure of Plant Communities*, Princeton University Press, Princeton, 1988.

Tilman, D., Biodiversity: population versus ecosystem stability, *Ecology*, 77, 350, 1996.

Tilman, D., Downing, A. and Wedin, D. A., Response: does diversity beget stability?, *Nature*, 371, 114, 1994.

Tilman, D. and Downing, J. A., Biodiversity and stability in grasslands, *Nature*, 367, 363, 1994.

Tilman, D., Knops, J., Wedin, D., Reich, P., Ritchie, M. and Siemann, E., The influence of functional diversity and composition on ecosystem processes, *Science*, 277, 1300, 1997.

Titus, B. D., Roberts, B. A. and Deering, K. W., Nutrient removals with harvesting and by deep percolation from white birch (Betula papyrifera [Marsh.]) sites in central Newfoundland, *Canadian Journal of Soil Science*, 78(1), 127, 1998.

Toman, M. A. and Ashton, P. M. S., Sustainable forest ecosystems and management: a review article, *Forest Science*, 42(3), 366, 1996.

Tongeren, J. W. v., *From SNA to environmental accounting*, United Nations Statistical Office, Washington, D.C., Documented prepared for a meeting of the expert group on SNA coordination (World Bank/UNSO), 3–7 December, 1990.

Trabaud, L., Fire as an agent of plant invasion? A case study in the French Mediterranean vegetation., *Biological Invasions in Europe and the Mediterrancean Basin*, Castri, F. d., Hansen, A. J. and Debussche, M., Kluwer Academic Publishers, Dordrecht, 417, 1990.

Tree Farmer Association, *Towards the practice of sustainable forestry: a special addition*, American Forest Foundation, Washington, D.C., 1997.

Tritton, L. M., Martin, C. W., Hornbeck, J. W. and Pierce, R. S., Biomass and nutrient removals from commercial thinning and whole-tree clearcutting of central hardwoods, *Environmental Management*, 11(5), 659, 1987.

Tuan, Y.-F., *A Study of Environmental Perception, Attitudes and Values*, Columbia University Press, New York, 260, 1974.

Tyson, C. B., Broderick, S. B. and Snyder, L. B., A social marketing approach to landowner education, *Journal of Forestry*, 96, 34, 1998.

U.S. Congress Office of Technology Assessment, *Harmful non-indigenous species in the United States*, U.S. Government Printing Office, Washington, D.C., 1993.

Uhl, C., Jordan, C., Clark, K., Clark, H. and Herrera, R., Ecosystem recovery in Amazon Caatinga forest after cutting, burning, and bulldozer clearing treatments, *Oikos*, 38(3), 313, 1982.

Ulman, N., A Maine forestry firm prospers by earning eco-friendly label, *Wall Street Journal*, New York, A2, 1997.

Underwriters Laboratories, Internet Website, http://www.ul.com, 1998.

UNDP, *Human Development Report*, United Nations Development Programme, New York, U.S.A., 1990.

Unesco/UNEP/FAO, *Tropical Forest Ecosystems: A State-of-Knowledge Report*, Vendome Presses Universitaires de France, 1978.

Upton, C. and Bass, S., *The Forest Certification Handbook*, St. Lucie Press, Delray Beach, FL, 219, 1996.

USDA Forest Service, *Forest Resources of the United States*, Forest and Range Experiment Station, Fort Collins, CO, General Technical Report, 1994.

Valerio, J., Salas, C. and Castillo, M., *Informe final de proyecto comportamiento del bosque natural despues del aprovechamiento forestal*, Instituto de Tecnologico de Costa Rica, Cartago, Costa Rica, 99, 1995.

van Miegroet, H. and Cole, D. W., The impact of nitrification on soil acidification and cation leaching in a red alder ecosystem, *Journal of Environmental Quality*, 13, 586, 1984.

Van Miegroet, H., Homann, P. S. and Cole, D. W., Soil nitrogen dynamics following harvesting and conversion of red alder and Douglas fir stands, *Soil Science Society of American Journal*, 56, 1311, 1992.

Vaux, J. J., The regulation of private forest practices in California: a case of policy evolution, *Journal of Forest History*, July, 128, 1986.

Vermeij, G. J., An agenda for invasion biology, *Biological Conservation*, 78, 3, 1996.

Vernet, J.-L., Man and vegetation in the Mediterranean area during the last 20,000 years, *Biological Invasions in Europe and the Mediterranean Basin*, Castri, F. d., Hansen, A. J. and Debussche, M., Kluwer Academic Publishers, Dordrecht, 161, 1990.

Viana, V., Certification: a Southern perspective, *Understory*, publication of the Good Wood Alliance (289 College St., Burlington, VT 05401. USA; telfax: +1 802 862 4448; email: warp@together.net), http://www.goodwood.org/, 4(3), 1994a.

Viana, V., *Subsidios para a elaboracao de criterios e indicadores de sustentabilidade florestal na bacia amazonica. Taller regional para definir criterios y indicadores de sostenibilidad del bosque Amazonico, Trapoto, Peru*, 23–25 December, 1995.

Viana, V. M., *Certification of forest products as a catalyst for change in tropical forest management in timber certification: implications for tropical forest management*, Yale School of Forestry and Environmental Studies, New Haven, CT, 1994b.

Viana, V. M., Ervin, J., Donovan, R. Z., Elliott, C. and Gholz, H., *Certification of Forest Products: Issues and Perspectives*, Island Press, Washington, D.C.; Covelo, California, 1996.

Vincent, J. R., The tropical timber trade and sustainable development, *Science*, 256, 1651, 1992.

Vincent, J. R., Gandapur, A. K. and Brooks, D. J., Species substitution and tropical log imports by Japan, *Forest Science*, 36(3), 657, 1990.

Visser, R., New Zealand forestry and the forest code of practice, *Forest codes of practice*, Dykstra, D. P. and Heinrich, R., FAO, Rome, 49, 1996.

Vitousek, P. M., Biological invasions and ecosystem properties: can species make a difference?, *Ecology of Biological Invasions of North America and Hawaii*, Mooney, H. A. and Drake, J. A., Springer-Verlag, New York, 163, 1986.

Vitousek, P. M., Biological invasions and ecosystem processes: towards an integration of population biology and ecosystem studies, *Oikos*, 57, 7, 1990.

Vitousek, P. M., Gosz, J. R., Grier, C. C., Melillo, J. M., Reiners, W. A. and Todd, R. L., Nitrate losses from distrubed ecosystems, *Science*, 204, 469, 1979.

Vitousek, P. M. and Matson, P. A., Disturbance, nitrogen availability, and nitrogen losses in an intensively managed loblolly pine plantation, *Ecology*, 66(4), 1360, 1985.

Vitousek, P. M. and Walker, L. R., Biological invasion by Myrica faya in Hawaii: plant demography, nitrogen fixation, and ecosystem effects, *Ecological Monographs,* 59(3), 247, 1989.

Vitousek, P. M., Walker, L. R., Whiteaker, L. D., Mueller-Dombois, D. and Matson, P. A., Biological invasions by *Myrica faya* alters ecosystem development in Hawaii, *Science,* 238, 802, 1987.

Vogt, D. J. and Edmonds, R. L., Nitrate and ammonium levels in relation to site quality in Douglas fir soil and litter, *Northwest Science,* 56(2), 83, 1982.

Vogt, K. A., Asbjornsen, H., Ercelawn, A., Montagnini, F. and Valdes, M., Roots and mycorrhizas in plantation ecosystems, *Management of soil, nutrients and water in tropical plantation forest,* Nambiar, E. K. S. and Brown, A. G., ACIAR — Australian Centre for International Agricultural Research, Canberra, Australia, 1997b.

Vogt, K. A., Asbjornsen, H., Grove, M., Maxwell, K., Vogt, D. J., Sigurdardottir, R. and Dove, M., Linking social and natural science spatial scales, *Integrating Landscape Ecology and Natural Resource Management,* Jianguo, L. and Taylor, W. M., Cambridge University Press, Cambridge, 1999a (In Press).

Vogt, K. A., Gordon, J. C., Wargo, J. P., Vogt, D. J., Asbjornsen, H., Palmiotto, P. A., Clark, H. J., O'Hara, J. L., Keaton, W. S., Patel-Weynand, T. and Witten, E., *Ecosystems: Balancing Science with Management,* Springer-Verlag, New York, 1997a.

Vogt, K. A., Grier, C., Meir, E. and Edmonds, R. L., Mycorrhizal role in net primary production and nutrient cycling in Abies amabilis ecosystems in western Washington, *Ecology,* 63(2), 370, 1982.

Vogt, K. A., Palmiotto, P. A., Tyrrell, M., Vogt, D. J., Patel-Weynand, T., Fanzeres, A., Larson, B., Wargo, P., Cuadrado, E., Johnson, K. H. and Coty, J., The Legacy Framework: a new integrative ecosystem framework to evaluate trade-offs and risks of natural resource uses, *Ecosystem Health,* 1998.

Vogt, K. A., Publicover, D. A., Bloomfield, J., Perez, J. M., Vogt, D. J. and Silver, W. L., Belowground ground responses as indicators of environmental change, *Environmental and Experimental Botany,* 33(1), 189, 1993.

Vogt, K. A., Vogt, D. J., Boon, P., Covich, A., Scatena, F. N., Asbjorgsen, H., O'Hara, J. L., Siccama, T. G., Bloomfield, J. and Ranciato, J. F., Litter dynamics along stream, riparian, and upslope areas following Hurricane Hugo, Luquillo Experimental Forest, Puerto Rico, *Biotropica,* 28, 458, 1996.

Vogt, K. A., Vogt, D. J., Boon, P., Fanzeres, A., Wargo, P., Palmiotto, P. A., Larson, B., O'Hara, J. L., Patel-Weynand, T., Cuadrado, E., Berry, J. and 1999, *A non-value based framework for assessing ecosystem integrity,* UDSA, Forest Service, Pacific Northwest Research Station, Portland, OR, General Technical Report, 1999b.

Waddington, D. G., Wildlife-associated recreation in the U.S.: results from the 1996 national survey of fishing, hunting, and wildlife-associated recreation and what it tells private land owners, *Natural Resources Income Opportunities for Private Lands Conference.,* Kays, J. S., University of Maryland Cooperative Extension Service, Hagerstown, Maryland, 1998.

Wade, M., Predicting plant invasions: making a start, *Plant Invasions: Studies from North America and Europe,* Brock, J. H., Wade, M., Pysek, P. and Green, D., Backhuys Publishers, Leiden, 1, 1997.

Walker, B. H., Biodiversity and ecological redundancy, *Conservation Biology,* 6, 18, 1992.

Walker, L. R. and Vitousek, P. M., An invader alters germination and growth of a native dominant tree in Hawaii, *Ecology,* 72(4), 1449, 1991.

Walker, T. D. and Valentine, J. W., Equilibrium models of evolutionary species diversity and the number of empty niches, *American Naturalist,* 124, 887, 1984.

Wardle, D. A., Changes in microbial biomass and metabolic quotient during leaf litter succession in some New Zealand forest and scrubland ecosystems, *Functional Ecology,* 7, 346, 1993.

Wardle, D. A. and Nicholson, K. S., Synergistic effects of grassland plant species on soil microbial biomass and activity: implications for ecosystem-level effects of enriched plant diversity, *Functional Ecology,* 10, 410, 1996.

Wargo, J., *Our Children's Toxic Legacy: How Science and Law Fail to Protect us from Pesticides,* Yale University Press, New Haven and London, 380, 1996.

Waring, R. H., Imbalanced forest ecosystems: assessments and consequences, *Forest Ecology and Management,* 12, 93, 1985.

Waring, R. H., Nitrate pollution: a particular danger to boreal and subalpine coniferous forests, *Human Impacts and Management of Mountain Forests*, Fujimori, T. and Kimura, J., Forestry and Forest Products Research Institute, Ibaraki, Japan, 1987.

Warkotsch, P. W., Engelbrecht, G. v. R. and Hacker, F., The South African harvesting code of practice, *Forest codes of practice*, Dykstra, D. P. and Heinrich, R., FAO, Rome, 75, 1996.

WCED (World Commision on Environment and Development), *Our Common Future*, United Nations, New York, 1987.

Weber, M. G., Methven, I. R. and Van Wagner, C. E., The effect of forest floor manipulation on nitrogen status and tree growth in an eastern Ontario jack pine ecosystem, *Canadian Journal of Forest Research*, 15, 313, 1985.

Wedin, D. A. and Tilman, D., Species effect on N cycling: a test with perrenial grasses, *Oecologia*, 84(4), 433, 1990.

Weetman, G. F., A forest management perspective on sustained site productivity, *The Forestry Chronicle*, 74(1), 75, 1998.

Wells, T. C. E., Botanical aspects of conservation management of chalk grasslands, *Biological Conservation*, 2, 36, 1969.

Weste, G., Impact of Phytophthora species on native vegetation of Australia and Papua New Guinea, *Australasion Plant Pathology*, 23, 190, 1994.

Westman, W. E., Managing for biodiversity: unresolved science and policy questions, *BioScience*, 40(1), 26, 1990.

White, T. C. R., *The Inadequate Environment: N and the Abundance of Animals*, Springer-Verlag, Berlin, 1993.

Whittaker, R. H., Evolution and measurement of species diversity, *Taxon*, 21, 213, 1972.

Whittaker, R. H., Levin, S. A. and Root, R. B., Niche, habitat, and ecotope, *American Naturalist*, 107, 321, 1973.

Wiegart, R. G. and Owen, D. F., Trophic structure, available resources and population density in terrestrial versus aquatic ecosystems, *Journal of Theoretical Biology*, 30, 69, 1971.

Wiens, J. A., Spatial scaling in ecology, *Functional Ecology*, 3, 385, 1989.

Wijewardana, D., Caswell, S. J. and Palmberg-Lerche, C., Criteria and indicators for sustainable forest management, *FAO XI World Forestry Congress*, Antalya, Turkey, 13–22 October, 1997.

Wilk, R. R., *Household Ecology: Economic Change and Domestic Life Among the Kekchi Maya in Belize*, The University of Arizona Press, 1991.

Williams, C. C. and Haughton, G., *Perspectives towards sustainable environmental development*, Avebury, Aldershot; Brookfield, USA; Hong Kong; Singapore; Sydney, 200, 1994.

Williams, M., *Americans and their forests: a historical geography.*, Cambridge University Press, New York, 1989.

Williams, R. A., Voth, D. E. and Hitt, C., Arkansas NIPF landowners' opinions and attitudes regarding management and use of forested property, *Symposium on Nonindustrial Private Forests: Learning from the Past, Prospects for the Future*, Minnesota Extension Service, Washington, D.C., 1996.

Williamson, M., *Biological Invasions*, Chapman & Hall, London, 244, 1996.

Wilman, I., Expecting the unexpected: some ancient roots to current perceptions of nature, *Ambio*, 19(2), 62, 1990.

Wilson, D. J. and Jefferies, R. L., Nitrogen mineralization, plant growth and goose herbivory in an Arctic coastal ecosystem, *Ecology*, 84(6), 841, 1996.

Wilson, E. O., *Biodiversity*, National Academy Press, Washington, D.C., 1988.

Winters, R. K., *The Forest and Man*, Vantage Press, New York, Washington, Hollywood, 1974.

Woodmark, *Responsible Forestry Standards*, The Soil Association Marketing Company Ltd. — Responsible Forestry Programme, Bristol, U.K., 41, February, 1994.

Woods, P., Effects of logging, drought, and fire on structure and composition of tropical forests in Sabah, Malaysia, *Biotropica*, 21(4), 290, 1988.

World Bank, *The forest sector: a World Bank policy paper*, World Bank, Washington, D.C., 1991.

World Bank, *Toward gender equality: the role of public policy (an overview)*, World Bank, Washington, D.C., 1995.

World Bank, *From plan to market: world development report*, World Bank, Washington, D.C., 1996.

WRI, *Tropical forests: a call for action*, World Resources Institute, World Bank, United Nations Development Programme, Washington, D.C., October, 1985.

WRI, *World Resources, 1994–95*, World Resources Institute, New York, 1994.

Wyman, R. L., Soil acidity, moisture and the distribution of amphibians in five forests of south central New York, *Copeia*, 1988(2), 394, 1988.

Yin, X., Perry, J. A. and Dixon, R. K., Influence of canopy removal on oak forest floor decomposition, *Canadian Journal of Forest Research*, 19, 204, 1989.

Young, A. and Mitchell, N., Microclimate and vegetation edge effects in a fragmented podocarp-broadleaf forest in New Zealand, *Biological Conservation*, 67, 63, 1994.

Youngberg, C. T., The influence of soil conditions, following tractor logging, on the growth of planted Douglas fir seedlings, *Soil Science Society of America Proceedings*, 23, 76, 1959.

Zapata, J. B., The Central American process of sustainable development, *FAO XI World Forestry Congress*, Antalya, Turkey, 13–22 October, 1997.

Zimmerer, K. S., Soil erosion and social discourse: perceiving the nature of environmental degradation, *Economic Geography*, 69(3), 312, 1993.

Zimmerer, K. S., *Changing Fortunes: Biodiversity and Peasant Livelihood in the Peruvian Andes*, University of California Press, Berkeley, xi, 1996.

Zinkhan, F. C., *Timberland Investments: A Portfolio Approach*, Timber Press, Portland, OR, 1992.

Zonneveld, I. S., Principle of bio-indication, *Environmental Monitoring and Assessment*, 3, 207, 1983.

Zonneveld, I. S., Scope and concepts of landscape ecology as an emerging science, *Changing Landscapes: An Ecological Perspective*, Zonneveld, I. S. and Foreman, R. R. R., Springer-Verlag, New York, 1990.

Index